AF351841

HARVEST HORIZONS

Perspectives on Contemporary Agrarian Experience, Art, and Literature

Richard D. Scheuerman

Foreword by Alexander C. McGregor

TRITICUM PRESS
Pasco, Washington

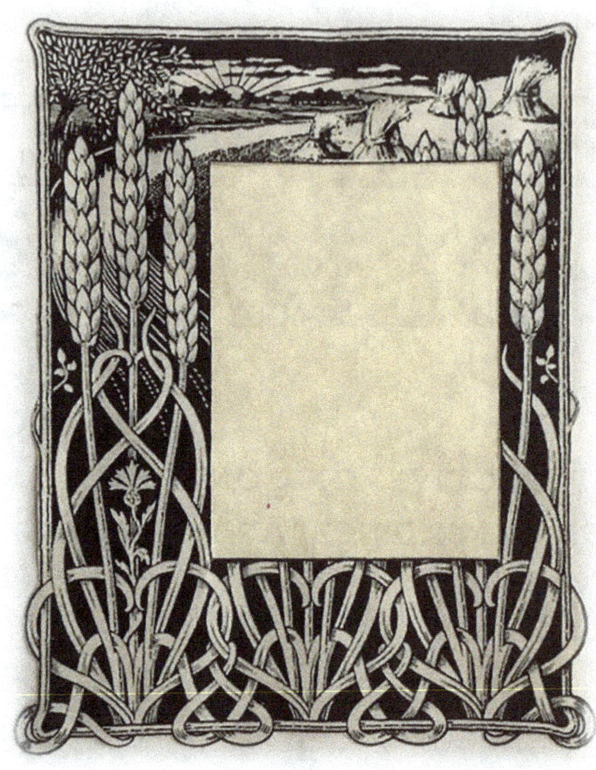

Above: Alfred Vivian, Arts and Crafts Bookplate Design (1907)
Columbia Heritage Collection

Cover: Jim Gerlitz, *Palouse Colony Harvest* (2016)

Chapter sickle and sheaves epigraphs woodcuts by Edgar Moore (c. 1920);
italicized epigrams are titles of representative art works.

HARVEST HORIZONS

Essays on Agrarian Themes in Western Art and Literature

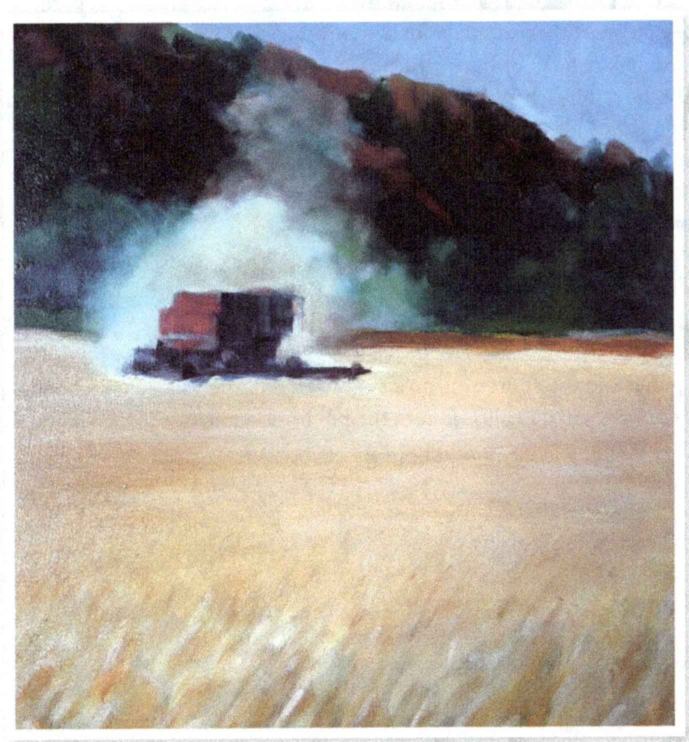

Richard D. Scheuerman

Foreword by Alexander C. McGregor

HARVEST HORIZONS
Perspectives on Contemporary Agrarian Experience, Art, and Literature

This is a publication of Triticum Press
9903 Coronado Dr
Pasco, Washington 99301

Design: Carol O'Callaghan, Triticum Press
Cover Background Image of grain: Depositphotos

Royalties from this publication support non-profit agrarian arts and related educational programs. Images in this book are either in the artist's collection, in the public domain, or shared via the Fair Use Act which supports dissemination of knowledge in scholarly endeavors.

PAPERBACK ISBN: 979-8-9891343-2-8
EBOOK: 979-8-9891343-3-5
HARDBACK: 979-8-9891343-8-0

Library of Congress Control Number: 2024942000

For Dan Birdsell and Dave Wells,
and in memory of
Arden Johnson and Louise Braun—
Teachers Extraordinaire

Other Titles in the Harvest Project by Richard Scheuerman
(Copies and Information available through www.triticumpress.com.)

Hallowed Harvests: Agrarian Depiction in the Bible, Literature, and Art to Early Modern Times

Harvest Hands: Reapers and Threshers in American and European Modern Art and Literature

Harvest Heritage: Agricultural Origins and Heirloom Crops of the Pacific Northwest
(with Alexander C. McGregor and John Clement)

Published with support from
The Schneidmiller Endowment for Palouse Regional Studies
The McGregor Company
and the
Anson Mills/Carolina Gold Foundation

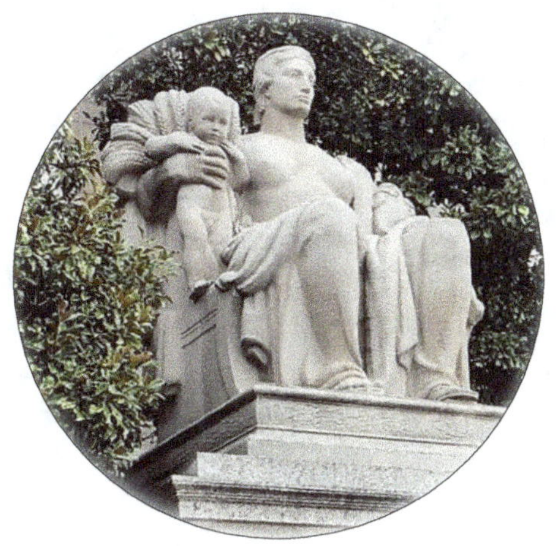

Heritage—Allegorical Matriarch with Child and Grain Sheaf (1935)
National Archives Entrance
James E. Fraser (designer) and Edward H. Ratti (sculptor)
Inscription: _The heritage of the past is the seed that brings forth the harvest of the future._
Columbia Heritage Collection Photograph

[S]peak to the earth, and it will teach you; ….
Does not the ear test words as the palate tastes food?
Wisdom is with the aged, and understanding in length of days.

—Job 12:8, 11-12

Harvest since time immemorial was understood in ritual terms as the
principal duty in humanity's relationship with Mother Earth for the
perpetuation of life. This was essentially the purpose of existence . . .
and the sacred nature of work in the fields long imparted a marked
artistic-spiritual dimension.

—J. Katarzyna Dadak-Kozicka

CONTENTS

LIST OF COLOR GALLERY PLATES

(following page 70)

FOREWORD

A thousand hills lay bare to the sky, and half of every hill was wheat and half was fallow ground; and all of them, with the shallow valleys between, seemed big and strange and isolated. The beauty of them was austere, as if the hand of man had been held back from making green his home site, as if the immensity of the task had left no time for youth and freshness. . . . The scene was heroic because of the labor of horny hands; it was sublime because not a hundred harvests, nor three generations of toiling men, could ever rob nature of its limitless space and scorching sun and sweeping dust, of its resistless age-long creep back toward the desert that it had been. Here was grown the most bounteous, the richest and finest wheat in all the world. Strange and unfathomable that so much of the bread of man, the staff of life, the hope of civilization. . . .

PIONEER COLUMBIA PLATEAU WHEAT FARMER RALPH SNYDER commanded my youthful attention when he first told me that Zane Grey likely composed these lines for his novel *The Desert of Wheat* while residing in our area during the sweltering summer of 1917. Although better known for stories of cattle drives and frontier lawmen, Grey wove a classic tale of farmland experience during rural America's momentous transition from work horses and mules to steam engines and gas-powered threshers. Ralph especially piqued youthful listeners' interest when he mentioned that the famed author boarded in our little hamlet of Hooper's nondescript white clapboard hotel—now called periodically from retirement to host groups visiting from other parts of the country for art workshops and writing seminars.

Despite my tiny hometown's isolation and a 1950s population numbering perhaps fifty inhabitants, my companions and I were impressed by our elders' abiding values of hard work and self-reliance informed with appreciation for history and the arts and humanities. We admired my older cousin Maurice McGregor's autographed copy of H. L. Davis's *Honey in the Horn*, paintings of farm and ranch life hung in neighbors' homes by regionalists like Anine Harder and my aunt, Ruth Burden. Framed prints like Millet's evocative *The Gleaners*, contributed by local subscribers to *The Gentleman Farmer* and *Collier's*, hung in the Hooper General Store.

Our wheat ranch was located near the larger rural community of LaCrosse on the western periphery of the Columbia Plateau's famed Palouse Country—a wondrous, mysterious region of steeply undulating hills still famed as the world's highest yielding dryland grain region. Geographer-philosopher Yi-Fu Tun has written, "Awareness of the past in an important element in the love of place," and that such appreciation is very different from the romantic notions of rural life often held from a distance. My mother, Norma Gantzer McGregor, came from Minnesota to put down roots here in the 1940s and found that rural life in the largely uninhabited western Palouse took some getting used to. Her father-in-law told her, "If you stay long enough, you'll come to love those basalt cliffs and the rolling wheat lands." And so she did, bringing with her a love of reading and culinary traditions that enriched the lives of a large circle of family and friends.

Residents of our rural community formed a colorful tapestry of cultures and experience depicted in many of the images included in this remarkable collection of agrarian art and literature. Mother was a tireless prime mover in the early 1970s organization of the Whitman County Historical Society and composed promotional material about farm history to encourage efforts to preserve a priceless legacy of regional oral histories, country life literature for rural libraries, and threatened agrarian architecture. She found inspiration for her work in the writings of many authors featured in this collection, and was fond of quoting from Gerald Brenan's *Thoughts on a Dry Season*: "We do not completely die if we have loved, for by the act of loving some part of us passes into loved person or object. Trees, hills, fields, dear ones, and our country. In the act of loving these we give some portion of ourselves. . . ."

In addition to farmers and ranchers like my grandparents, Ralph and Aimee Snyder, and Dave Carter, whose ancestors had come to area in territorial days from eastern states, others like Emile Morod and Hans Kahler ventured here from the other side of world to labor in the fields and ranges of the American West. Emile had been orphaned at an early age and had left the Basque region of France for fear of impending war in 1914 and the prospect of life in a reformatory. Hans emigrated with other German farmers from the Volga steppes seen in Ivan Shishkin's canvases where they had lived since the days of Catherine the Great. A longtime and dependable member of the McGregor harvest crews, Hans regularly stirred us into action after substantial early morning July breakfasts by announcing, "Vee got to get a harvest-ting!" And out the door he'd go. Long-suffering Lenora Barr Torgeson worked for eighty-seven consecutive twelve-hour days during one harvest season to fill the appetites of some one hundred workers. Such labor vividly brings to mind women's stories of harvest has recounted in Lois Phillips Hudson's 1932 semi-autobiographical novel, *The Bones of Plenty*.

I admired the abilities of these men and women, many with little formal education, to

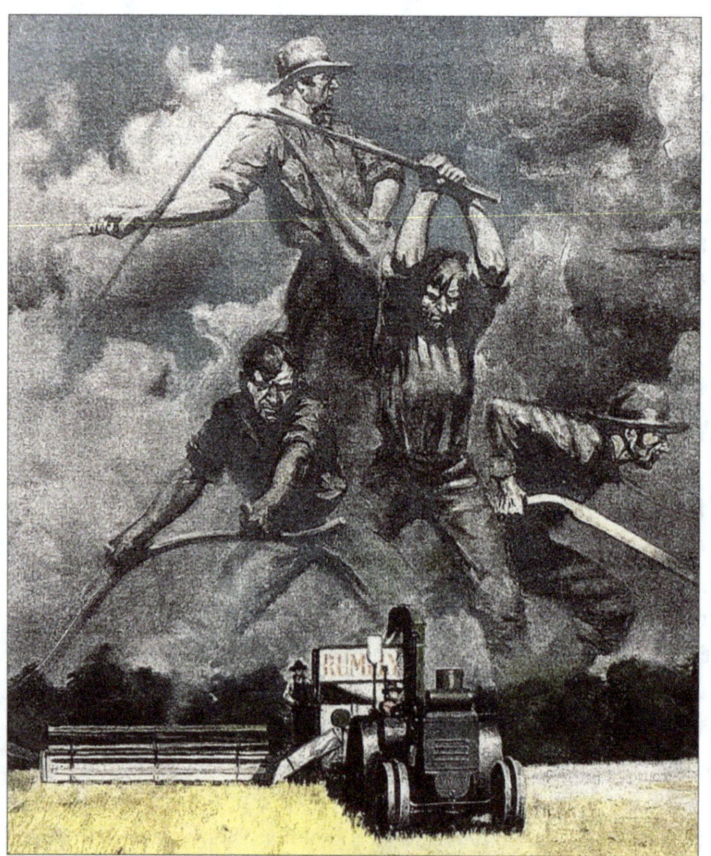

L. M. Collins Hardware & Machinery Calendar Cover Art (c. 1920)
Fairfield, Washington
Dennis Solbrack Collection

quote lines at length from a poem by Longfellow or Whitman while "rogueing" rye from stands of barley, or surveying Zane Grey's fields of "richest and finest wheat in all the world." These strong-willed, opinionated individuals taught us to appreciate art and music, and with the teachers in our country school introduced us to great authors like Willa Cather and T. S. Eliot whose rural themes gave special meaning to our own near experiences. They also pointed out more obscure works we found fascinating and available through the weekly Whitman County Bookmobile visits like "Uncle Dan" Drumheller's *Thrills of Western Tales*, and Kirby Brumfield's wonderfully illustrated *This Was Wheat Farming*. In later years Northwest regionalists Nona Hengen and Sherryl Evans, two of the region's foremost painters of pioneer farm scenes, created dramatic works that foster appreciation for the sacrificial labors of men and women during *The Horse Interlude*. The expression is the title of Thomas Keith's fascinating 1990 book of harvest history and the era of America's transition to modern agriculture.

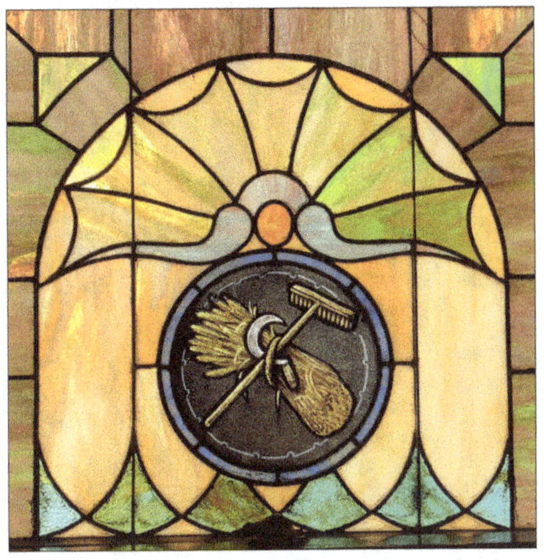

Sheaf and Sickle "Bread of Life"
Stained Glass Window (c. 1920)
United Methodist Church; LaCrosse, Washington
Columbia Heritage Collection Photograph

Richard Scheuerman has examined a vast realm of agrarian literature found throughout the American West and wider world that informs interpretations of renowned paintings like Monet's Wheatstack series, van Gogh's vivid Grainfields collection, and the evocative harvest canvases of Millet and Breton. Much of this writing and art depicts the stolid nobility of rural life amidst trying circumstances and brings to mind moral obligations of stewardship and benevolence. These varied influences through art, literature, and country living have contributed through the years to efforts by my colleagues and family farmers to promote local food relief throughout the country in "contemporary gleaning" efforts like Second Harvest, and to support regional and state arts and humanities programs. Fostering a greater sense of heritage and public service has led to renovation of space for art galleries and museums in rural communities as well as for local and county library expansions. Great art and literature encourage these important initiatives and renew agrarian vitality in this day of flat worlds and virtual connections. Seeking insight and inspiration from the images and stories in *Harvest Horizons* and other resources in Richard's remarkable decade-long "Harvest Project" series wonderfully serves these worthy endeavors.

Alexander C. McGregor
Pullman, Washington
January 31, 2024

PREFACE

The idea of a study on agrarian art first occurred to me in September 2013 after seeing a century-old postcard of Canadian-American illustrator Robert Atkinson Fox's luminous *After the Harvest* (c. 1910, PLATE 1) at the Washington State Fair in Puyallup. I did not buy it, but the scene stirred something deep within and kept appearing in my mind's eye. Late summer and fall festivals across the country are heirs to medieval post-harvest celebrations that hearkened back to even earlier times. The word entwines Latin *feria*, days off for the buying and selling of wares at a "fair," with *festum*, a religious holiday and feast. Like these bygone special occasions, present-day festivals and fairs provide momentary separation from daily life for entertainment, consumption, competition, and exhibition rooted in pastoral provision. Many people have memory of a fair visit that sparked a special interest or festival attendance that led to an important spiritual realization or social transition. Others have found connection with some deep feeling amidst the cacophony of commercial sights, sounds, and smells that stretch back to communal ancestral ways woven deeply into the soul. For whatever reason, the Fox harvest image broke through the nebulous boundary of time and place and stayed with me.

I had no intention of acquiring images or prints like I had seen at the fair until I made a random café stop a few months later during meetings at the Washington State History Museum in nearby Tacoma. While awaiting lunch fare across the street, I noticed a tattered array of magazines from the 1930s and '40s strewn about the tables including large-format copies of *Fortune*. I later learned they have long been popular among collectors for their rich colors and high-quality paper stock. At my table was the October 1935 issue featuring a serigraph cover by Ernst Hamlin Baker of heads of grain in rich golds and browns silhouetted against a combine reel and header-tender. I also remember that the sandwich I enjoyed while perusing was made of thick-cut bread cut from a hard-crusted artisanal loaf and that I could choose salads made with pastas of various shapes, buttery croissants, delicately layered baklava, and whole-grain muffins. While paying my bill, I explained to the proprietor my special interest in the cover and for a few dollars he kindly parted with it. On my stroll through city's museum district later that day I passed along a monumental exterior wall mural being painted in sea and earth tones, *Working Forward, Weaving Anew* (2017, PLATE 2), by Esteban Camacho Steffensen and Jessilyn Brinkerhoff. The painting shows a young Puyallup Indian girl plaiting a basket foundation from grain stalks and cedar bark.

These happenstances got me to think seriously about agrarian art and the anonymous souls who inhabited farmsteads across the country, many now abandoned, like those in Fox's painting and more recent depictions. Certainly there are other farming tasks essential for putting bread on the table like cultivating, seeding, weeding; and grains and breads are among many foodstuffs. Other labors and crops have also been variously depicted in art and literature but my investigations were fenced by reasonable limits. Likewise, I am mindful of the valued contributions of Hispanic- and Asian-Americans to Northwest agriculture. Their association with production of fruit, vegetables, and specialty crops involves important aesthetic considerations not specifically addressed here. The fates of families and entire nations have substantially depended for generations upon the year's culminating grain harvests. These souls left the kind of roadside evidence commonly encountered in rural areas—rusting threshing machines and abandoned combines, derelict wooden "crib" elevators, and country cemeteries with gravestone sheaf carvings. Art and literature help provide rescue from oblivion, and I began collecting affordable vintage prints and period novels about farm life to better understand the experience. Subsequent investigation turned up harvest scenes by Peter Helck and other prominent artists that had been featured in *Fortune*, *Country Life Magazine*, and other periodicals.

One discovery led to another with relevant acquisitions that contributed to this study including seventeenth- and eighteenth-century Labors of the Months harvest woodcuts, a Rembrandt-Watelet counterproof print of *Landscape with Hay Barn and Flock of Sheep* (1660), and age-stained engravings from James Thomson's *The Seasons* (1726). This in turn introduced me to stirring images in Robert Bloomfield's *The Farmer's Boy* (1800), André Theuriet's incomparable *La Vie Rustique* (1896), and lithographs from *A Natural History of New York* (1851) by James Hall, and Thomas Craven's *A Treasury of American Prints* (1939). Little by little, I became fascinated with harvest themes in a broad art-historical context, and the availability of print forms for tactile chronological documentation that came to form the Columbia Heritage Collection of Agrarian Art.

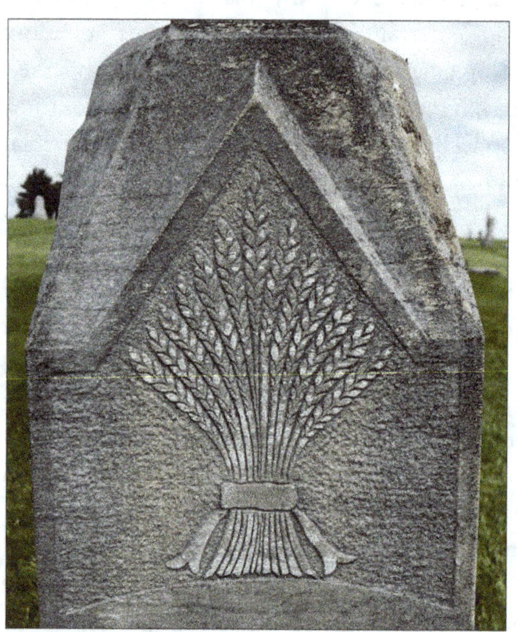

Grain Sheaf Headstone Carving (1908)
11 ½ x 8 ¼ inches
Mountain View Cemetery; Endicott, Washington
Columbia Heritage Collection Photograph

Prints of various kinds in Western art have constituted a prolific if fragile form ever since Europeans first appropriated papermaking and block-cutting from the Chinese in the fifteenth century. Masterful European woodcut prints first appeared in Italy and Germany from artists like Andrea Mantegna (c. 1431-1506) and Albrecht Dürer (1471-1528), followed later in the century there and in France with painstaking intaglio line engraving on iron and copper. Sixteenth-century Dutch printmakers like Hieronymus Cock (1518-1579) perfected shadowing techniques that provided depth to landscapes and other subjects, while engravings in the 1600s and 1700s are characterized by more ornamental design and the introduction of mezzotint for tonal shadings.

Eighteenth-century English engraver William Wynne Ryland (c. 1738-1783) developed stipple etchings as a variant of French engraving methods by using dots as well as lines for rich tonal variations. Italian Francisco Bartolozzi (1727-1815), who learned the process in England, became the most famous practitioner of stipple and was the first to use color with the process. Many of his magnificent creations were first published by John Boydell (1719-1804) in London, a skilled engrav-

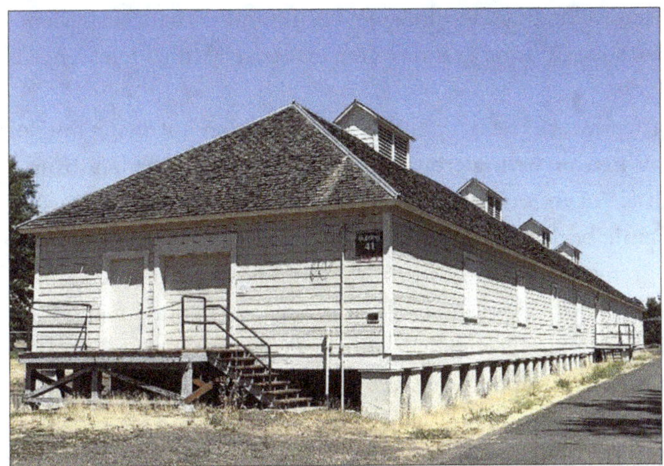
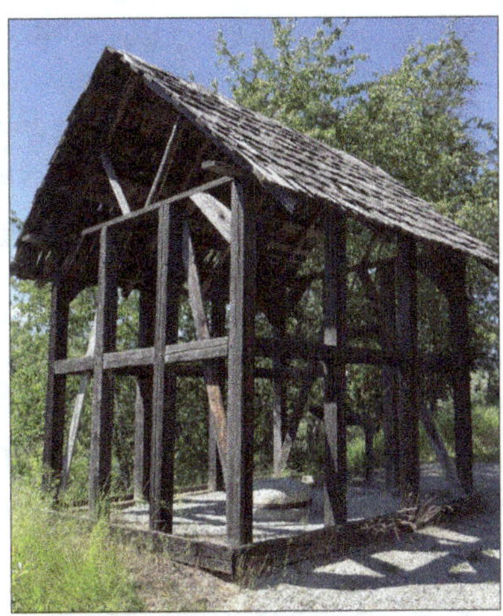
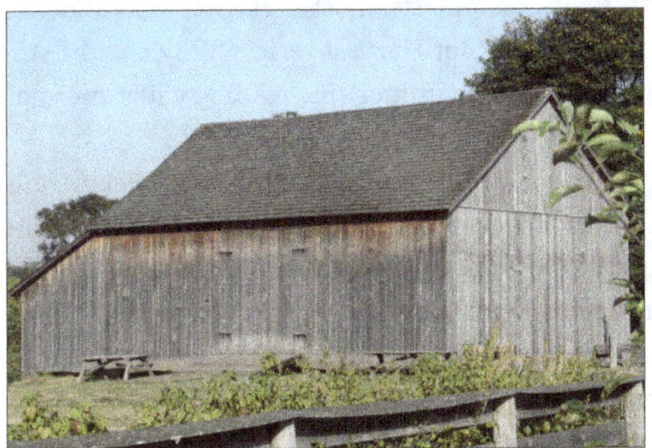
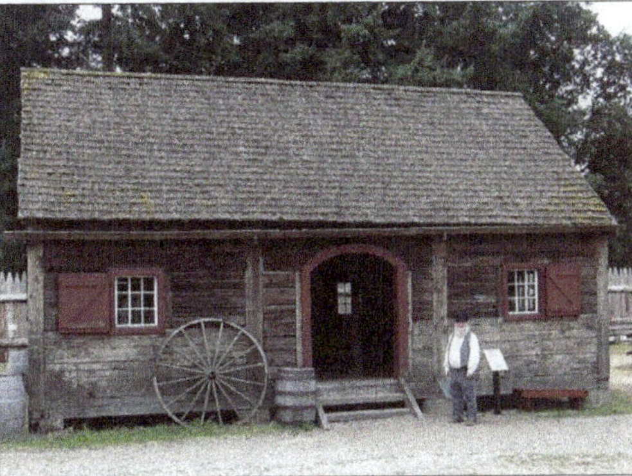

Clockwise from top left: Ft. Walla Walla Granary (1888); Cataldo Mission Gristmill Restoration; Cataldo, Idaho (c. 1845); Ft. Nisqually Granary—Washington's Oldest Wooden Structure (1843), Tacoma, Washington; Manson Threshing Barn, Champoeg National Historic Site; Champoeg, Oregon (1862) Columbia Heritage Photographs

er himself, and collected by connoisseurs of fine art in early America like George Washington and Thomas Jefferson. The advent of lithography in the eighteenth century led to detailed color reproductions of fine art, and nineteenth-century printmaking innovations included application of steel coating to copperplate to make more impressions possible. The rural landscapes of British artist John Constable (1776-1837) and engraver David Lucas (1802-1881) are considered the epitome of tonal mezzotint. Modern printmaking has featured mixed media and digital processes.

Being ignorant of many things fortunately does not disqualify for setting out on a mission of mind and many miles. Indulgent artists, authors, and curators kindly enabled my inquiries. Their guidance led to consideration of broader themes of social change, technical innovation, and unintended consequences depicted in art and literature since ancient times. A troubling modern cultural trend has been separation between those with narrow technical and intellectual specializations, and some who might lack expertise but dismiss knowledge as irrelevant or misguided. Still others are greatly occupied with material values suggested in the art of Andy Warhol and minimalist fiction of superb contemporary authors like Mary Miller. The young protagonist in her road novel of vacuity, *The Last*

Days of California (2014), is indifferent to the countryside while recounting endless banal experiences at generic motels and franchise restaurants. Cultural anomie often associated with adolescence is harder to excuse among parents and elders.

Harvesting in its various aspects of teamwork, reward, and frustration is an apt metaphor for the research and writing of this book. When surveying crops, for example, farmers benefit from a familiarity with landscapes I could not claim in terms of background for this study. Few brushes with art history swept across my academic background. Country folk also know that harvest can be a haphazard undertaking. Various grains mature at different rates, and any single crop ripens unevenly depending on seeding time and angle of exposure to the sun. My investigation had similar twists and turns. I benefited from life in a farm family with ancestors having pursued that way of life as far back as we know, and have attempted to interpret expressions of agrarian experience through the centuries by investigating some of its artistic and literary masters—van Gogh and Willa Cather, Thomas Hart Benton and Edvart Rølvaag, and many others. But I mostly wanted to enjoy the beauty and meanings of profound works that bring forth their creators' steadfast presence. Why did van Gogh devote his last fevered months to painting the magnificent Harvest series? Why could I still name the main characters in Rølvaag's *Giants in the Earth* after summertime reading a lifetime ago? Why do the lyrics of Sting's "Fields of Gold" often play in the mind?

This study evolved slowly and a complete list of those who provided valued assistance in this project would fill a harvest truck driver's logbook. I especially thank kindred and generous spirits Mark and Carol Moseman of David City, Nebraska, who have significantly contributed to public and scholarly recognition of agrarian art as an aesthetic genre. I also thank Clifford and Lee Ann Trafzer, Jack and Ann Dorsey, Vicki Broeckel, and Katherine Nelson for informing my understandings. Charles and Rose Ann Finkel helpfully encouraged my studies at Seattle's Frye Art Museum, and I thank Jo-Anne Birnie Danzker and Cory Gooch at the Frye for permission to reproduce works by Lhermitte and other painters of agrarian scenes from their priceless collection. Arthur and Marie Ellis and Michelle Beauclair assisted in translations of documents important for this study. Artist-author Nona Hengen is one of the country's leading painters of draft horse and yesteryear farm scenes like her Palouse Wheat Harvest series. She has long shared her knowledge of agrarian art and literature with me and others in her family homestead studio near Spokane, Washington, and popular books of regional fiction.

New York writer Amy Halloran turned my attention to rural themes in the literature of Herman Melville, and also made possible memorable explorations of the Yale Center for British Art. We benefited there from guidance by Director Amy Meyers regarding masterful harvest scenes by George Stubbs and Samuel Palmer, and landscapes by John Constable. Rosemarie Oehler Adcock attended the American Academy of Art in Chicago under Eugene Hall, an apprentice of Russian painter Alexander Zlatoff-Mirsk. Hall himself had been a student of the nineteenth-century Russian Itinerant luminary Ilya Repin. Rosemarie also directed me to Monet's breathtaking Grainstack series at the Chicago Art Institute. Contribution of her monumental canvas *Ruth and the Barley Harvest* (2014, PLATE 3), winner of the 2016 Year of Jubilee Mission Exhibition in Chicago, for this study is a gesture of uncommon dedication and generosity. Her art speaks to the relevance of the ancient theme of graceful provision. In kindred spirit Seattle artist Jim Gerlitz created a series of masterful illustrations for my recent book, *Hardship to Homeland* (2018), and painted *Palouse Colony Harvest*

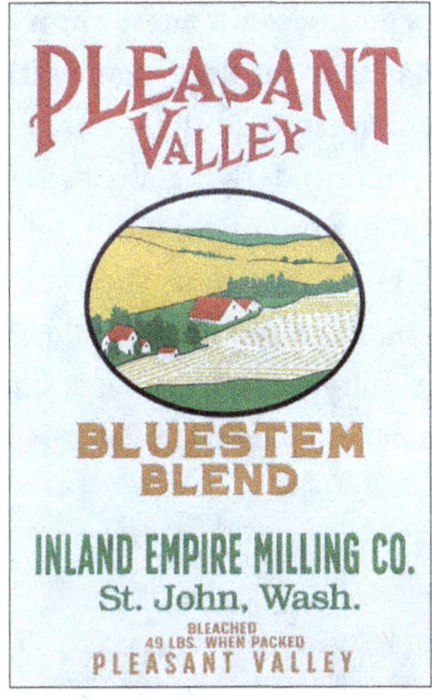

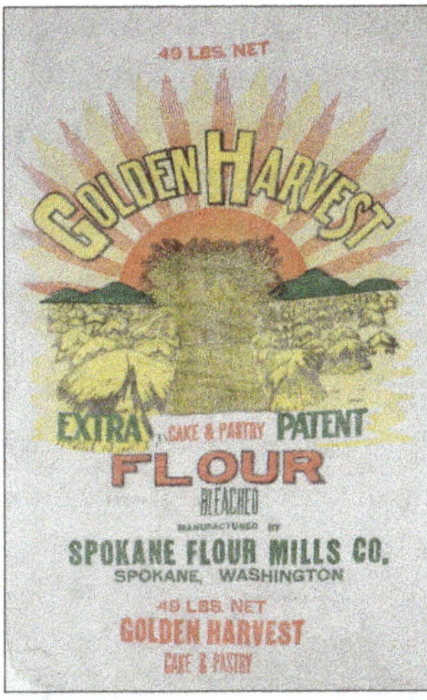

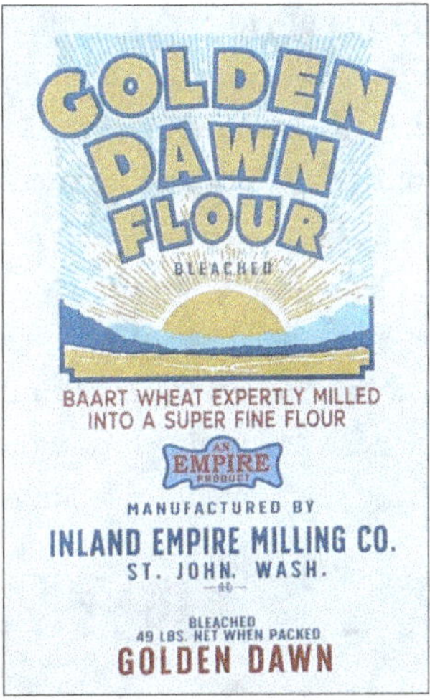

Left and right: Heritage Grain Product Line Commercial Art (c. 1940)
Inland Empire Milling Company; St. John, Washington
Center: Golden Harvest Flour Sack (c. 1915)
Museum of Arts and Culture; Spokane, Washington

(2017, PLATE 4) on a theme of modern bounty to commemorate the 125th anniversary of the farm's founding. The significance of this historic location to our families is described in these pages.

Alexander McGregor, president of The McGregor Company farm supply business, is a historian and author of numerous books and articles on regional agricultural history. Our families' longtime association goes back to the 1950s when my father was among the firm's first employees. I have vivid boyhood memories of annual visits to tiny Hooper, headquarters of the McGregor Ranch and clan, for their festive summertime picnics that featured games, races, and fascinating tours of rural memorabilia in the Hotel Glenmore and Hooper General Store. These events were successors to Chautauquas, Fourth of July celebrations, and other summertime gatherings of the early settlement years when local literary societies debated matters with contemporary relevance like, "Resolved, that modern inventions are detrimental to the laboring class." Through collaboration on a coauthored work, *Harvest Heritage: Agricultural Origins and Heirloom Crops of the Pacific Northwest* (Washington State University Press, 2014), Alex reacquainted me with regional agrarian themes in novels, short stories, and poems by Zane Grey, H. L. Davis, Edwin Markham, and other writers.

John Clement is a longtime professional landscape photographer from Richland, Washington, and honoree of the Professional Photographers of America International Hall of Fame. He has accompanied me to galleries throughout the United States, Europe, and Russia to document many of the paintings featured in this study. A selection of recent Northwest harvest scenes from his considerable archive is included in this series. In forays through pastoral place and time, Alex, John, and I have viewed and documented many ordinary things like sickles, scythes, and flails made and sold throughout rural America. The decades have transformed them and the harvest machinery that replaced them into wonderful objects of meaning, innovation, and beauty.

Color images of John's "Northwest Drylands" photographs like *Bringing in the Sheaves* (2010) show the influence of two prominent American watercolor artists whose works he has closely studied since starting his career in the 1970s—Winslow Homer and Andrew Wyeth. Although the substantial portion of Homer's paintings depict realistic Eastern landscapes and ocean scenes, Impression-istic views like *Schooner at Sunset* captured John's imagination just as they inspired a generation of modern American artists like Wyeth and his father, Nathaniel. Clement studied the watercolors of the younger Wyeth and learned that the drier Pennsylvania prairie and underlying abstractions in paintings like *Christina's World* held lessons in originality for photography of the arid Columbia Plateau grainlands. His unpeopled landscapes typically feature evidence of humanity's waning pres-ence—dilapidated barns and fences, retired farm machinery, and fields of maturing grain. John's ideas about the "saturating luminosity" of dawn and dusk suggest affinity with the nineteenth-cen-tury American Luminists and pioneers of color photography whose detailed agrarian views beneath soft, hazy skies engender feelings of melancholy and meditation.

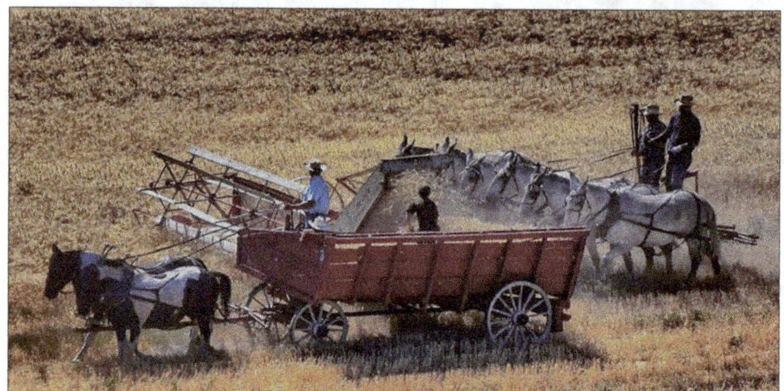
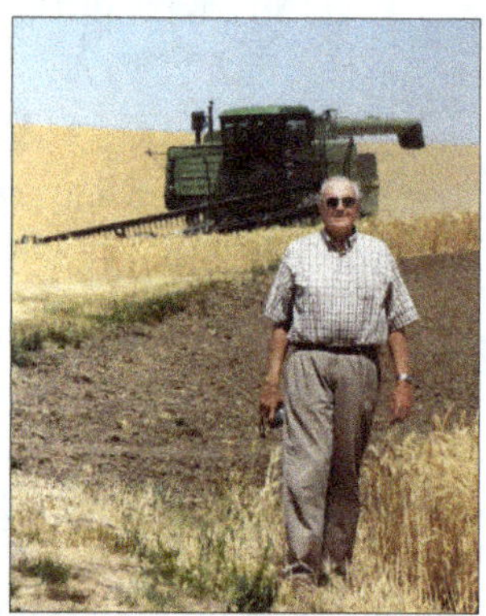

Left: John Clement, Palouse Empire *Threshing Bee* (2011)
Right: Bill Smick at Spillman Agronomy Farm Harvest, Washington State University, Pullman (1989)

Jim Howell's family has sold harvesting equipment in the West for a half-century and supplied helpful information and publications on the history of McCormick-International Harvester and the J. I. Case Threshing Machine Company. Longtime president of Inland Empire Milling Al Schauble, also a native of St. John, kindly shared from his considerable knowledge of regional grain processing and related commercial art since the company was founded by his grandfather in 1919. Stan Riebold of Winona has long been active in the Palouse Empire Threshing Association and fielded many of my questions about steam-threshing and horse-powered headers and wagons. Claiming that he was born a half-century too late, Stan and other members of the group rekindle the spirit of bygone har-vests in an annual September fairground threshing bee that celebrates the spirit of cooperation, labor, and pioneer innovation. Brothers Arvin and Paul Edstrom, sons of Endicott harvest equipment in-ventor Arthur Edstrom, kindly explained their father's remarkable work and other aspects of design and farm mechanization. My late Shetland Islander friend René Peschel recurrently shared from his vast knowledge of Old World farming to aid in understanding what is depicted in agrarian art from medieval and early modern eras. George "Judd" Robinson introduced me to Amish steam threshers of the Kenton, Ohio, area for whom religious faith and communal labor with draft animals remain vital ways of life.

I acknowledge the valued contributions Daniel Arp, Dean of Agricultural Sciences at Oregon State University; Shelley Curtis, art historian and curator of OSU's acclaimed Art About Agriculture

Collection of Fine Art; Ryan Hardesty and Anna-Maria Shannon of Washington State University's Museum of Art; Vivian Ross at the National Agricultural Museum Center for Rural Art in Bonner Springs, Kansas; Mark Moseman and Mark Mohler, Bone Creek Museum of Agrarian Art, David City, Nebraska; and Judy Mansfield, Museum of Art, Springville, Utah. Others who shared valued original research and translations include Judith Gray, American Folklife Center, Library of Congress; Brent Shaw, Princeton University; David Ruark, Eastern Washington Agricultural Museum, Pomeroy; James Payne and Groover Snell, Ft. Walla Walla Museum, Walla Walla; and Kenneth Robison, Montana Museum of Agriculture, Ft. Benton. I also thank Juleigh M. Clark, John D. Rockefeller, Jr. Library at Williamsburg, Virginia; Daniel Kempker, Chicago Art Institute and Ryerson Art Library; Wendi Sott, The Museum of Russian Art, Minneapolis, Minnesota; Cheryl Strichik, Monthaven Arts & Cultural Center, Hendersonville, Tennessee; and Rudolf Stock, Laubach Museum, Germany. My work also significantly benefited from published studies by art historians Liana Vardi, State University of New York at Buffalo; Richard Brettell, University of Texas—Dallas; Charles Eldredge, University of Kansas—Lawrence; and British historians George Fussell and John Barrell. USDA senior staff members Thomas Hoffman and Lillie Brady provided most valued access to and commentary on the department's substantial art collection.

The Schneidmiller Endowment for Palouse Regional Studies at Washington State University in Pullman and Anson Mills/Carolina Gold Foundation, Columbia, South Carolina, supported research for this study. I thank my longtime friend Gary Schneidmiller for his encouragement of my cultural interests ever since we resided at Washington State University's FarmHouse Fraternity in Pullman in the early 1970s, and Anson Mills founder Glenn Roberts for recurrent discussions on agrarian history as rich as Carolina red earth. Trevor Bond and Cheryl Gunselman at WSU's Holland/Ter-

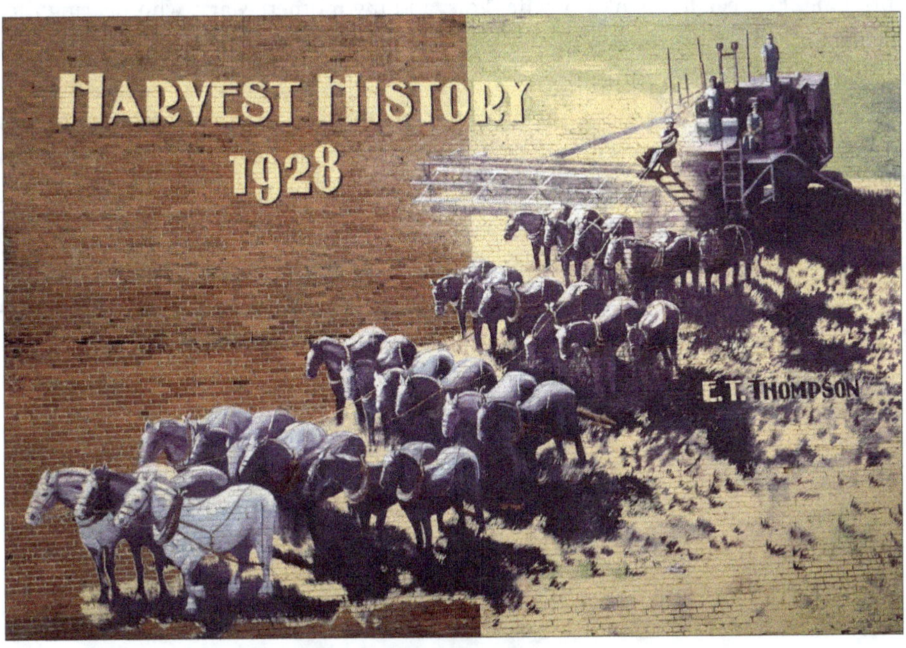

Sharon Bogen, *Harvest History 1928—E. T. Thompson* (2006)
American Legion Building Mural; Connell, Washington
Columbia Heritage Collection Photograph

rell Libraries' Manuscripts, Archives, and Special Collections have long provided valued assistance to my research endeavors. I thank Triticum Press founder/editor Carol O'Callaghan for her abiding encouragement and shaping of this volume and others in the Harvest Project Series into published form. Distribution of this book by Washington State University Press marks one of many titles on which I have been privileged to work with Linda Bathgate, Ed Sala, Kerry Darnell, Caryn Lawton, and Laurie Brown. Access to the Washington Association of Wheat Growers history files and art collection in Ritzville was kindly provided by Michelle Hennings and Trista Crossley.

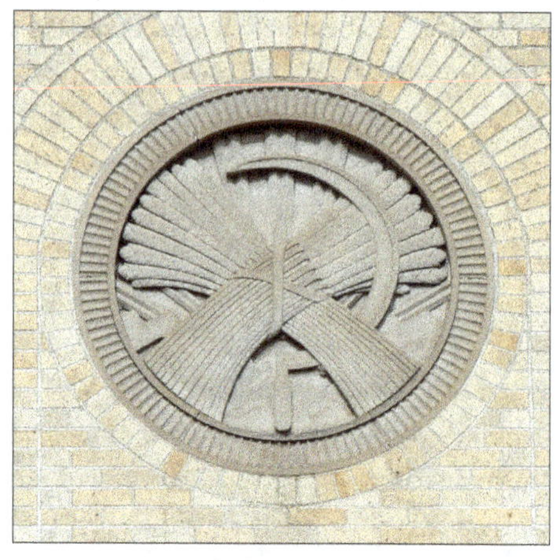

Sheaf and Sickle Art Deco Sculpture (1941)
Adams County Courthouse
Ritzville, Washington
Columbia Heritage Photograph

I owe special gratitude to the late farmer-scholar William Schmick who introduced me in the summer of 1989 to agronomists Stephen Jones, Steve Lyon, and the magnificent world of heritage grains at Washington State University's Spillman Farm near Pullman. I also thank Bruce LePage, Pat and Daryl Kleweno, Dennis Solbrack, Dwight Gibson, Mike Lust, Bill Woodward, Colette De Phelps, Jennifer Saunders, Jennifer Kilmer, Gary and Sandy Matthews, and Dale Daniel for their special contributions to this project. My brother, Don Scheuerman, and sisters, Debbie Wolfe and Diane Smith, have joined me for much of this journey and I am grateful for their abiding support and recollections of our shared rural upbringing. I especially thank my wife, Lois, and my mother, Mary, who nearing the century mark remains a most capable editor. They have sustained my far-flung travels to fields real and imagined that are explored in these pages.

Progressive change to promote well-being of the countryside and future generations can be unwisely limited by amnesia as well as nostalgia. Amnesia forgets about cultural legacies bequeathed by ancestors and society, while nostalgic appeals to life in an idealized past overlooks the sobering challenges of such times. We remember places, mark lines and verses, and appropriate elders' counsel for personal guidance, social cohesion, and to safeguard natural resources for future generations. To these ends the study and appreciation of agrarian art and literature remain vital endeavors for clarification of truth and experience. They provide visibility and voice for millions of self-reliant souls who have labored anonymously from seedtime to harvest throughout the ages to provision humanity and tell of creation and conscience.

Richard D. Scheuerman
Richland, Washington
April 1, 2024

INTRODUCTION

Association of the term "agrarian" (from Latin *agrarius*, "relating to a field") with agriculture, rural experience, and art derives from ancient Rome. In 133 BC the Roman tribune Tiberius Gracchus's controversial *Lex Sempronia Agraria* ("Agrarian Laws") attempted to redistribute public lands to the poor. The related word "pastoralism," from *pastoralis*, "relating to a shepherd," denotes tending livestock in open spaces but is also used more broadly in aesthetic portrayals of country life. The genre casts rustic folk in idyllic landscapes where they generally affirm the carpe diem tradition of romance and tranquility. Virgil's first-century BC *Georgics* owes its title to Greek *geōrgos*, "farmer," though the English noun form, "georgic," does not appear until the sixteenth century. Europe's population at that time reached approximately 100 million of whom some 95% inhabited rural areas. The peasantry had known little more than subsistence or abject poverty in the wake of epidemic disease, high incidence of war, and stratified social hierarchies that substantially precluded economic improvement. A typical eighteenth-century village in England would have numbered about three hundred souls with perhaps two-thirds subsisting as laborers, servants, the elderly, and other unpropertied poor. Only about one-third of the villagers were families of farm tenants and small freeholders.

An egalitarian sense of agrarianism appears in the eighteenth-century writings of Thomas Jefferson about yeoman farmers who transform the American wilderness for independent living and the foundation of freedom and democracy. Jefferson's erstwhile Revolutionary War friend Tadeusz (Thaddeus) Kościuszko returned to his native Poland homeland and led the 1794 Kościuszko Uprising against the imperial forces of Russia and Prussia. Devoted to democratic principles, the

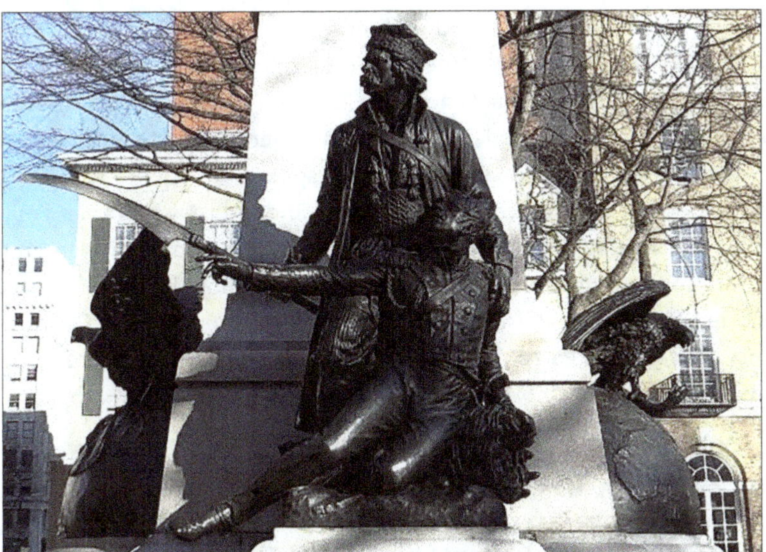

Antoni Popiel, Tadeusz Kościuszko and Peasant Scytheman Statue (west side, 1910)
Lafayette Square Historic District; Washington, D. C.
Columbia Heritage Collection Photograph

charismatic agrarian apostle's "scythemen" fought under the crimson *Zywia y Bronia* (They Feed and Defend) banner emblazoned with a golden sheaf. During the continental 1840s social revolutions, impoverished peasants in many parts of Europe famously wielded scythes and flails against their oppressive landlords. (Widespread failure of European grain and potato harvests significantly contributed to the era's social and political instability.) Association of the term agrarian with economic and political struggle endured into modern times.

In *Agrarianism in American Literature* (1969), M. Thomas Inge identifies key tropes of agrarianism to include reciprocity (mutuality of healthy rural communities), romance (redemption through natural harmonies), and religion (farmer reliance upon God and nature). These elements have long been expressed in art and literature that inform present considerations of rural challenge and environmental sustainability. To others, like novelist-historian Saul Bellow, the rural American experience "has had a long history of overvaluation," with notions of self-reliance and fairness mixed with considerable unhappiness, alienation, and provincial pride. Historian Richard Hofstadter further observed that despite the Jeffersonian yeoman-citizen ideal, American farmers generally prized personal commercial interest above rural community wellbeing and conserving the soil upon which they depended. Hofstadter identifies two phases in the nation's fundamental shift from subsistence agriculture to a market economy. These developments, argues Hofstader, were abetted by farm mechanization and rural literacy. The first phase involved federal support under Washington and Jefferson for frontier roads and national expansion, followed by the Lincoln era Homestead and Morrill Acts, first transcontinental railroad, and creation of the Department of Agriculture. These perspectives reveal the enduring contradictions of America's rural experience and appear in generous works of art and literature with depictions of isolation and melancholy as well as optimism and muscularity.[1]

Emmanuel Leutze's grandiose mural *Westward the Course of Empire Takes Its Way* (1861), painted for the U.S. Capitol within a year of these legislative achievements, presents a romanticized conception of a unified nation's Manifest Destiny. Settler hardship is juxtaposed against the dawn of progress and medallions beneath the main painting show farm and mining tools. Criticisms of agrarian moral commitments by lifelong urban academicians like Bellow and Hofstadter, who presumably benefited from rural provision, eschew factors like isolation and backbreaking labor that affected pioneer self-sufficiency. Several cultural-political trends emerged in late nineteenth-century American history that significantly influenced discourse on agrarian aesthetics and politics. Traditional social structures and values confronted the rise of cities and industry. Agrarian Populism appeared in 1880s and '90s America with political leaders like William Jennings Bryan and Thomas Watson advocating foreclosure reform and federal regulation of monopolistic railroad and grain storage rates. These efforts led to the establishment of the National Grange of the Patrons of Husbandry, Farmers' Alliance, and People's Party "Populists." Walt Whitman lauded the prospects for a Midwest of prosperity and diversity in his short poem "The Prairie States" (1881):

> *A newer garden of creation, no primal solitude,*
> *Dense, joyous, modern, populous millions, cities and farms,*
> *With iron interlaced, composite, tied, many in one,*
> *By all the world contributed—freedom's and law's and thrift's society,*
> *The crown and teeming paradise, so far, of time's accumulations,*
> *To justify the past.*

Much has been written about rural America's thrift and parochialism with limits on personal freedom. Currier & Ives's idealized depictions of country life in views like *The Wheat Field* and *The Farmer's Harvest Home* (c. 1860) and *Harper's* woodcut illustrations obscure social and racial inequalities, burdensome toil, and rural isolation. Similar charges have been made against twentieth century American Regionalists like Thomas Hart Benton and Grant Wood who were products of Whitman's "newer garden of creation" and whose agrarian scenes generally depict stolid, hardworking Whites. But vast agrarian America has been a place of many places. Paintings by Benton's and Wood's fellow Regionalist, Missouri native Joe Jones, also show Black fieldhands and less favorable interpretations of farm mechanization. Historian John Lauck observes that for all its imperfections the Midwest has been among the most democratic and socially progressive sections of the nation. Rural America has been home to flourishing public schools and land-grant universities, and bequeathed to the citizenry a spirit of Lincoln-era egalitarianism that still motivates. Twentieth-century repression of progressive farm labor and reform movements and nativistic abuse of rural minorities found willing acolytes throughout the United States but did not arise from the countryside. Presidential candidate Woodrow Wilson's efforts as a progressive visionary were undone by the country's entry into World War I in 1917 and embrace of industrial favoritism, postwar global trade, and attacks on American labor.[2]

Country Life Progressivism had widespread support throughout the West as social reformers, church leaders, and educators concerned about rural stagnation joined to implement "modern" solutions. These efforts led to the 1914 Smith-Lever Act for educational extension to rural counties through Department of Agriculture partnerships with state universities and for implementing cooperative commodity marketing strategies and training in farm mechanization. Each of these movements drew reaction. Populist candidates initially found voter support in the countryside but came to be associated with socialist agendas; Neoclassical Arcadians were dismissed for overlooking the gritty realities of rural poverty; and some farmers viewed USDA "efficiency planning" as an assault on time-honored custom and lifeways.[3]

"Purpose of Existence"

The ubiquity of people's familiarity with agrarian scenes, labor, and traditions throughout the world since time immemorial is evident in a wide range of artifacts, art, and literature. This vast realm of evidence has rendered aesthetic interpretations of harvest in greatly varied ways. The longstanding popularity of the harvest theme from ancient to modern times, throughout both East and West, has contributed works that range from sublime to exceedingly hackneyed. Yet these attest in the main to a conviction that beauty, cooperative endeavor, and remedies to cultural and environmental threats are moral imperatives. Cultural anthropologist J. Katarzyna Dadak-Kozicka observes that since time immemorial harvest "was essentially the purpose of existence," and that field labors had a latent contemplative and spiritual dimension commemorated through art and rituals that long formed the basis of culture. The original meaning of culture is the tending of crops, from which the terms agriculture and cultivate are derived, as well as the cognate word coulter, or blade of a plow or seed drill. Only with the deepening influence of liberalism and individualism in the eighteenth century did culture more broadly manifest itself as both personal expression and commodity in ways less connected to agrarianism.

In many postmodern circles culture is characterized by materialism, fantasy, and spectacle. But a century ago journalist Alfred Henry Lewis offered a sobering practical corollary to humanity's ultimate agricultural dependency: "There are only nine meals between mankind and anarchy." The unprecedented pace of social change since industrialization has shifted populations from the coun-

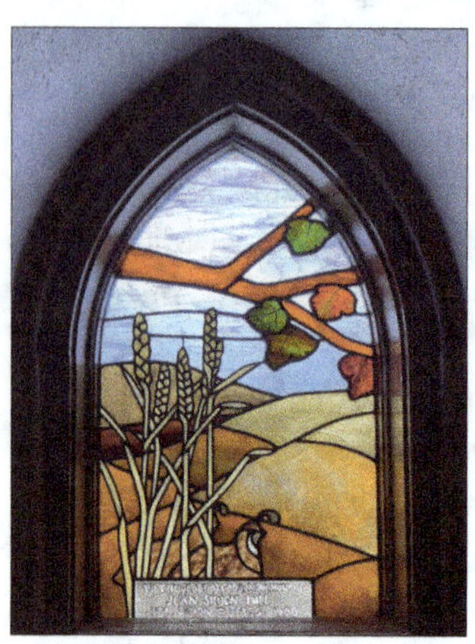

Grain and Branches
Stained Glass Window (1986)
Redeemer Episcopal Church
Pendleton, Oregon
Columbia Heritage Collection

tryside to cities and distanced human connections to nature. For many generations farm work has required intimate knowledge of natural systems and long hours of hard physical work whether using human, animal, or mechanical power. These demands have fostered improved tillage methods to increase crop yields and ingenious labor-saving inventions. But such developments have inexorably if irregularly distanced societies from their fundamental reliance on the wellbeing of the land.

The term "harvest" has often been invoked as a quaint synonym for agrarian bounty or some distant ingathering of crops. Throughout the course of civilization, however, harvest has determined sufficiency or want, been the subject of endless anxious speculation throughout the seasons, and in many times has been a matter of life or death. *"Give us this day our daily bread."* British scholars note significant social dislocation and political instability associated crop failures in England (e.g., 1481-1482, 1555-1556, 1596-1597), which were usually caused by late rains and resulted in yields of less than 50% of normal production. Periodic "harvest dearths" of such magnitude have been a significant factor in human migrations. In modern times nations have established storage facilities and enacted multilateral policies to ensure food supply resiliency. Yet annual harvests remain the heartbeat of national economies in the twenty-first century and are increasingly at risk from food processor consolidations, climate change, and degradation of vital soil biomes.[4]

In continental and global contexts imperialism originated, and endures, in the quest for the most coveted natural resource—harvested foods. Various ideologies have been formed since ancient times to justify the conquest. Substantial Roman grain ships transported wheat from Egypt, North Africa, and Sicily; medieval European traders tapped the fertile Great Hungarian Plain, Rhine-Mosel Valley, Paris Basin, and Ukraine's "Black Earth" district. Rural colonizers in the modern era transformed the American heartland, Argentine pampas, and western plains of Australia's New South Wales. Populations of many contemporary societies are preoccupied with various commercial and secular endeavors and take a dependable and diverse food supply for granted. But this confidence belies serious risks, and public concern has been expressed in recent sustainability movements and examinations of exploitive geopolitics.

Harvest Labors

The long days of summer grain harvest begin before dawn with a serene vitality soon witnessed at first light as described by Nebraska author Willa Cather: "Every morning the sun came up a red ball, quickly drank the dew and started a quivering excitement in all living things. In great harvest seasons . . . , the heat, the intense light, and the important work in hand draw people together and make them friendly. Neighbors helped each other to cope with the burdensome abundance of nourishing grain; women and children fell to and did what they could to save and house it."[5] Labor traditions and rituals bequeathed from ancient and medieval times changed significantly with the advent of post-World War II technologies but still requires fundamental responsibilities for individuals, families, and communities. Period literature, art, diaries, and harvest lore passed along through oral histories attest to the season's abiding significance.

For generations the harvest of cereal crops throughout the world has marked a momentous time that marshalled individuals of all age and backgrounds who gathered for a series of essential endeavors that were generally ordered by gender, experience, and status (i.e., landowner vs. hired laborer). Although work rituals and laborer hierarchies have been influenced by changes in technologies from the eras of sickle and scythe to reaper and combine, four core tasks generally separated by gender have long defined harvest labor. Responsibilities typically, but not exclusively, taken by males has included reaping, carting, threshing, and storage of grains. (Before the advent of mechanical reaper-binders in the early nineteenth century, young women commonly followed hand-reapers to gather the rows of cut stalks into bundles.) When combines appeared in the late 1800s, reaping and threshing took place in a single operation without need for binding.

In early medieval times carting involved transport of grain stalks from the field on racks and wagons to a barn where hand-threshing by flails could take place during fall and winter. Carters, later known as header-box drivers and by other terms, hauled reaped stalks to steam- and combustion engine-powered stationary threshers made by manufacturers like McCormick, Deere, Case, and Advance-Rumley. Today's carters are among the busiest field workers who transport threshed grain from high-capacity combines on bank-out wagons to large tandem-trailer trucks. Proper grain storage has long been essential to protect the harvest. This task has required other workers and record-keeping to safeguard delivery of bulk grain for farm storage or to immense "country castle" elevators where care is taken to protect the crop from damage by vermin, moisture, and malfeasance. For each of these roles, physical strength and mechanical skills are core elements of the harvester work ethic that are expressed in period agrarian art and literature.

Female harvest labor traditions and rituals consisted of four central roles essential to the season's success—provisioning, cooking, boarding, and childcare. Women generally supervised provisioning of the farm larder as garden and orchard produce were gathered and preserved by canning and in earthen root cellars. In the hilly Palouse these structures were commonly attached to the house by a porch. A farm couple's older daughter was often assigned to supervise garden weeding and watering by younger members of the family while older boys joined their fathers in the field. Women also milked cows, collected eggs, and tended livestock during harvest as well as other times of the year, and procured groceries from local stores. Long hours of intense labor generated enormous appetites for harvest crews that could numbered from eight to two dozen men. One of the surest ways to keep reliable workers was to ensure all were well fed with plenty of fresh meat and potatoes, vegetables, and fruit. Whether in farmhouse kitchens or portable cookshacks with wood-fire stoves, women toiled from predawn hours to feed the workers up to five times each day—breakfast by 6 AM, mid-morning lunch break, dinner at noon—perhaps the most memorable in harvest folklore, afternoon lunch break, and supper after 8 PM. The men sometimes ate in shifts on expandable tables inside,

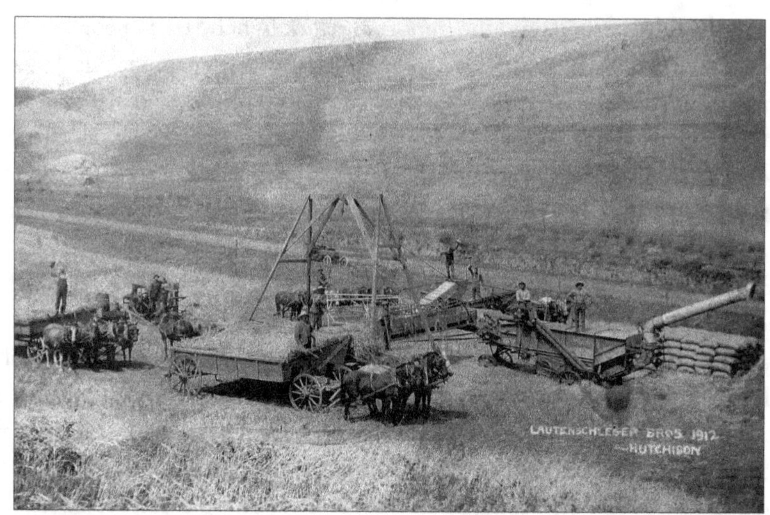

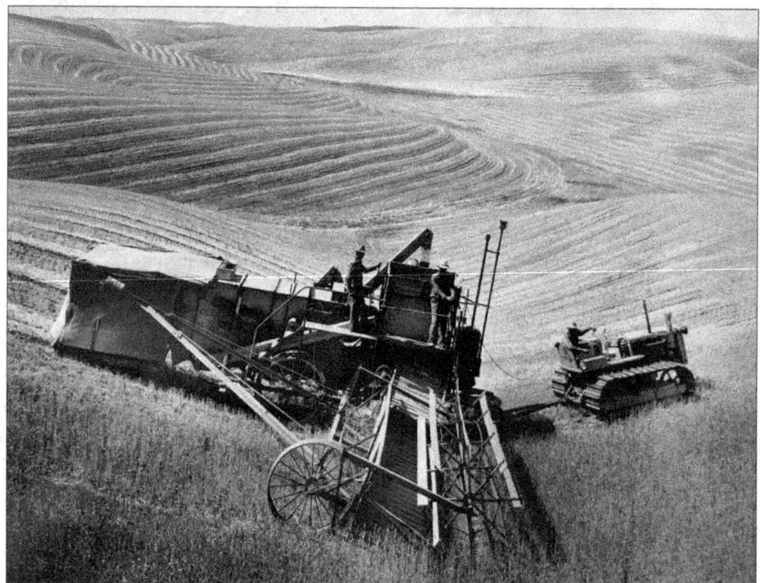

Above: (Philip and Adam) "Lautenschlager Bros." McCormick Header and Case Thresher Outfit (1912)
Between Endicott and St. John, Washington; R. R. Hutchison Photograph
Below: Adam P. Morasch John Deere Pull-Combine and Caterpillar Tractor (1941)
Between Endicott and LaCrosse, Washington; Columbia Heritage Collection
(The lower photograph was featured in Wayne G. Broehl, Jr.'s comprehensive history,
John Deere's Company: A History of Deere & Company and Its Times **[1984]).**

picnic-style on makeshift tables outside, or on long narrow benches and tables in the cookshack behind screened windows.

Meal quality and portions were a source of considerable pride to farm wives and other women who assumed this responsibility and routinely served food on regular dinnerware. Willamette Valley pioneer farm wife Mathilda Siegmund Jones recalled late nineteenth-century harvest fare preparations: "Almost an entire mutton was used daily, besides ham, bacon, and quantities of eggs. Bread was baked twice a day, nine large loaves at each baking. Ten or more pies were baked daily for both breakfast and dinner, and dozens of cookies for supper. There was fresh applesauce or can fruit for every meal and Royal Anne cherries canned in half-gallon jars. . . . Two kinds of pickles, potatoes, gravy, and another vegetable at each meal with butter, plenty of milk and coffee, cream, sugar, and often a salad of cabbage, lettuce, or potatoes."[6]

In addition to these responsibilities, women lodged threshing crews by making clean spare rooms with beds available to workers in the main house, cool basements, or nearby bunkhouses. Itinerant workers sometimes packed bedrolls to sleep in barns or outside. Wash basins, towels, and pitchers of lemonade with fruit pulp and coffee pots were regularly available for field workers. Younger girls were often assigned some of these ancillary tasks and also cared for smaller children under the watchful eye of mothers and older women. Grant Wood's painting *Dinner for Threshers* (1934) and Ben Shahn's many Depression era harvest dinner photographs depict the importance of women's contribution to threshing labors and the fostering of rural hospitality and accomplishment.

Harvest as Aesthetic Motif

Art is in the soul of the creator before it reaches the eye of the beholder, and reflects profoundly personal responses and representations of ideas, external influences, and inner experience. In some cases these depictions promote a prevailing civic virtue, while in others they suggest a vague oth-

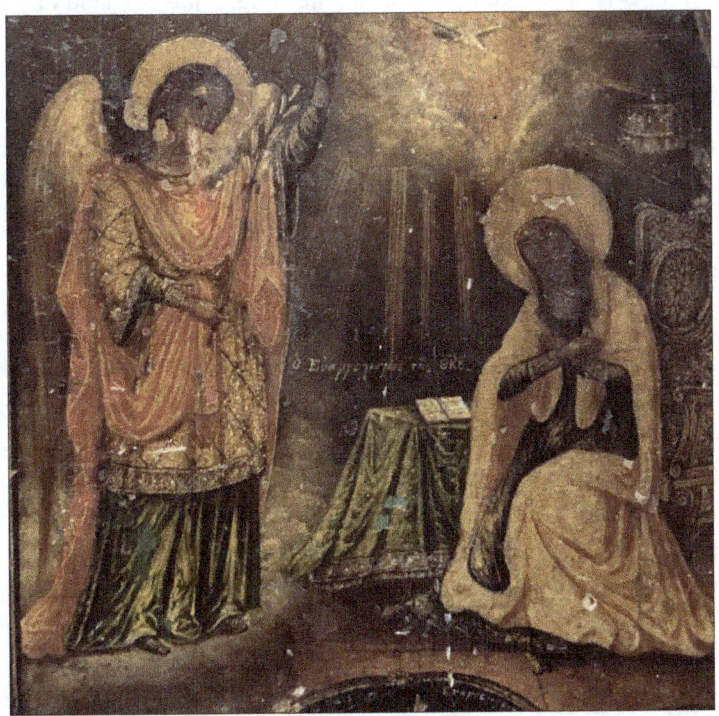

The Annunciation Icon (detail, c. 1805)
(Archangel Gabriel holding a cluster of grain before the Virgin Mary—Luke 1:26-28)
Sitka Centennial Visitors Center Museum; Sitka, Alaska
Columbia Heritage Collection Photograph

er-worldliness associated with ill-defined shapes, and described as visionary, or mystical. But these latter terms are derived from words that mean "clear-eyed," and "to reveal," as in great art's expression of a truth or sensibility. Since ancient times the authors of vernacular literature have been called "seers" for composing stories in various forms, rhymes, and meters as a bridge between traditional knowledge and prophetic wisdom. Throughout time artists and writers have risked deep personal exploration and public expression of understandings in a wide range of forms and styles that reach beyond literal depiction. Many experienced extremely difficult circumstances and were often unacknowledged in their own day. In an age when artists and writers seek formal study to achieve proficiency in skill and technique, cultural historian Jacques Barzun observed that true greatness also involves solitary toil, extraordinary self-confidence, and recurrent disappointment. Farmers have also long identified with such traits.[7]

Artists and authors create various expressions of reality and imagination for the sake of the works themselves as well as for pure enjoyment and seeing anew. Aesthetics also relate spiritual and moral truths that evoke a range of emotional responses variously considered beautiful, disturbing, ambiguous, and transcendent. (In the Eastern Orthodox tradition of icon painting beauty is transcendence, a contemplative "window" to spirituality that can be experienced if not rationally explained. Creativity is understood as fundamental to the divine image as related in Genesis 1.) At its heart, these are expressions of love, or reactions to its absence. The peculiar shapes and shades of abstract art in idiosyncratic combinations of commonplace objects may express universal themes about life and struggle, or inner perceptions with no point of external meaning. Louis Menand's monumental studies of contemporary culture explores the cynicism seen in some prominent modern art and literature, and the clownishness of creators disillusioned by post-war experience and commercial demands of self-promotion. These considerations relate to Max Weber's notion of modernity as disenchantment in the face of global threats to societal wellbeing and ambivalence over national claims of progress. With these changes have come new settings for public art beyond traditional religious and cultural institutions to include structural murals, field installations, and an array of electronic domains. Art has never felt so physically scattered while serving such varied purposes.

The commercial "culture industry" has compromised many selfless quests for truth and beauty, while endless foreign conflicts, growing economic disparity, and climate disruption replace vision with disillusion in some post-modern writers like Charles Simpson. Agrarian perspectives, which may be far removed intellectually from sophisticated cultural conversation or gallery receptions for disparate canvases of cork and trash, offer rich humus for inquiring and troubled souls. Many works by agrarian artists and authors make essentially conservative cases on behalf of vulnerable lands and people. They serve as mirrors to view judgments on culture and nature in times of unprecedented change and uncertain futures. To self-taught African-American Alabama painter and sculptor Charlie "Tin Man" Lucas, who cites the influence of his country blacksmith great-grandfather, art "is about building a legacy of knowledge." Degas sought a "well-ordered whole" in which no stroke on the canvas would appear accidental, while Baudelaire considered artfully crafted writing an efficacious force of highest spiritual power.[8]

Great art of any age is authentic and autonomous and can present multifarious perspectives about truth and beauty on the same painted surface or within a single story or musical composition. Many artists' or authors' intentions are provisional given the primacy of a creational journey above any destination. To the extent some finished works come to represent something essential and enduring, they may also be considered classic. Baroque master Pieter Bruegel's ambitious *The Harvesters* (1565) presents a grand and complex scene of toiling reapers and exhausted mealtime workers against a Belgian village background crowned with a church steeple. (Bruegel created several large-format versions of the same scene with subtle differences in each painting.) In works like *Wheat Fields* (c. 1670), Dutch Golden Age artist Jacob van Ruisdael used agrarian landscapes to explore the dynamics of depth and surface, and light and color for grand views that impart humility in the wake of sublime creation. The small ancillary forms of laborers and livestock that painters call staffage reveal much about agricultural history.

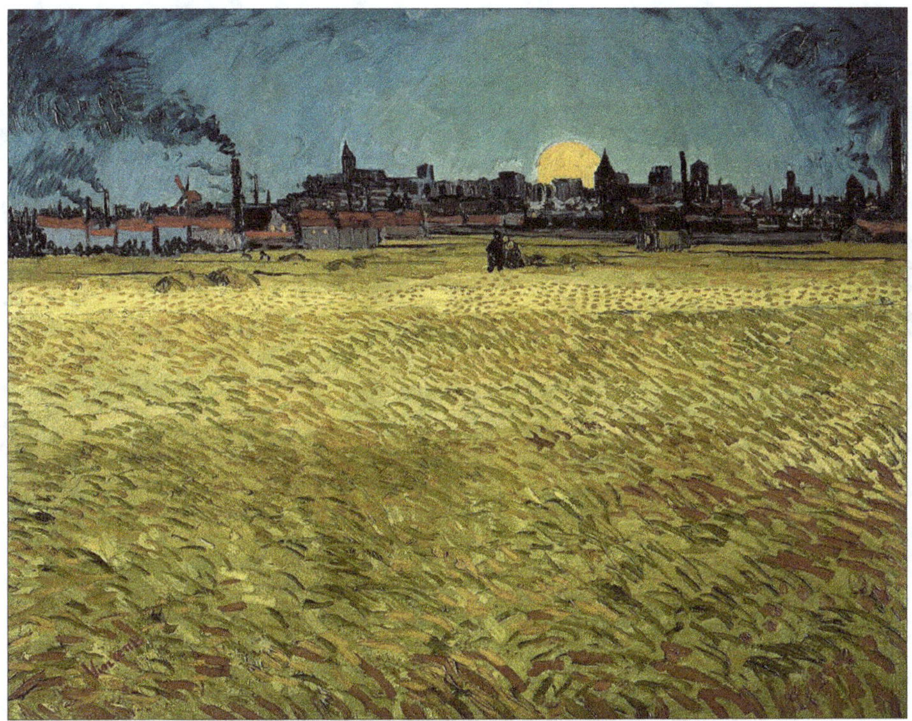

Vincent van Gogh, *Summer Evening* (1888)
Oil on canvas, 30 x 36 ⅓ inches
Kunstmuseum Winterthur, Switzerland

Harvest field labors foster myriad memorable sensory experiences that artists, authors, and composers have long sought to express—vigor and exhaustion, solitude and chaos, boredom and beauty. Winslow Homer's *The Dinner Horn* (1870), popularized as a contemporary illustration by *Harper's Weekly*, depicts a young woman summoning distant harvesters and suggests respite, refreshment, and a flavorful ample meal. American composer Charles Ives's unconventional "Song for Harvest Season" (1894) includes the sharp report of similar sounding horns along with the thrum of scythe-like percussion. Anton Chekhov's evocative novella *The Steppe* (1888) renders the reaper brigade's rhythmic "vzhee, vzhee" that would be drowned out today by the baritone roar of a diesel combine engine.

The dramatic range of spectacular harvest color is expressed with remarkable precision in one of van Gogh's many letters to painter Émile Bernard. Writing from Arles in southeastern France in June 1888, he tells of reveling in the fields "like a cicada" and includes a sketch for what would become another masterpiece in his harvesttime series, *Summer Evening* (1888). "... [T]he wheatfields have all the tones:" van Gogh informs his young friend, "old gold, copper, green gold, red gold, yellow gold, green, red and yellow bronze."[9] To this palette might well have been added other metallic tones van Gogh was known to use for dramatic landscape highlights and the distinctive sheen of final grainfield ripeness— lead white and chrome yellow. His genius for daring color combinations is seen in *Summer Evening* features his characteristic bold pairings of golds with raw sienna, red ochre, and Prussian blue.

Variety in Harvest Expression

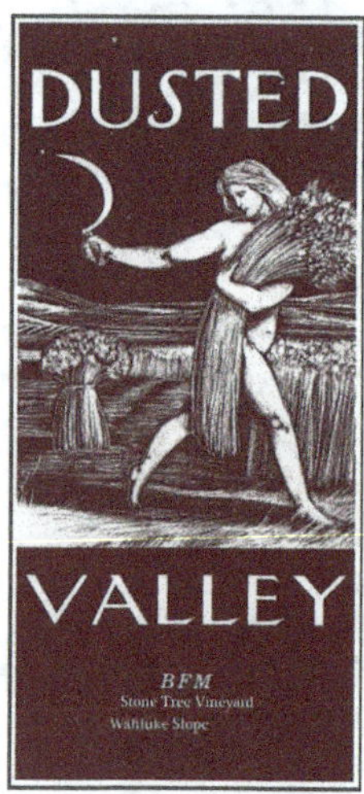

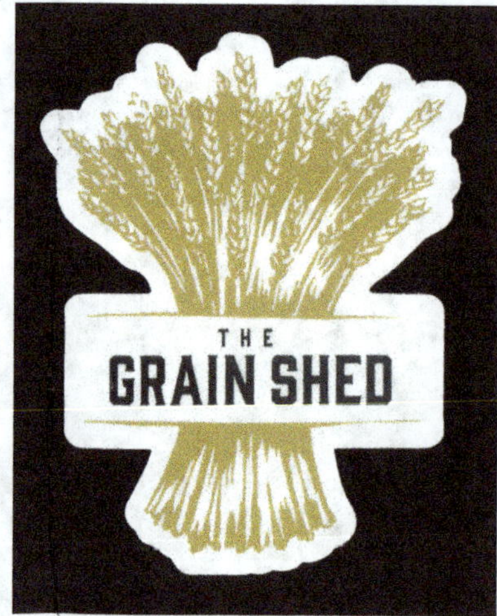

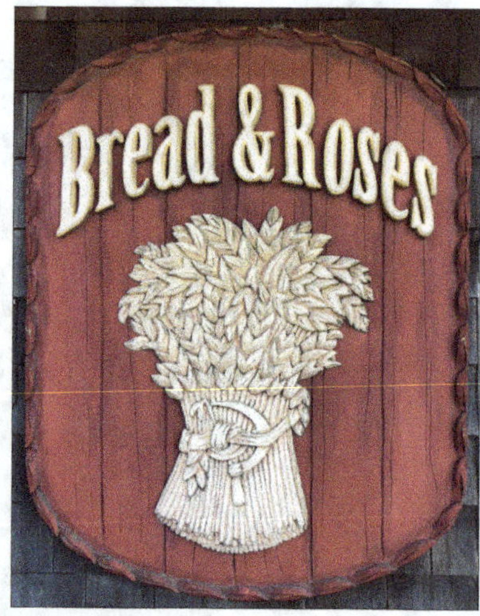

From left: Squire Broel, Dusted Valley Ceres Winery Banner
(after Rockwell Kent; Walla Walla, Washington
The Grain Shed Bakery and Taphouse Logo; Spokane, Washington
Bread & Roses Bakery Sign; Yachats, Oregon
Columbia Heritage Collection Photographs

Thomas Hardy's *The Mayor of Casterbridge* (1886) relates the numbing conditions of rural labor in Edwardian England and also questions assumptions of modern industrial progress. Currier & Ives are known for their colored lithographs depicting sentimental views of country life tacked to the parlor and bedroom walls of clapboard farm homes that bore little resemblance to the idealized scenes. But prints in the series like Fanny Palmer's *The Farmer's Harvest-Home* (1864) also offer commentary on aspirational values from the time and provided opportunity for female artists to participate in the arts and enterprise.

In rare instances rural laborers with aesthetic inclinations managed to find advocates for their abilities who made possible their publication and participation in art exhibitions. Many of German Romanticist Caspar David Friedrich's spectacular nineteenth-century landscapes would likely have been lost, or never created, had it not been for the unexpected encouragement of Romanov tutor Vasily Zhukovskly that Nicholas I become his patron. Northampton peasant poet John Clare published a collection of poignant verse in *A Harvest Morning* (1820) that captures joys and miseries of country life reflecting his own experience. Often first scribbled on paper scraps and even bark, Clare's writings contain vivid descriptions of English landscapes and fieldworker conversation that give authentic voice to the brigades of reapers and binders seen in paintings like *The Harvesters* by Peter De Wint (1784-1849) that he so admired.

Despite blindness and village obscurity, nineteenth-century Russian blind "poor girl poet" Domna Anisimova (b. c. 1810) dictated lyrical verse on harvests and rural life to the local priest after he had read poems, some coincidentally also by Zhukovsky, to his reclusive parishioner. Rare advocacy from generally class-conscious provincial officials eventually led to academy patronage for Anisimova. Behind the pathetic figures in Jean-François Millet's *The Angelus* and *The Gleaners* (1857) to those who pitch barley bundles in John Steinbeck's *Of Mice and Men* (1937) are the poor and downtrodden who have endlessly toiled in anonymity through the ages for subsistence and others' wellbeing.

The reciprocating influences of agrarian art and literature offer important understandings to a contrasting complex of cultural ideas in a day when humanity wields capacity for irreversible environmental damage that threatens food resources long taken for granted. Restorative influences evident in agrarian aesthetics include fulfillment and routine in rural labor, individual and cooperative endeavors, the facts and fictions of life on the land, and consideration of technological impacts. These topics warrant study in a day of increased distance between rural landscapes and societies that ultimately depend on their bounty. While traveling in Singapore, for example, composer Joseph Curiale, for example, viewed a telecast about the farm country of America's Midwest. The segment included verse from Pulitzer Prize-winning Nebraska poet Ted Kooser about meadowlarks, hollyhocks, and hands "gliding lark-like over the wheat." After his return to the States and a visit to Nebraska, Curiale moved there from Los Angeles to find inspiration for his acclaimed compositions

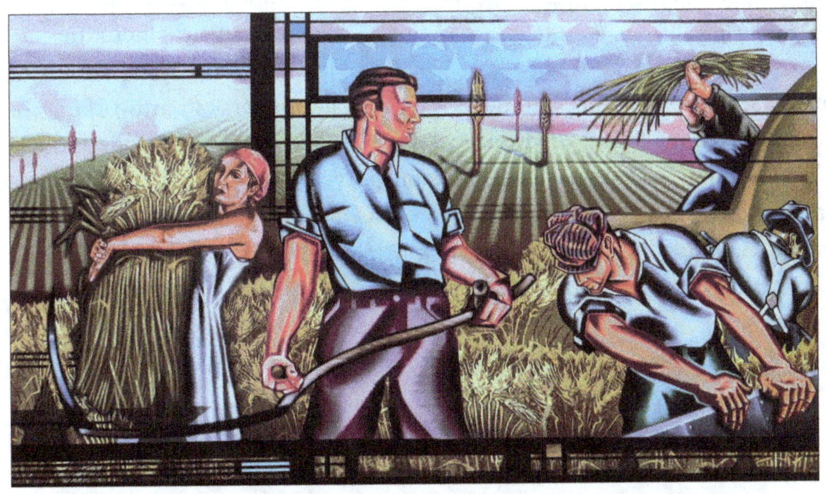

Martin Webb, after Mark McIntosh, WPA-Style Harvest Mural (2012)
Fabric print of acrylic mixed with sand, 9 x 16 feet
BJ's Restaurant; Redmond, Washington
Columbia Heritage Collection Photograph

"Awakening" (1997) and "Prairie Hymn" (2001). More recently botanist Ive De Smet from Ghent University teamed with Belgian art historian David Vergauwen to formulate iconographic analysis of ancient crops using period depictions in painting and sculpture. Their studies have revealed important insights about physical changes and distribution patterns of grains in ways that have long influenced consumption and appreciation for rural landscapes.[10]

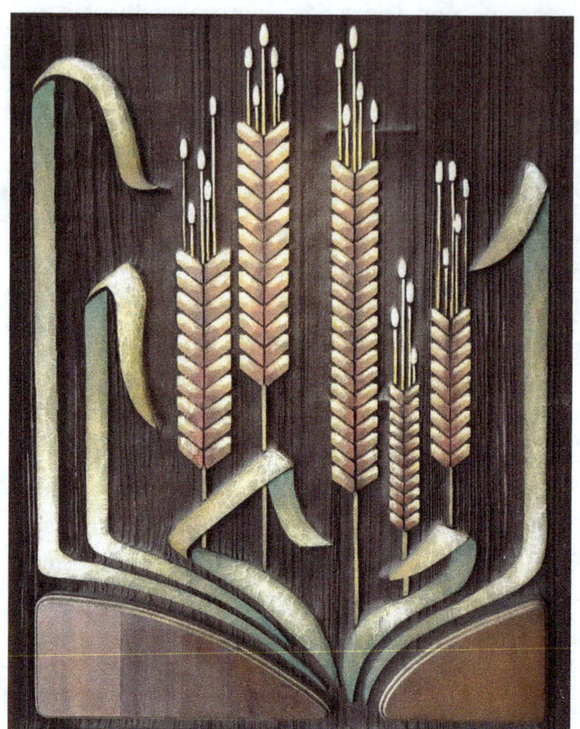

Pietz-McClaskey Woodshop
Stylized Grains Cedar Sculpture (c. 1975)
Red Lion Hotel; Pasco, Washington

Artists and writers of agrarian themes present an ever-unfolding legacy of understanding, beauty, and mystery to foster human flourishing against threatening forces of ignorance and disregard of nature. Alienation from land and place contributes to alienation from one another and from community. Yet progressive social change and mechanization have also improved myriad aspects of country life. Early modern poet Stephen Duck expressed unvarnished truths of mundanity and rural toil in *The Thresher's Labour* (1730) just as British innovators were on the cusp of developing the first mechanical harvesters that greatly reduced demands for grueling physical labor. These innovations led to modern combines that can be appreciated for beauty in form and function as have sickles and sheaves since time immemorial.

Agrarian art, literature, and music are rich in complexity and provide ongoing conversations on shared experience and the rescue of vast numbers of hardworking souls from historical erasure. Nineteenth-century stories by Anton Chekhov like *The Steppe* and verse by Italian Symbolist poet Giovanni Pascoli serve both this purpose and another that characterizes great literature: fostering appreciation for the simple gifts of life amidst little noticed daily experience—idle conversation among family and friends around a wholesome meal, the silent splendor of a harvest moon, or a grain sheaf's pleasing shape.

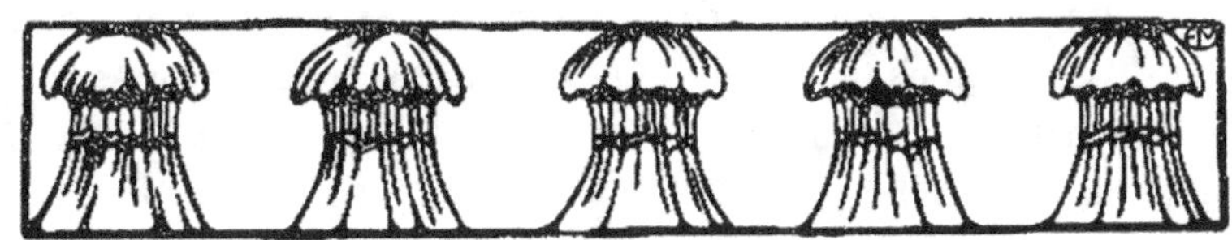

I

The Bread and the Wine
GRAIN SCYTHES AND ALTAR CLOTHS

In a weathered outbuilding on our small palouse country farm in southeastern Washington, my parents kept a substantial scythe in semi-retirement that leaned against the wooden wall. A small sickle with a fluted blue handle hung nearby on a rusty nail. The larger implement, pronounced *sigh*, appeared only when barnyard grass and weeds needed to be cleared, or when our father determined that my brother and I had unnecessary time on our hands. We didn't have a powered sickle-bar mower like some of our neighbors, though a dilapidated reaper-binder from our grandfather's day rested steadfastly in the low branches of a hillside cherry tree just above our house. The scythe's smooth, serpentine handle seemed taller than I stood so was awkward to hold. But Dad could adjust the weathered grips, which he called nibs, and put a keen edge on the sword-like blade with a whetstone. Then might come a refresher lesson in swath and rhythm before he would send one of us out to battle against formations of wheatgrass and Jim Hill mustard. "There's a right way, and a wrong way," was Father's familiar refrain while demonstrating many tasks like this one for our rural edification. As if performing some peculiar country dance, he showed us how to keep the ancient tool's heel down with each measured step to avoid sticking the blade's point into the ground.

The ancient hillside contraption married to the cherry tree had a wooden reel, drive chains, and faded green sprockets. It was an object of endless boyhood fascination. Dad had come of age during the Great Depression and recalled scenes of reapers and grain shocks dotting the countryside like those captured by famed Farm Security Administration photographers Arthur Rothstein and Dorothea Lange in the 1930s. Our binder's corroded carcass had been abandoned long before Dad took over the farm in the 1940s, but he somehow mastered the elegant art of crafting decorative sheaves. I remember well his patient evening lessons at the kitchen table teaching us how they were made. After careful arrangement of small, tightly tied clusters of stalks and bristled heads, the brittle forms formed up against each other. This eventually created a beautifully proportioned convex mass that crowned the column of lustrous, upright stems that was slightly tapered at the waist. We entered a sheaf each year in the Palouse Empire Fair where it stood proudly with similar arrangements of club wheat, bushy oats, and bearded barley. To this day the fair still features grain sheaf and "Wheat King" kernel-quality competitions along with folk art displays of country quilts and seed mosaics. I remember the sheaf's pleasing symmetry, and other arrangements of grain festooned with ribbons and placed in glass vases to adorn our church sanctuary in the fall. Whitman County's Cheryl Es-kelson was crowned Washington's first "Wheat Queen" in 1966 by the state Association of Wheat Growers as a promotional distinction for young women that harkened back to the Harvest Queens of Early Modern Europe.[1]

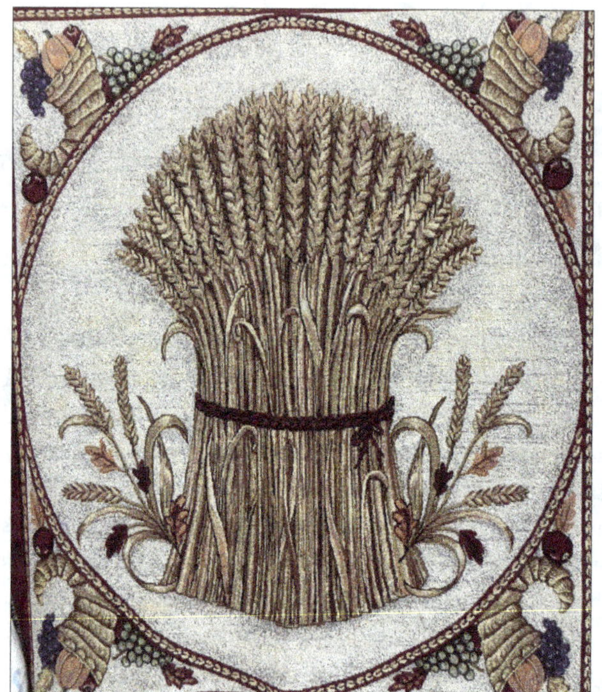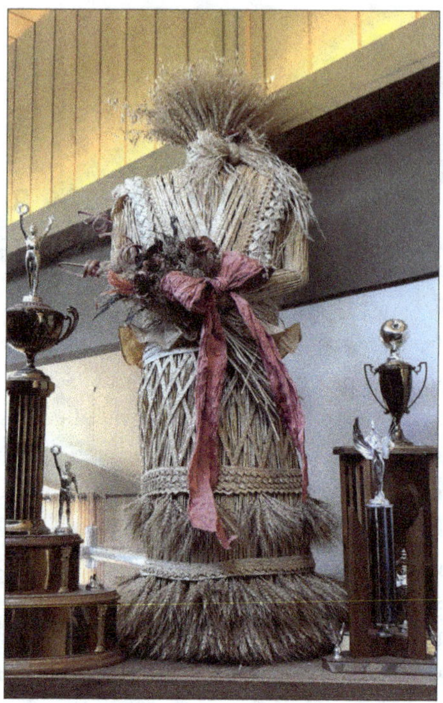

Left: Pataha Flour Mill Exhibit Tapestry (c. 2000)
Garfield County Fair, Pomeroy, Washington (2023)
Right: Aileen Warren and Jackie Penner, *Wheat Lady* "Corn Dolly" (1997)
Washington Association of Wheat Growers; Ritzville, Washington

In those bygone times Harvest Queens customarily carried an elaborate "Corn Mother," or "Maid-en," fashioned from the last sheaf and festooned with colorful streamers and ribbons. In parts of northern England and Scotland the honored last bundle was known variously as the Mell or Kirn (Corn) Mother, Child, or Doll (Irish "Churn"). Similar reference to the Rye Mother, Wheat Mother, etc., took place among many rural peoples in Russia, Poland, and elsewhere in the Slavic East. The designated "Harvest Lord" drove the queen and last sheaves home by wagon accompanied by the band of mirthful field workers. Traditions honoring the grain's vitality perpetuated the Old World craft of wheat-weaving with artfully twisted shapes are still featured at county fairs throughout America. Agrarian folklorist Rene Peschel traces the origins of Northwest wheat-weaving to the 1974 centennial commemoration of Mennonite immigration to Kansas from Russia. Midwestern Mennonite women wove mementos with Russian "Turkey" Red wheat and the following year Mo-ses Lake resident Phyllis Franz learned the skill from a Mennonite visitor the area and taught it to members of her fellowship and other friends. One of the most spectacular examples of the craft is the life-size "Wheat Lady" (1997) by Aileen Warren and Jackie Penner of Dayton, Washington. The pair wove a grain dress from 225 feet of wheat straw and embellished the effigy with over 900 hand-tied decorative knots and 2500 heads of wheat.[2]

Our Eastern European immigrant elders recalled *Kirchenweih* (church consecration) services in which villagers took part in the *Kerb* (*Kärwa, Kerwe*) "fun-fairs" lasting several days following harvest that featured feasting and youthful revelry. The affair was not generally condoned by village pastors and priests, so churches and schools were off limits to the merry-makers. Festivities usually commenced after worship on an October Sunday as residents gathered in the village square or before the town hall to march down streets singing *Gasse Lieder* (street songs) on their way to appointed residences that hosted musicians with horns, fiddles, drums and melodious hackbrett (hammered dulcimer). Traditional galloping dances, waltzes, and mazurkas commenced amidst enormous spreads of succulent dishes including recently harvested foods, roasted and smoked meats, cabbage salads, rich egg-cake desserts with almonds and spices, and drink.

The ancient *Kerb* tradition harkens back to pre-Christian times and derives its name from ancient Indo-European *kerp-*, "to gather, harvest." Observance was perpetuated among many of the 27,000 Germans who immigrated to Russia's Volga region in the late 1700s during the reign of Catherine the Great. Her landmark 1763 "Manifesto of the Empress" promised that colonists would have, "the right . . . of establishing annual fairs according to their discretion." A generation later, thousands more Germans with similar customs settled in the Ukraine and Caucasus regions. Following months of arduous field labors, villagers of all ages—but especially young people, eagerly awaited *Kerb* to receive the season's wages, settle cash rents, perhaps buy something new, and enjoy long-awaited celebration. Typical Russian-German "festival cakes" prepared for such occasions included *Zucherkuchen*, a sugar and cinnamon cake often cut into bars, *Flammkuchen mit Apfelschnitz*, a thin crust tart topped with apple slices, and spiced pumpkin *palatchinta* turnovers.

Approximately 20,000 first- and second-generation Germans from Russia had settled in America's Pacific Northwest between the 1880s and 1920, and variations of these dinners and delicacies continue in our communities. Endicott historian Anna Weitz recalled century-old tales of harvest festivities at the Volga German Palouse River colony several miles north of town: "The families had a grand harvest celebration with a feast and music and singing and plenty of refreshment. . . . No wonder they lived so long!" Farm families also gathered grain and flowers into festive arrangements to decorate local churches.[3]

Universal Symbols

Many years later when encountering writings by nineteenth-century British author-artist John Ruskin I read about innate affinity to grain sheaves and clusters of flowers as universal symbols of beauty, peace, and plenty. This, in turn, led to encounters with notable works of art featuring sheaves and stooks by van Gogh (for whom the sheaf was a symbol of infinity), Early American architect Samuel McIntire, German Expressionist Marc Franz, American landscapist May Lydia Ames, and many others. Modernists Milan Konjović and Fedja Soretić of Serbia's fertile Vojvodia region considered grain shocks to be icons of primitive, regenerative cultural strength in a land where traditional methods of harvest continued throughout most of the twentieth century.

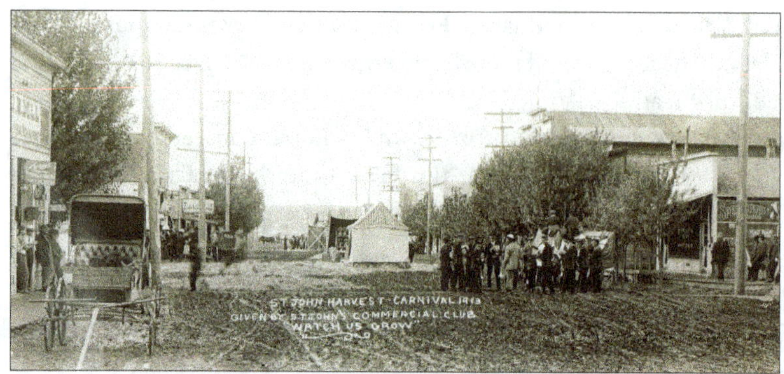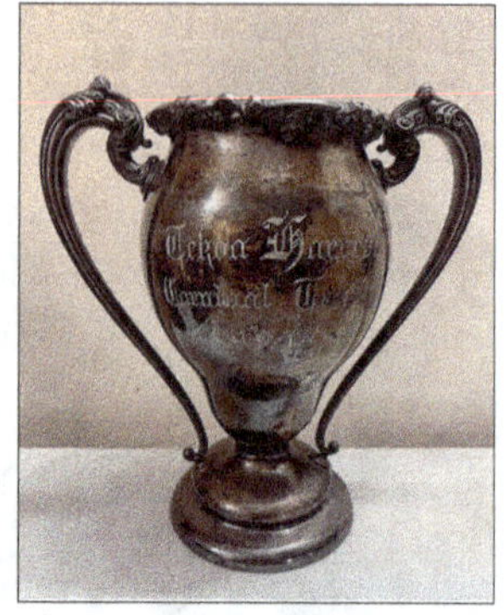

Left: *St. John Harvest Carnival, 1913* **(Whitman County Library Heritage Collection; Colfax, Washington)**
Right: Tekoa Harvest Festival Trophy—Champion Draft Stallion Royal Graham (1912)
Roger and Sheri Hager Collection

Austrian-born contemporary Canadian artist Armand F. Vallée depicted the grandeur of harvest on the Alberta prairies in paintings of loose brushwork that also honor these iconic forms (see PLATE 5). Son of a Jewish cantor in pre-World War I Galicia (now western Ukraine), Ben-Zion Weinmann immigrated to the United States and became a member of "The Ten" avant-garde Expressionists. Unlike his New York contemporaries, however, Weinmann created recognizable figures and considered his work a religious experience confronting the mysteries of existence. His Biblical Themes portfolio (1951) features engravings like *Joseph's Dream* ("Behold, we were binding sheaves") and Boaz and Ruth ("And it came to pass at midnight") that feature sheaves and stars as divine symbols of provision and connection. In Catholic, Orthodox, and Protestant traditions sheaves of wheat are commonly featured in print, sculpture, and stained glass to represent the Eucharistic Host.

A key figure in the development of the Arts and Crafts Movement that emerged in mid-nineteenth-century Britain, John Ruskin wrote that highest intrinsic value in art, as with excellence of life in general, consists of beauty combined with benefit. Mere fashion apart from function is irresponsible since time and attention are precious in life. In describing this two-fold function of quality, Ruskin singled out a sheaf of wheat as exemplar. Modernist sculptor Tracy Lindner, raised in rural south-central Montana, used leather, resin, and fiberglass to create graceful *Wheat* (2007) consisting of suspended forms of immense grain heads. "Wheat is one of the most durable crops that survives wind, drought, wildlife, [and] humans . . . ," she observes. "The field survives as a unit, yet the individual plants must endure much strife." Lindner's installation invites consideration of the farming's significance and inevitability in the circle of life.[4]

Possessing the most complex genome of any organism on earth with fifteen billion DNA letters, wheat itself is a mystical product of nature that serves practically as staff of life with a one seed yielding a thousand golden kernels from twenty or more stalks in a single plant. A bushel basket weighing some sixty pounds could hold one million of the densely nutritious nuggets. "Wheat is almost sacred for the great service it yields," wrote nineteenth-century art critic Philip Hamerton in the context of feeding the world's masses as a single bushel of wheat makes about ninety loaves of bread. Gathering and binding grain into the form of a sheaf is a distinctly human creation of stylistic beauty. Sheaves feed both body and soul. This understanding, the "cultivated" knowledge of Hamerton's mentor Ruskin, esteems grain as both symbol and resource that "binds" society like sheaves through a social contract. Farmers provision consumers whose existence ultimately rests upon agriculture, and they reciprocate by supporting producers. In Early America sheaf carvings were emblems of prosperity and hospitality and a familiar motif in the new nation's popular Federal Style, which in turn borrowed Greek and Roman architectural elements to commemorate the new nation's democratic origins. The sheaf was a common decoration for bakery signs and smaller household items like wooden butter molds and earthenware bread trays. For all these reasons our kitchen table was more than somewhere to partake of Mom's satisfying meals. It also served as a place to pass down an ancient craft and teach lessons of patience amidst happy conversation between parents and children.[5]

My boyhood sheaf-building forays fostered a special fellowship with the past as well as opportunity for Dad to point out yet again the "right way and wrong way" of doing small things. I could not know at the time that in faraway Belfast poet Seamus Heaney (1939-2013) had begun giving voice to rural experience in his first collections *Eleven Poems* (1965) and *Death of a Naturalist* (1966). Heaney's lyrical description of "The Barn," a short selection from the latter, might well have been about our own with reference to grain sacks, cobwebs, and musky darkness. His evocative recollections in "The Harvest Bow" (1979) ascribed special meaning to my father's quiet interest in imparting sheaf craft. The five-stanza poem is a visionary tribute to love between father and son that tells about traditional "golden loop" wheat-weaving deftly plaited by "fingers somnambulant." The small bow that had once festooned his farmer-father's lapel evoked memories of tranquil moments and rustic beauty as if in a Dutch Masters' still life. The intimate remembrance stirs Heaney to observe in the final stanza, "The end of art is peace." He was preoccupied at the time with Northern Ireland's "Troubles," while my late 1960s adolescent world was regularly rocked with news about Vietnam and campus riots. Against the pains of past or present, a sheaf or object crafted from grain stalks can mysteriously serve as a token of solitude and devotion.[6]

Located between the small rural communities of Endicott and St. John, Washington, our home place was deep in the Pacific Northwest's Palouse Hills—a sinuous grainland of labyrinthine swirls

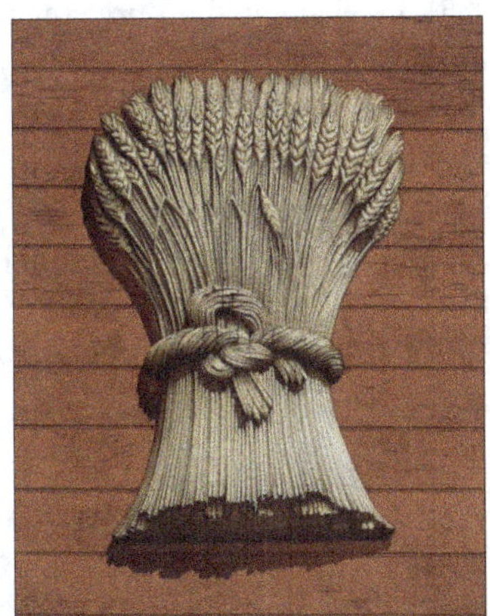
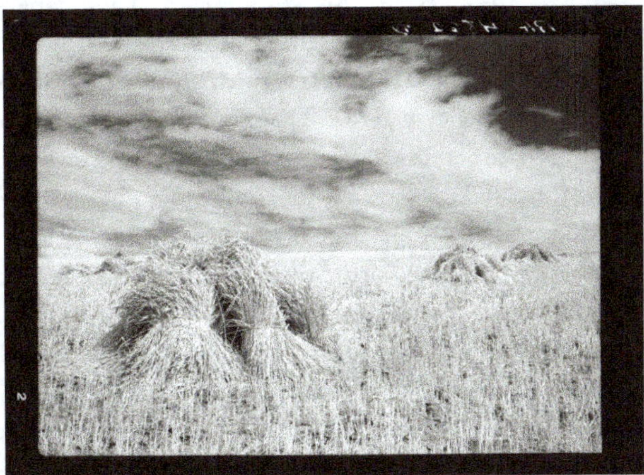

Left: Robert Pohle, *Sheaf of Wheat Shop Sign* (c. 1935, after a Rhode Island shop sign by W. Clarke Noble) Watercolor and pencil on paper, 16 ¼ x 12 ¼ inches; National Gallery of Art
Right: Arthur Rothstein, *Shocks of Wheat*; Whitman County, Washington (1936) Prints and Photographs Division, Library of Congress

and whorls resembling a deific thumbprint when viewed from high above. These hills of fabled loessal fertility, nicknamed the "Tuscany of America" in recent national press accounts, also produce legumes and oilseeds and are a favored destination for landscape photographers from around the world. Tending the dryland terrain could itself qualify as artistry writ large. Farmers have long used plowshares like trowels to form braided earthen furrows, and sickled combine headers for brushlike cropland contours. Rural Indiana columnist Rachel Peden recognized the phenomena in patterns of the rolling grainfields of her native Monroe County: ". . . [A]rt exists from the beginning of land tilling. There is hardly anything more artistic than a well-tilled, fertile, disciplined field at any time. When freshly plowed, brown and shining; or when newly worked down, fertile, full of promise and waiting to receive the seed; or when the first green sharp-pointed sprouts make dotted lines across the field, almost but never quite meeting the horizon, and always beckoning on. . . ."[7]

Younger farmers in our vicinity spoke admiringly of an elderly neighbor whose tightly disced corner turns on steep sidehills formed series of near perfect circles as if finely furrowed penmanship drills. (He also proudly preserved a World War I era green and orange bulk grain wagon used just one harvest season before replacing it with a gas-powered truck.) Hamerton would have approved of his abiding commitment to free-range cattle and to share the fields with native creatures. On countless boyhood drives to town we passed this orderly farmstead that was flanked on the west side of the road by an insequent stream that hosted a substantial stand of lush cattails. Vast numbers of chattering red winged blackbirds and furtive pheasants made their home among the lush green and tawny rushes. Every fall the children of visiting friends and relatives delighted in gathering the exploding brown heads to throw at each other. The prospect for such fun ended in the 1970s when the longtime steward passed away and the property was sold to someone who lived elsewhere. The meager brushland was soon plowed under and seeded to grain that yielded poorly except in silence.

Hamerton's musings a century earlier on rural change in England and France alerted him to concerns over stewardship of land in the wake of industrialization absentee farm management. In an essay titled "The Effects of Agriculture Upon Landscape" (1885), he elaborated on preference for spaces of "pure Nature" while acknowledging the value of flour and loaves. "We like to be rich, we like our land to be fertile, and yet 'a rich and fertile vale' may be destitute of beauty," he wrote. The unfortunate tendency had been the removal of scenic groves and confinement of livestock in stalls. Hamerton mourned the loss of agrarian coexistence with the natural world as seen in the destruction of copses and wastes unfit for cultivation. Pulitzer Prize-winning novelist Marilynne Robinson grew up in the 1960s playing among the cattails in the shallows of northern Idaho's Lake Coeur d'Alene. "We live at the Earth's scale," she writes in the essay, "A Theology of the Present Moment," which should instill a sense of humility when one ponders the vast universe. "Evening and morning, seedtime and harvest, it shapes time for us." Robinson's "theology of the present moment" prizes the marvel of consciousness—perhaps singular in the universe, for appreciation of every tree, herb, and blossom, like those so long protected by our farmer neighbor.[8]

Some years after the assault on the blackbird sanctuary, I returned to our hometown to serve as public school teacher and administrator. One day our irrepressible art instructor Arden Johnson insisted that we take all middle school students to the area's last undisturbed stand of cottonwood and hawthorn. The area covered perhaps an acre and was surrounded by grainland along Union

Flat Creek several miles south of town. Arden argued that the fieldtrip—one of many adventures in "experiential learning," would provide instructional grist for studying art and literature as valuable as any textbook. Following the adventure with yellow buses navigating the soft terrain and youngsters examining stalks and bark, we returned to school where they committed their impressions to drawings and stories. One who had fared poorly in academic work submitted an essay with descriptions and illustrations so fulsome that a teacher called home to confirm authenticity. Next day the paper was read aloud as the finest from the class. Thinkers like Hamerton, Robinson, and Mrs. Johnson have long understood that threatened natural areas not only complement agrarian landscapes but stir minds to consider other important values like imagination and creation care.

"The Unembodied Call of a Place"

Pioneer Palouse farmers devised a peculiar system of "back-landing" for horizontal tillage around hills to reduce the strain of implements on draft animals and to reduce the persistent problem of soil erosion on vulnerable slopes. Palouse topography resisted rationalization imposed by cadastral survey grids though landscapes gradually yielded in the wake of mechanized equipment. Maneuvering a twenty-ton combine across an undulating slope of swaying grain beneath migrating cloud shadows imparts a sense of wonder at any age. But the machine's operator, aided by hydraulic steering and an array of monitoring instruments, must concentrate on what Palouse Country photographer Bill Woolston terms a "choreography of . . . maximum efficiency" to guide each full swath to minimize any waste of time and fuel. Unique views of tillage and harvest patterns are apparent in aerial vistas—Garrison Keillor's "God's eye views" that depict fertile constellations akin to starlit skies. Swiss-born photographer American Georg Gester devoted three decades to documentary aerial photography of farmland through all seasons as a tribute to this grand art form. The colorful images of grainlands and other terrain in his masterful volume, *Amber Waves of Grain: America's Farmlands from Above* (1990), appear as a grand palimpsest depicting the textured relief of earth and crops artfully fashioned by field implements and combines. Northwest landscape photographers John Clement, Alison Meyer, and Dean Davis have contributed significantly to this genre with their images of Columbia Plateau harvests and other field operations throughout the year.

With only 320 acres, our farm was small even by 1960s standards. My boyhood friends and I spent considerable dusty time in the fields with our fathers riding seed drills and other equipment as if carnival rides that we soon learned to operate. During my teen years we leased an additional 700 acres of exceedingly steep ground about eighteen miles south near Colfax. Farming the "Foley Place" meant recurrent travel over gravel and dirt roads until our landlord died unexpectedly. The arrangement abruptly ended and we experienced some distress. Like some other small farmers in the vicinity, Dad maintained economic viability from supplemental income, which he earned by selling crop insurance and as a rural mail carrier. In this way they embodied the entrepreneurial spirit of Early American yeomanry to a much greater extent than what Thomas Jefferson had envisioned. Despite noble intent, the ideal of self-sufficiency from the land alone was as irregularly present in the new nation as in the Midwest following passage of the Homestead Act a century later. Making a living on a half-section in many parts of the country in our day remained challenging and households often engaged in trades of all kinds to make ends meet—carpentry and engine repair, sewing and baking.

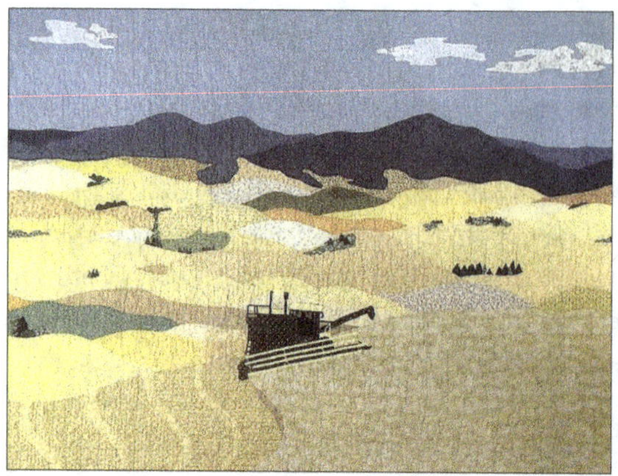

Karen Schoeflin Hagen, *Palouse Puzzle* (1992)
Jig-saw Puzzle Quilt, 105 x 82 inches

Author-agronomist Jack DeWitt recalls 1950s youthful rural experience in the Wal-la Walla Valley: "As soon as I was old enough to hang on, I spent hours riding on the hood of Dad's old tractor, my legs hanging over the sides of the motor as if I was riding a horse. . . . I watched how he did everything—how he turned corners, how he tilled sometimes around and around, sometimes back and forth. . . . I would get in Dad's old truck and, again in my mind's eye, take a load of grain to town, shifting like I'd watch him do."[9] In these ways farm youth learn of changes on the land through the seasons, and valuable lessons of self-reliance. Shouldering adult responsibilities while driving harvest truck remains an important rite of passage for many rural youth that requires respect for a handful of core field rules: no parking behind another vehicle, check engine fluids daily, no backing trucks downhill. Scooping up any spilt grain and never driving into the standing crop went without saying. The combine driver establishes a relationship with the machine known through touch, sight, and sound. Safe and effective operation requires regular checks of myr-iad chains and sprockets, and belts and pulleys that drive shafts, augers, grates, and other equipment that must be properly lubricated and tightened.

As with other cultural rituals, much of the har-vest experience is solitary and happens in wide open spaces, requires guidance of elders, and brings de-manding challenges overcome through emotional and physical struggle. Occasional mishaps still hap-pen—a dented fender, empty gas tank in a remote field location, or getting stuck on a sidehill. But in the end the responsible initiate earns meaning-ful if informal membership into the harvest crew. The work also brought material reward. Temporary farm permits are allowed in many farm states for drivers as young as fourteen. We apparently had a bumper crop when I was about that age because I remember Dad presenting my brother and me with a hundred dollars each at the end of a long harvest run—my only memory of ever being compensated. I could scarcely fathom such wealth.

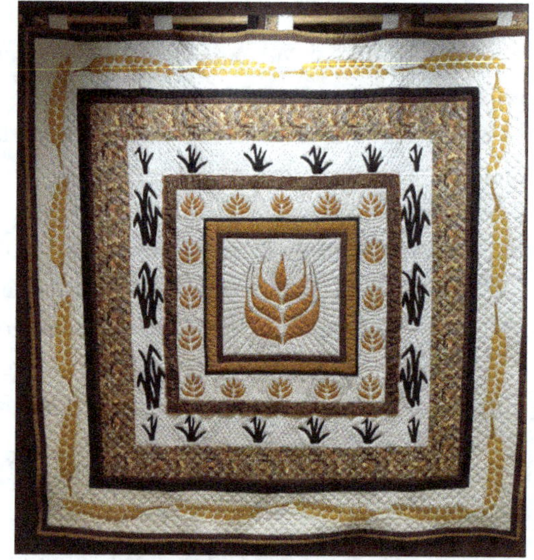

Palouse Piecemakers
Commemorative Quilt (detail, 1991)
Association of Wheat Growers
Ritzville, Washington

Small spreads like ours held in common with farm families most everywhere peculiar toponymns necessary for communication about what had happened long before, or what needed to be done with scythe, hoe, or fence. The far northwest corner of our property was the exceedingly steep and deep

Huvaluck (Hessian dialect for "Oat Hole," from *Hafer*, German "oat") and nearby were Windmill Hill and Spud Draw. For three generations our clan had planted potatoes in what seemed to be an endless furrow of the latter during the week of Good Friday to be dug in September and placed in a cluster of burlap gunny sacks destined for the root cellar. The head of Spud Draw led over a narrow ridge aptly named The Saddle that joined two broad hills. With so many landmarks on that tussled half-section, there could be no doubt where we might be directed to attack a purple-topped thistle patch or rendezvous with a truck at harvest time to unload grain from the combine.

Over time I came to better understand my father's and grandfather's familiarity with the land's every twist and turn in terms that defy analytics. Nature writer Barry Lopez characterizes such natural relationships in *Home Ground* as "the unembodied call of a place, that numinous voice" when a hill, draw, or untilled sod-patch wildlife sanctuary seems to become "something that knows we are there."[10] This mystical relationship fosters deep connection with landscapes that predisposes some to feel a kinship of peace and familiarity with certain places. I would also come to sense that the undulating Hessian Vogelsberg district in central Germany, and the steeply rolling *Bergseite* of southern Russia's black earth Volga region where my paternal ancestors dwelled long ago. People have perceived these mystical forces since time immemorial, and they continue to be felt. The Lake District sustained Ruskin and Wordsworth like Suffolk enlivened Constable, and the Great Plains stills feeds souls as it did for Willa Cather and Hamlin Garland.

Mom's family hailed from Norway where our kindly, indulgent Great Aunt Mary had been born in 1882. She had been raised near the village of Gol in Norway's scenic Hallingdal Valley northwest of Oslo where the family eked out a hardscrabble existence on a small farm and dined on Nordic peasant fare—boiled mutton, carrots and peas in milk gravy, cottage cheese, and black coffee. Because she never married, Aunt Mary volunteered her time in later years to help out farm family relatives like us with harvesttime cooking and she introduced us to the delights of buttered *lefse* and rice pudding. She told us about the small herd of cattle and goats they tended in the Old Country, and of the wild berries and mushrooms gathered on the higher slopes.

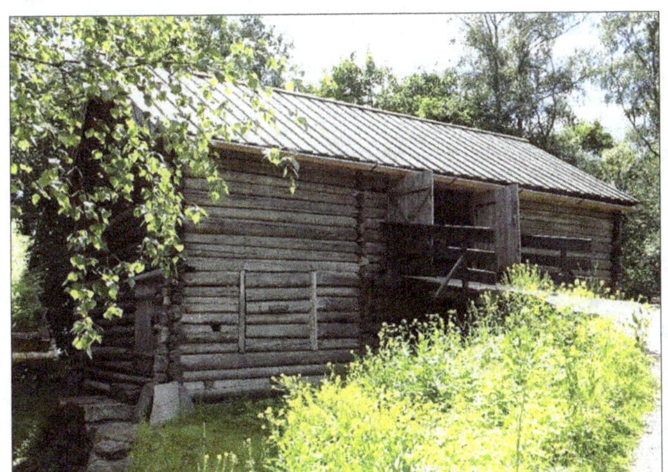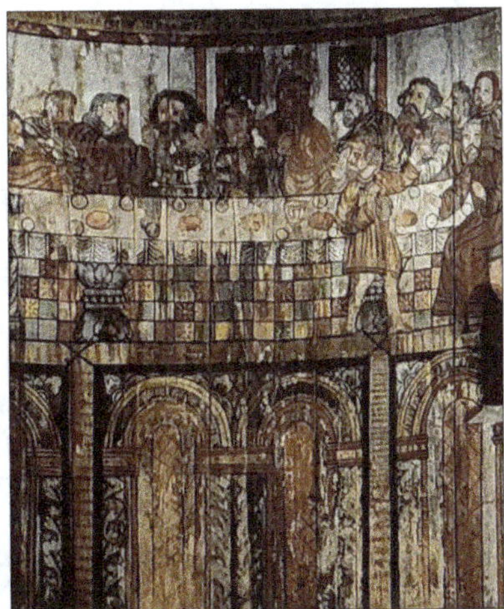

Hallingdal Threshing Barn and Gol Stave Church Altar
Norsk Folkemuseum, Oslo (2017)
Ruth Knight Pagac Photographs

Several wooden stave churches dating from medieval still existed at the time in the vicinity of Gol. The distinctively capped structures featured steeply sloped roofs and flanking prows of fearsome carved animals said to protect them from evil spirits. Many Norwegian legends tell of enchanting fairy-like Nokkens and malevolent trolls who sometimes camouflaged themselves as corporeal grain shocks, and our spinster aunt told of peculiar happenings on the dark slopes where they lived. In the early 1880s, Norwegian artist Theodor Kittlesen made the first of his many folktale illustrations and became one of the leading contributors to the Asbjørnson & Moe standard edition of Scandinavian myths. His *Grain Stooks in the Moonlight* (1900) shows a field of tied grain poles like Aunt Mary described that appear to conceal the trolls who would appear after dark. Dating to about the year 1200, Gol's principal stave church and several local farmsteads attracted the architectural interest of Norway's King Oscar II. In 1881 and 1882 the buildings were carefully taken apart and reassembled with others on the grounds of his summer residence, the Bygdøy Royal Farm near Kristiana (Oslo). This location became the world's first open-air museum and living history farm. The magnificent church features an immense altar painting of the Lord's Supper (c. 1650) showing Jesus' distribution of bread and wine to the Disciples.

Citadel of Understanding

The base of the broad hill with the entangled cherry tree reaper that rose westward from our Palouse Country home was steep enough to accommodate a spacious root cellar attached to our ramshackle porch. The cherry tree and nearby crab apple exploded in fragrant white blossoms in the spring, and Mom made delicious golden jelly from the small, colorful fruit. The two trees were positioned so close to the fence that moving through them when pulling wide tillage implement with the tractor involved split-second maneuvering of twists and turns. Dad could have easily removed the gnarled pair but protected them out of regard for their seniority and abundant yield. We also safeguarded the dark purple lilac bushes his mother had planted years earlier behind the house and found other isolated patches of blue iris and hollyhocks across the fields that marked where neighbors long since gone had once lived. Years later I encountered Montana poet Grace Stone Coates's (1881-1976) "Portulacas in the Wheat" (1919) in which she recalls a childhood harvest time memory of field flowers familiar to many rural folk:

My mother told me they were portulacas
Gone wild, once planted by an
   earlier tevnant.
No! They were rich enchantment,
   silken flame,
A whole new continent in Fairyland!
That timeless, golden afternoon I held
Grave converse with my Fellow of the Sun,
Companionship beyond the need of words.

Deep in the sun-drenched wheat, content,
   I heard
The whirring binders drop their tawny loads
Nearer and nearer, clanking nearer still;
A pause, a question, then my father's voice,
Abrupt, imperative, "Swing out, I say!
The child shall have her flowers!
   Swing around!"[11]

Our cavernous hillside cellar had been dug not long after our Russian-born German ancestors began arriving in the Palouse in the early 1880s. We were warned to stay out due to threat of collapse

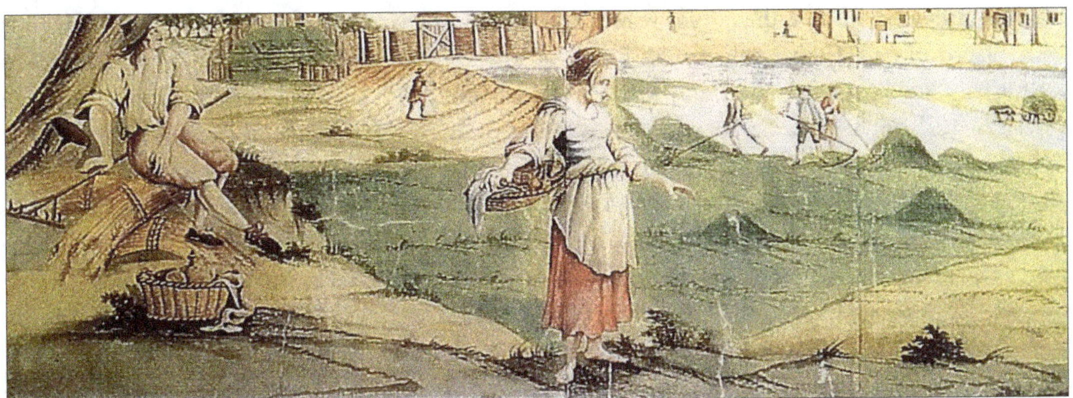

Volga German Scything Scene (1768 map illustration detail)
Columbia Heritage Collection

from the weight of the ground above. Old timers called such places *zemlyanki*, Russian for "earth-homes," which likely were successors to the medieval German *Grubenhäuser* ("dug-houses") our forebears had fashioned in Hesse. Our paternal ancestral and agrarian roots could be traced to the hilly slopes of the Vogelsberg district north of Frankfurt a. M. Count Berthold of Nidda had transferred farmland in the area to the Order of St. John Hospitallers and in the mid-1700s our ancestors, Hartmann and Elizabeth Scheuermann, tended the *Johanniterhof* in nearby village of Unter Lais. In the aftermath of the Seven Years War in Europe (1756-1763), the couple and their seven children joined the epic eastward trek eastward with other peasant immigrants who sought better lives in Russia. They established 104 farming villages along the lower Volga River near Saratov including our ancestral village, Yágodnaya Polyána (Berry Meadow). Many spent the first winter in the primitive *zemlyanki* earth-homes.

When *usu Leut* ("our people") first settled in the Pacific Northwest, some inhabited such dens again until proper family houses could be built. Our Scheuerman great-grandparents first lived with relatives in "the colony," a scenic quarter-section of Palouse bottomland and rugged basaltic bluff that had been established as a commune by the half-dozen immigrant family founders. Today the place is known as Palouse Colony Farm and produces heritage grains including legendary "Turkey Red" and "Scotch Fife" (both actually native to Ukraine), progenitors of most modern American bread wheats, as well as historic White Lammas. Named for the August 1 Lammastide (Loaf-Mass) celebrated throughout medieval England as the ceremonial commencement of harvest, White Lammas was introduced by the Hudson's Bay Company in the 1820s as the Pacific Northwest's first cereal grain. (Later introductions of the same variety from Australia were widely raised throughout the region as Pacific Bluestem.) Descendants of Palouse Colony founder families safeguard treasured photo albums that document early farm life along the river in a genre sometimes called "unintentional art." Protected through generations between heavily embossed faux leather covers, these archives represent important visual documentation with scribbled titles showing everyday rural life—men working in the fields, women preparing meals and tending gardens, children milking cows and riding horses.

The young are disposed to explore, so in our youth we ventured inside the cellar periodically to see dimly lit evidence of life from yesteryear. As in Russia, this is where our elders stored their substantial garden produce in the fall for winter and the coming year. Small green *arbuza* watermelons were placed in barrels or crocks of salt brine for a tangy sour treat. Cucumbers were also picked in this way, and crocks of sauerkraut were typically layered with carrots and apples, along with the prized *dawza* cabbage hearts. Potatoes, carrots, and parsnips were generally stored in burlap sacks beneath braided onions hung from the ceiling. An exceedingly dusty *pituvfka* kitchen worktable stood against one wall of our cellar with two rounded flour bins suspended beneath, and some old glass canning jars remained on wooden shelves. Most had long been empty, but some may once have held grain safeguarded by our grandparents. Perhaps these were remnants of the *Hirsche Brei* ("millet porridge" in Hessian dialect) ingredients that Dad remembered eating with honey as a boiled wheat breakfast mush in Depression days. I learned later that some of the immigrants had brought ancient landrace grain seed from Eastern Europe. Varieties were named for colors of kernels and brush-like beards like Turkey Red and Odessa White wheats, purple Egyptian Hulless barley, and Green Russian oats. Perhaps the cellar had also served as our elders' seed vault.

Grandpa was a citadel of understanding. I listened with fascination to his and other elders' stories about cutting grain by scythe and reaper, forming shocks of ten to fifteen upright bundles, pitching them into wagons, and threshing the stalks in stifling summer heat with a skilled team of workers. Their exhausting harvest labor required their tending header-box wagons, derrick table and Jackson fork, and a mammoth steam-powered threshing machine and engine. Their tales imparted appreciation for meaningful, co-operative work carried out throughout the year in the great outdoors. I also marveled at Grandpa's and our Norwegian-born Aunt Mary Sunwold's ability to quote Victorian countryside verse at length— ". . . where ill thoughts die and good are born, / Out in the fields with God," by poets like Elizabeth Barrett Browning and Alfred, Lord Tennyson. In a day before hard-drives and cloud storage, country folk commonly fashioned substan-

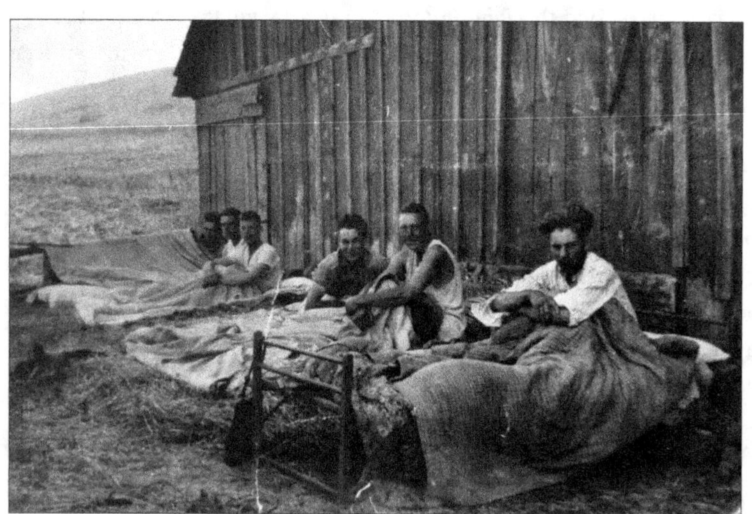

Palouse Colony Harvesttime Scene (c. 1935)
Gayle Schoeflin Collection

tial scrapbooks of nostalgic rural poems clipped from the pages of regional papers and magazines. Tucked within the pages of Aunt Mary's extensive collection are poems like "The Old Log Barn" contributed by lesser known or anonymous authors. Their only published work may have been on the page of a Depression era issue of the *Spokesman-Review* or *Country Life Magazine*. Such lines foster imagination of their lives when passing places where such recounted events might well have happened.

. . . [T]here is the oldentime thrashing floor,
Where busily moved our feet.
To handle the hay or the bearded sheaf
Or winnow the golden wheat.
For the merry old times that I sported there,
The song that I sung in my play.
Hath left recollections within my heart
That never will fade away.
They are gathering the fruit of the
*   plenteous year,*
In granary and spacious mow,
And the laborers shout of the harvest home
Is floating around me now.[12]

Once at my Grandfather Scheuerman's home in town I asked about harvest in bygone days. He retrieved his leather-cased sack-sewing needle—still razor sharp after many years in retirement, and an old photograph mounted on dark gray hardboard from his bedroom closet. I instantly recognized Grandpa as a young man standing under the wooden derrick clasping the handle of a pitchfork and flanked by a horse-pushed McCormick header and

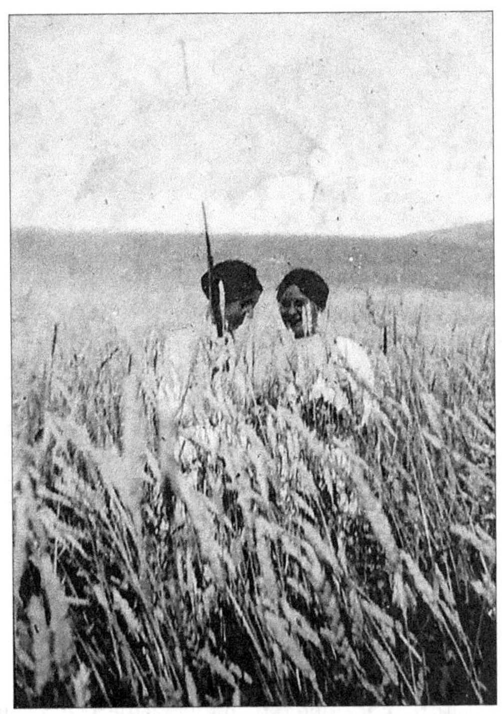

Palouse Colony Harvesttime Scene (c. 1935)
Gayle Schoeflin Collection

steam-powered Case threshing machine. He then patiently described the role of each member of the substantial crew and introduced me to terms like derrick table, header-puncher, hoe-down, and other agrarian vernacular from the steam threshing era. Many farm families treasure such panoramic views today, and I have unrolled many that stretch as wide as a kitchen table. Grandpa delighted in relating tall tales of bygone August "thrashin' weather" happenings like when the Moore brothers threshed more than a thousand sacks of grain in a single day during the same harvest season pioneer Endicott photographer R. R. Hutchison took that picture. He also recalled the bumper crops of 1908-1910, and how Black field hand Otis Banks could lift a 120-pound sack of wheat with his teeth.[13]

A gradual transition from steam-powered stationary threshers to gas-powered tractor-pulled combines took place from about 1920 to 1940. Among Grandpa's most poignant stories was how one the father of one of my schoolmates had come to the community as a young man to work in harvest for a prominent local farmer. In the midst of steam-threshing operations the landowner attempted to push a tangled mass of grain into the feeder mechanism that struggled with menacing movable claws to draw in the stalks. Somehow he became snagged and the thresher began pulling him into its howling maw. Young Joe Bensel sprang up and pulled the man to safety. Although the leg had to be amputated, the hired hand's quick action had saved his life. In an enduring gesture of gratitude, the landowner allowed Joe to farm the place for the rest of his working days. As a young man, my wife's uncle, Chuck Pugh, worked as a "mule-skinner" in the summer of 1939 on the Jordan ranch west of Endicott near Winona. This demanding task required driving some twenty-seven mules or horses

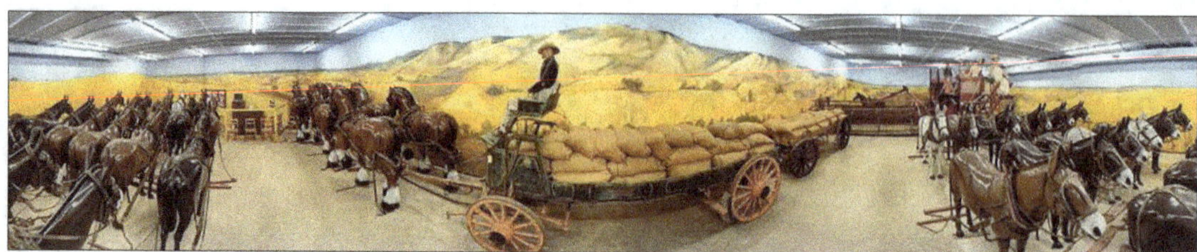

Carol Poppenga, Walla Walla Harvest Mural (1999)
Courtesy Fort Walla Walla Museum; Walla Walla, Washington

using jerk-lines to two lead animals while perched atop the hurricane seat of a wooden ladder that extended from combine high above the horses. (Drivers routinely kept an arsenal of pebbles, dirt clods, or small green apples nearby to prod any dithering animal.) While sweeping across a hillside the harness tongue pulling the machine broke and Pugh crashed down into the trace chains of the stampeding animals. The fall shattered his leg and required two years of rehabilitation.

One of the most impressive museum harvest exhibits anywhere is Fort Walla Walla's life-size 33-mule team "Shenandoah hitch" pulling a 1919 Harris combine. The Agricultural Exhibit Hall display is surrounded by an enormous sun-drenched harvesttime mural by local artist Carol Poppenga showing the combine and teamsters in action. One aspect of the era's transition to agricultural mechanization as expressed in art, literature, and memoir was the rapid obsolescence of the faithful creatures considered by many farmers to be almost extensions of the family. Recalling his affinity to mules, veteran Walla Walla, Washington, area wheat rancher and area historian Carl Penner (1889-1983), who helped design the Walla Walla exhibit, claimed, "[I]f you'd treat 'em halfway right and try to pet him a little bit and curry him nice and give him plenty to eat and a little bedding they soon learned who was boss."

Up to three-dozen animals were joined in complex hitches to pull the mammoth threshing machines, and drivers prided themselves in skillfully maneuvered serpentine swaths around hills and straight-as-an-arrow strips along fence-rows and flatlands. The advent and affordability of gas-powered tractors meant dependable mechanized alternatives to "bullwheel-driven" pull-combines and stationary threshers that also required considerably less care than animals weighing up to a ton or more. Stockton, California, manufacturer Benjamin Holt patented the first "caterpillar" cleat-crawler tractor in 1907 and introduced gas engines to separately power combine cutting and threshing mechanisms. Penner was wistful in recollecting the transition: "You hated to quit the mules. They was just some life. You know you get attached to a mule the same as you do people. . . . But you couldn't get anybody to drive the mules. . . . The Caterpillar [tractor] came along and you bought a Caterpillar." Hames and harness were hung in dusty corners of vacant barns as farmers built machine sheds with shelves for socket sets and spark plugs.[14]

The Desert of Wheat

Our small hometown of Endicott had been platted in 1880 as one of many rural trading centers and grain stations along the Columbia & Palouse line of the Oregon Railway & Navigation Company. The Oregon Improvement Company, an O.R.&N. Co. land development affiliate, established the region's first demonstration farm adjacent to the community to show prospective settlers Palouse

Country agricultural potential. In the early 1890s, the O.R.&N. Co. brought the Pacific Northwest's first header-and-thresher "sidehill combined harvester" to Endicott where a vast crowd watched the Holt mechanical marvel drawn by a column of horse teams navigate the steep slopes surrounding the town. One of the most widely circulated postcards of the time, the Spokane Card Company's "Combined Harvester Scene" (1906), shows five wood-frame Holt "Oregon Special" sidehill combines pulled by thirty-three-horse teams for a total of 165 animals on a Dry Creek Gulch farm near Walla Walla. Use of stationary threshers continued to dominate for several decades until the advent of self-propelled models, but early combine manufacturing businesses were established in Harrington for the Dunning-Erich Harvester and for Moscow's popular Rhodes Combine Harvester and Idaho National models.[15]

World War I era threshers on the Columbia Plateau are prominently featured in Zane Grey's 1919 novel, *The Desert of Wheat*, illustrated by German-American artist W. H. D. Koerner. Community elders from Hooper and Connell to Moses Lake and Ritzville told tales for years about the bestselling author's visit to their communities in the summer of 1917 to gather background material for the book. How glorious to first encounter a notable published work about our area. Although Grey's exact itinerary is unclear, the 2019 discovery of his picture album from the trip confirms his visit to area harvest fields. While Grey (1872-1939) is best known for idealizing the American frontier experience with larger-than-life characters, several of his novels venture beyond the unforgiving Southwest deserts to the farm country of Washington and Oregon. In contrast to Jean Toomer's favorable considerations of tension between the old and new cultural orders amidst a tumultuous era, Grey castigated grainland labor reformers in *The Desert of Wheat*. Socialist-leaning IWW. "Wobblies' are seen as unwelcome and dangerous interlopers who sabotage harvesting equipment and burn crops. Moreover, they seek to organize the migratory "bindle stiff" workforce on whom area landowners depend to take in the crop.

The novel opens with lines inspired by his summertime journey from Spokane across the ripen-

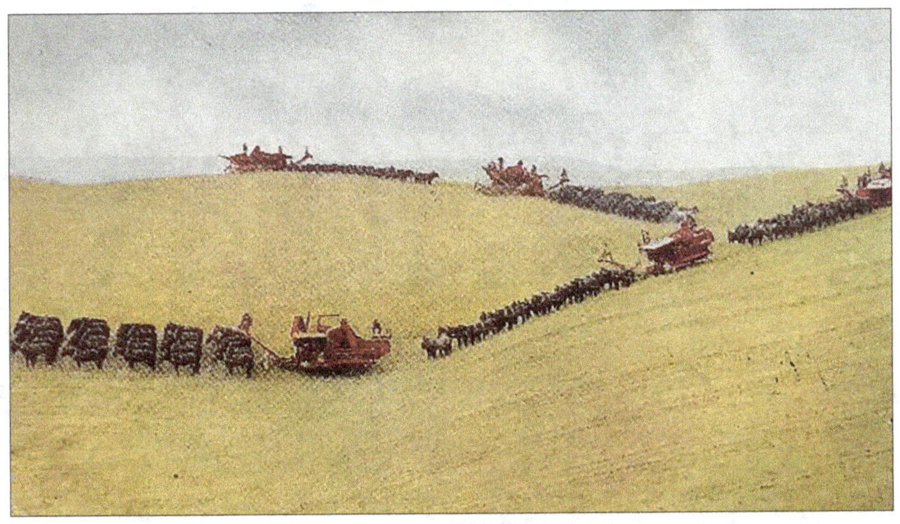

"Combined Harvester Scene" (Spokane Card Company, 1906)
Holt Combines near Walla Walla, Washington
Columbia Heritage Collection

ing grainlands: "Late in June the vast northwestern desert of wheat began to take on a tinge of gold, lending an austere beauty to that endless, rolling, smooth world of treeless hill. . . ." Through dialogue about Bluestem and Turkey Red wheats and rattling threshers, the story lauds the hard work and struggles of taciturn German immigrant farmer Chris Dorn in the nativist climate of World War I. Grey's principal characters often pause for imaginative reflection that reveals the author's melancholy preoccupation with interpersonal conflict and the impact of "progress" on the natural world's inherent beauty and mystery. Dorn's young adult son, Kurt, ponders the significance of change in his affairs and deeper meanings of land and life:

> *After a while the young man sat up and looked at the heavens, at the twinkling stars, and then away across the shadows of round hills in the dusk. . . . Where had the wheat come from that had seeded these fields? When the first and original seeds, and where were the sowers? Back in the ages! The stars, the night, the dark blue of heaven hid the secret of their impenetrableness. Beyond them surely was the answer, and perhaps peace.*
>
> *Material things—life, success—on this calm night lost their significance and were seen clearly. They could not last. But the wheat there, the hills, the stars—they would go on with their task. Passion was the dominant side of a man declaring itself, and that was a matter of inheritance; self-sacrifice, with its mercy, its seed like the wheat, was an infinite as the stars.*[16]

Although some critics panned Grey's plots as formulaic, British poet John Masefield, Ernest

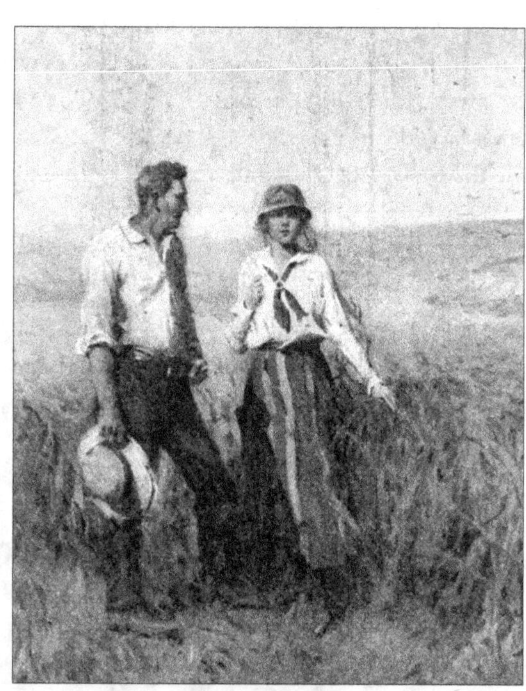

Left: Zane Grey, *The Desert of Wheat* Manuscript Opening Lines (1917)
Zane Grey Papers, Library of Congress
Right: W. H. D. Koerner, "The Undulating Sea of Wheat"
***Country Gentleman Magazine* (May 14, 1918)**

Hemingway, and Earl Stanley Gardner considered his writing of the highest order. Grey considered himself an author is the spirit of Romantics like Victor Hugo and Alfred Lord Tennyson who wrote adventurous quests in which the main character's values and courage overcome a host of challenges. From this classical genre Grey wove plots that portrayed the restorative effects of country life as expressed in the Early American literature of agrarian theorists Thomas Jefferson and Hector St. John Crevecoeur. The story still inspires. Seattle playwright Paul Lewis's two-act musical *Lost in the Hills* (2022) is adapted from Grey's 1919 farm novel. For melodic themes, Lewis credits the mystical influence of landscapes and rural communities across the Columbia Plateau and Palouse that combined with the novel's depictions of country life. Had the celebrated author been writing today, Lewis believes the book would have cast itinerant laborers in more favorable light. Grey's descriptions of the harvest spirit remain timeless, and his prose suggested vibrant images for songs by Lewis like "Harvest Day—Agrarian Reverie":

Bright sun!　　　　　　　　　*Sweat, shout!*
Chaff, dust!　　　　　　　　*Gripe, cuss!*
Hands have arrived　　　　　*Out to the harvest, one and all*
From miles around　　　　　*Wheat sheaves sway, fall!*
Wagons waiting　　　　　　*Into the grinding maw.*[17]
Ready to take their haul

The prose and poetry of Northwest author H. L. Davis (1894-1960) challenges categorization. Davis drew from the experience of his eastern Oregon upbringing, keen powers of observing nature, and the behavior of country folk to express unvarnished and universal rural reality. A series of his poems were published by *Poetry* in 1919, including "The Valley Harvest"

'Are the hollyhocks full bloomed?'
It is harvest then.
The hay falls like sand falling in a high wind

and "The Threshing Floor"

Yellow is where the threshing-floor is, and horses' hoofs
Beat the grain-heads into chaff; and cold wind
Stews chaff over the bushes and into the eyes

For this collection, Davis received the Levinson Prize and the admiration of Carl Sandburg. His verse is colorful while sometimes cryptic in relating beauties and hardships, and hopes and frustrations of farm life. "Poet of the Rockies" Thomas Hornby Ferril characterized his writing as the "low incandescence" of a "naturalist of transitoriness."

Davis's 1935 novel, *Honey in the Horn*, which received both the Harper Prize and a Pulitzer, presents a meandering coming-of-age tale in the Inland Northwest drylands with humor and humanity where the farm work and landscape known to orphan Clay Calvert shape his life as much as any person. Harvest labors of the time range from exhausting to exhilarating for the boy who learns lifelong lessons on the value of community and commitment:

The worst part of the job was the noise, which was incessant and deafening, and the chaff and dust, which worked into throat and clothes and skin and eyes and hair and itched an un-practiced man half-crazy. The best part was to look up, when the headers were making their round in the heat of the afternoon, and watch them climb up through the hot yellow wheat into the hot yellow sky and move in it, tall and scarlet and steady, turning their fan-arms in its yellow light for a couple of minutes and then wheeling and coming down again as if that had been nothing.[18]

Harvesttime Yesteryears

In remarkable documentary photography careers, Endicott's R. Raymond Hutchison (1887-1967) and Bill Walter (1915-1999), who resided in Colfax, Endicott, and St. John, captured over 100,000 images of Palouse area landscapes, people, events that spanned the century. Their archives of glass plates, film, and early movies, now at Washington State University, represent some of the period's largest collections by individual photographers. Hutchison's and Walter's commercial interests complemented lifelong passions for recording the region's heritage and scenic beauty during its transition from homestead to modern eras. While just a boy on the family farm near town, Hutchison carried his first Brownie on recreational trips to the Palouse River and into the surrounding fields. He captured men, women, and children working throughout the year—cultivating and harvesting, feeding threshing crews, freighting and storing grain, and tending to countless other ordinary tasks he understood to be significant to rural families. Hutchison's signature wide-format sepia scenes of stationary thresher and combine crews document their harvesttime labors using equipment manufactured by Holt, Case, Hunt, Best, McCormick, Deering, and Harris. He would go on to establish studios in nearby Pullman, Washington, and Moscow, Idaho, and was a prime mover in the establishment of the Inland Empire Photography Association.

The Palouse became synonymous far and wide with grain production and its unique topography attracted prominent photographers farther afield. While Seattle's acclaimed Edward Curtis (1868-1952) trained his lens on Columbia Plateau Indian chiefs and Southwest Native American harvesters, his brother, Asahel Curtis (1874-1941), documented farm life in the Palouse and throughout the Inland Pacific Northwest. Many of the latter's artful harvest scenes like *Combine-Harvester* (1915, PLATE 6) were used for postcards and promotional materials that ac-

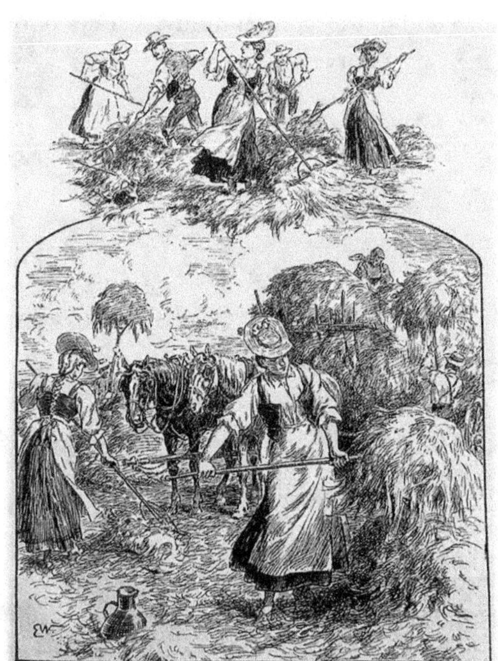

The Harvest woodcut, *Grosser Volkskalender des Kinkenden Boten* **(Lahr, Baden: J. H. Geiger, 1912) Columbia Heritage Collection**

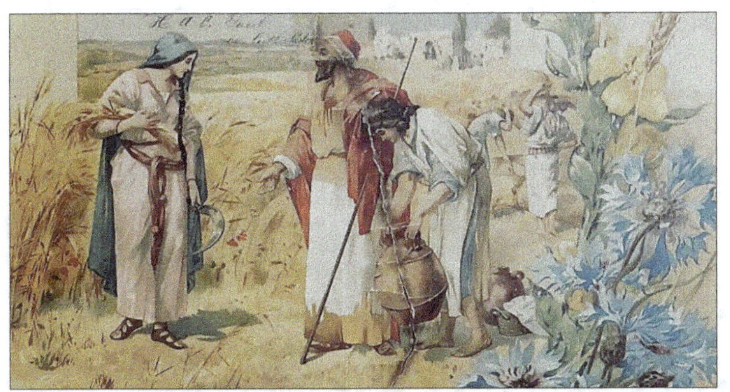

Ruth and Boaz Gleaning lithograph
Conrad and Emma Repp Marriage Certificate (1905)
Columbia Heritage Collection

quainted viewers throughout the country with the region's agrarian bounty and beauty. Among the books I recall in my grandfather's home were a German and an English Bible, and an enormous three-volume German New Testament commentary. He had also owned a 1902 copy of *Magner's Farm and Stock Book* with chapters on soil fertility, crop diseases, horse-shoeing, and numerous other subjects relevant to a suc-

cessful mixed farming operation. We kept the 900-page tome, which was probably handed down by his father, as a family heirloom. Many of our Russian-born community elders had such books and others like the red and gold clothbound folk calendars (*Volkskalendar*) that provided windows into a dimension they had inhabited in the Old Country. Questions about stories, sayings, and illustrations from these works provided an enriching supplement to formal studies at school.

Our father's most frequented volume may have been the weighty and thoroughly smudged parts-manual to our dilapidated International-Harvester Model 160 pull-combine. Though still in stolid service, the ancient machine bore an uncanny resemblance to Jack Dorsey's *Harvester at Rest* (PLATE 16). I felt a bit embarrassed in the 1960s days of efficient self-propelled machines operating in every direction that we still resorted to an exceedingly faded red Rube Goldberg dinosaur of howling sprockets and pulleys that lumbered behind our unmuffled "three-lunger" (cylinder) crank-start Caterpillar tractor. Somehow Dad managed to navigate these roaring monsters through swells of bristled golden fullness during our annual month-long grain harvest. But feelings of accomplishment swept across all the crew with the cutting of the final swath that vanquished any boyhood unease over equipment breakdowns, spilt grain, or other field mishaps. "No one should be deprived of harvesting," rural artist-folklorist Eric Sloane observed in his illustrated 1977 rural memoir, *I Remember America*. "Beyond the value of feeling the fruition of nature all about you, there is the satisfaction of beholding the results of your own efforts."[19]

Like many teens in wheat country, my brother, Don, and I started driving truck in the harvest field about age fourteen on special farm permits. This legalized our trips throughout the day to the Endicott and Thera elevators. We delivered grain loaded from the field into our pale red and blue '56 Chevy truck and rather older black Ford that turned heads with a throaty purr. The obligation came with explicit warnings about harvest time dangers—field fires, equipment collisions, and tragic combine tip-overs on steep Palouse hillsides. The latter claimed the life of more than one boyhood acquaintance. Dad roused us from adolescent slumber before dawn for Mom's fulsome breakfast of eggs, hashbrowns with onions, and homemade German *wurst* pork sausage. We then headed to the field as the sun transformed the western horizon from indigo to light blue. In the morning dampness we fueled and cleaned and greased equipment before the morning parade of combine and trucks to

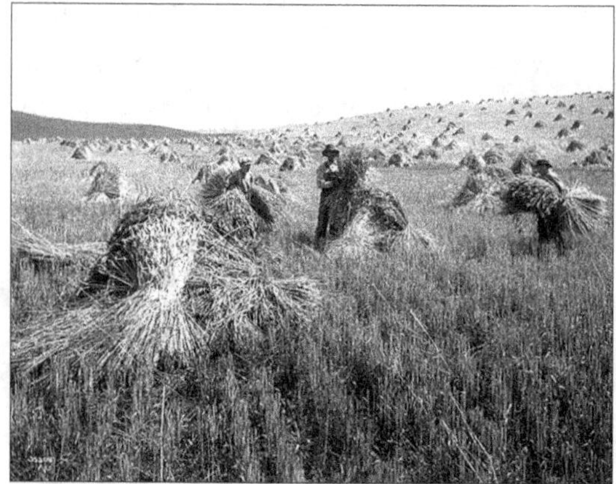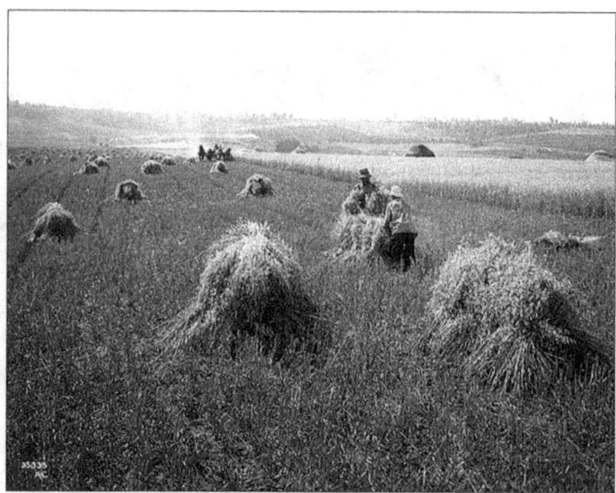

Left: Asahel Curtis, *Wheat Shocks in Field near Rosalia* (1916)
Right: *Shocked Oats in Field near Spokane* (1916)
Washington State Historical Society; Tacoma, Washington

a nearby uncut hilltop. Whenever possible we delayed harvesting of shaded draws where heavier soil and great moisture hosted denser grain until the dew had dissipated by mid-morning.

Even when the combine was too distant to see or hear, the hills were never lifeless. Coursing red-tailed hawks scanned the freshly cut stubble for mice that audibly scurried about, whirlwinds frequently punctuated midday breezes, and few days passed without seeing deer or coyotes in the distance. By late afternoon a vast layer of atmospheric chaff spread across the horizon to cast spectacular sunsets of glowing magenta. At least once every harvest such serenity was violently breached by the instantaneous roar of one or two fighter jets on a training mission flying so close to the hilltops that one could wave to the pilots. Periodic visits to the field by friends and city cousins provided welcome breaks in the daily routine of waiting for the several "dumps" needed to fill a truck and unload in Endicott's immense concrete grain storage elevator. Built in the late 1940s, the structure soars skyward some 150 feet and consists of a dozen interconnected exterior silos and several interior bins. The complex of elevators, augers, and pipes used to move incoming grain and fill railcars was capably handled in my day by foreman Harvey Blumenschein and a reliable crew of seasonal workers including lifelong friends of my generation like Steve Gerlitz and Mike Lust. Riding to the top of "Concrete" in the narrow cable birdcage afforded the finest view for miles around.

Mike went on to major in English in college and taught high school for almost his entire career in the rural northeast Washington community of Republic. But he reported year after year during summer vacation for harvest duty and reflected on the experience in numerous short stories like "Homage to Santiago," "The Inspiration of an Engine Roar," and "Russian Thistle"—named for the bane of many a dryland farmer. Mike's writing mingles routine happenings with meaningful considerations of place and purpose:

> There are two great sounds in these dusty elevators. The first is the sound of the leg when you push the button to turn it on. The tick, tick, tick of the cups sliding gently past the leg casing assures that all is well and the machines will continue raising the grain from the unloading pit to the top of the storage tanks. But the best sound of all is at the end of day when the leg motor is turned off. Quiet fills the dark air heavy with dust, and peace has come. . . .

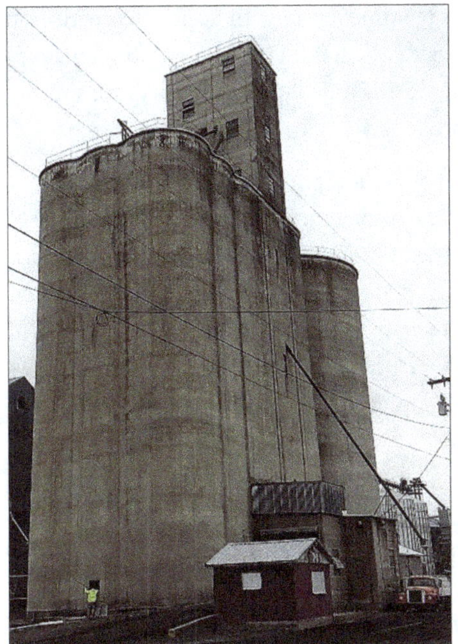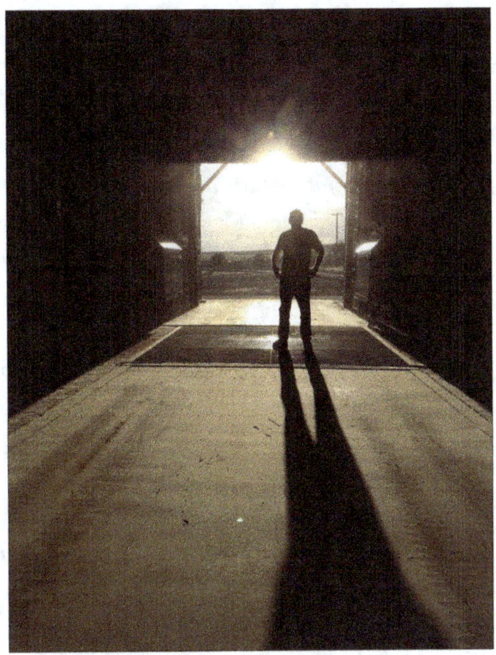

Endicott "Concrete" and Winona "Elevator Author" Mike Lust
Columbia Heritage Photographs

So I wonder. Is my managing the elevator an art? Rather, is the joy I sometimes bring suffi-
cient to actually entertain discriminating fans of elevator work?

The locals do occasionally show up when work is not bringing them, but I don't believe it is
to watch me work. They come mostly to visit which is even better. The job put me in the center
of the local harvest social scene. Now all the news from the drivers moved through me. How
good of a job am I doing? Does it approach sublime? I admit the thought is at best fanciful.
Still, let us appreciate this. I do my small part to put food in the stomachs of the masses. Can
little, unknown jobs be done beautifully? Hemingway wrote of a Hispanic man so great at his
craft that it stretched the imagination of the reader. Would Santiago be honored by the work
of this elevator man?[20]

As in Aesop's fable of the city mouse, the special aura of harvesttime also attracted relatives from
more distant places who came with their families each summer to renew the fellowship of kinship in
hot, dusty fields. The women labored throughout the day in sweltering kitchens to prepare sumptuous meals proudly served on grain-pattern dinnerware like popular Homer Laughlin Golden Wheat
and Kaysons Golden Rhapsody that were offered as promotions by local merchants.[21] In Steve Turner's poignant story of Columbia Basin wheat farming, *Amber Waves and Undertow* (2009), the Maryland native recounts his first experience driving harvest truck. Turner (1937-2012) had come west
in the summer of 1957 to earn money for college. Circumstances led him to the Bob Phillips place
near Lind in neighboring Adams County just in time to commence harvest. The Phillips crew ran
two pull-type International combines similar to ours and a John Deere self-propelled "pusher." After
a 5:30 AM breakfast and minimal instructions on operating the hoist and hi-lo axle range, the fore-

man issued brief orders about "loading on the go" to save time along with accident warnings before directing the new hire to a distant parking point alongside the field. Filled with apprehension about driving just inches away from a deafening beast and not losing any of the precious grain spilling onto the bed, Turner felt his heart race as the machine approached.

> *Now, the giant metal wheels were rolling, and every visible part on my side of it was moving, jittering, turning. . . . And I was mesmerized. Drive next to this wallowing monster, in a truck I hardly knew the width of? I was cranking the window up as fast as I could to block the sudden blizzard of chaff and dust blown by the no longer mysterious rear-end propeller. Flying straw was ticking hard against the hood and windows with the sound of a fast, forceful typewriter. . . . Sweating now in the sealed-up cab. Hands clenched on the steering wheel. Crazy to drive into this mess, but no choice. . . . [T]o monitor the ongoing dump through the rear window obviously required driving one-handed, one arm atop the back seat so I could look behind at the grain spouting down instead of forward toward my doom.*
>
> *. . . And then, as the elephantine parade of cat and combine plodded ahead, pulling away its noise and turbulent miasma, the big header reel came into view out the combine's other side. . . . With distance, as the harsh sound and blasting dust cloud moved away, the reaping took on a sedateness much more appropriate to the dignity of this ancient process than the fright I had just experienced.*[22]

Long Days, Endless Fields

The first stirrings of harvest in our part of the 1960s Palouse took place in late June. Longtime Endicott businessman and mechanic Conrad "Coonie" Fox began servicing his three McCormick 151 self-propelled combines to commence a six- to seven-week custom harvesting run. At the same time local Morasch cousin couples Leo and Alma—my future in-laws, and Carl and Emma set off on a similar route with their families in a brigade of three John Deere 95H combines, trucks, and cars laden with food and cooking supplies. Coonie recruited a dependable cadre of college students home for summer and other locals to man the annual brigade. While crops around home were still mostly green, the group lumbered nearly a hundred miles southwest to "Basin Country" in a parade that included the three red combines and two substantial International trucks that pulled a cluttered "trap-wagon" to hold tools, spare parts, and fuel, and a portable bunkhouse. Cutting commenced in the sweltering dryland and irrigated fields near Pasco and slowly moved eastward like a cavalcade of crawling crustaceans with the gentle gain in elevation and ripening crops all the way to Tensed, Idaho. The young men learned lessons about teamwork and self-reliance while contending with breakdowns, grain fires, and landowners zealous to procure every kernel. The crew covered 320 miles and worked seven days a week from dawn to dusk to earn combine operators twenty dollars a day; truck drivers drew fifteen. Artist Peggy Hanover captured the brigade in operation in a lurid harvest painting (PLATE 7) that seems to stir the heat and chaff of a Columbia Basin July afternoon.[23]

Considerable time was always available in the truck to think or read while awaiting the combine's return in a stifling draw bottom or on a breezy hilltop. Perhaps our mother's example had led us to be readers of paperbacks available on a large revolving wire rack at Jim Huff's local drug store. While my brother was attracted to Ian Fleming spy thrillers, this is where I first met writers of historical fiction like O. E. Rølvaag, Vilhlem Moberg, and James Michener. The Galilean archaeological dig in Michener's epic *The Source* (1965)—a thick book I thought would last all summer, acquainted me with Neolithic wadi life in the fictional village of Makor. Here the Ur family matriarch comprehends the value of planting grains for self-sufficiency while the men travel widely to hunt.

We grew up hearing many tales from Aunt Mary about our Sunwold great-grandparents on the Dakota and Montana frontier which eventually led me to Rølvaag's stirring and occasionally disturbing scenes in *Giants in the Earth* (1927). Great-Grandfather Andrew Sunwold had come to America about 1882 and settled on a small farm along the Missouri River near Washburn, Dakota Territory, before sending for Gertrude and young Mary. To make ends meet, taciturn Andrew also traveled as a steamboat woodcutter to fuel the vessels' seasonal runs to faraway Ft. Benton, Montana—westernmost extent of Missouri River steamboat traffic. This would be the genesis of family farming connections in the Flathead Valley. Billings artist Charles Fritz magnificently captured century-old Big Sky Country farm scenes in luminous paintings like *August Harvest* (1998), *Homesteader's Harvest*, and *When Toil Brings Bounty* (2010, PLATE 8). Describing the latter work, Fritz observes, "In the diaries of the homesteaders we often find the admirable who were motivated to gamble everything for an opportunity to prosper. They were . . . hard working, social, family-oriented people who loaded their wagons with farming tools and the family Bible. . . . We can read about their struggles and their successes, but most importantly, we can admire and emulate their perseverance."[24] Certainly Gertrude persevered as Andrew seems to have been gone for long periods. She used a rope to find her way from house to barn during blizzards, and nursed her husband back to health when he once returned so sunburned from days of prairie wander that she first mistook him for a stranger.

Rølvaag's Per Hansa and his wife, Beret, struggling against storms, locust plagues, despairing homesickness, and the mystical universe of Old World thought. The Hansas, in turn, led me to meet Karl Oskar and Kristina Nilsson in a subsequent summertime encounter with Moberg's magisterial four-volume Emigrant Series (1949-1959). The books dramatize the 1850s Swedish farmer immigrant saga of home building and barn raising, and planting and harvesting in Minnesota Territory. Experiences described on many pages reminded me of family tales my grandfather often spun about Palouse "sod-bustin'" days while riding in the harvest truck with us to see the hills of his youth:

> *He liked to sit at the window and look out at his fields; this was the land he had changed. When he came the whole meadow had been covered with weeds and wild grass. Now it produced rye, wheat, oats, corn, potatoes, turnips. The wild grass had fed elk, deer, and rabbits; now the field yielded so much there was enough for them as well as for other people.*[25]

Reflecting on those lines, I am mindful of the many creatures of horn and hoof that still roam the Palouse Hills if in diminished numbers from a century ago. However, I have not seen a jack rabbit near our home place in decades and the once commonplace barn swallows, yellow sulfur butterflies, and brown "miller" moths seem far fewer. The latter are a pest to grain growers as the hatchling cutworms feed upon emerging seedlings. But they are also an important food source for swallows and songbirds. As heralded in Rachel Carson's *Silent Spring* (1962), the resiliency of exquisitely complex ecosystems is not without limits, and relatively few conservation-minded farmers in our 1960s rural neighborhood established riparian greenbelts or planted highly erodible slopes with native forbs and grasses. On the other side of the world at that time in Ireland, young Francis Duggan watched the last vestiges of stationery thresher harvests and the ascendance of modern agricultural mechanization. Long since living in Australia, Duggan became an uncommonly prolific poet who speaks through folksy verse about benefits and loss through rural change in poems like "The Corncrake Heard No More" and "The Last Threshers."

> *Threshing day was a busy day in every farm yard*
> *And the neighbors came to lend a hand and the work at times was hard*

Like Carson, Duggan especially mourns damaging recent impacts on the natural world like the disappearance of corncrakes, chaffinches, and other wild creatures of his youth.[26]

Old Country Harvests of Sorrow

As young people do in many rural areas, I spent considerable time among community elders. While still in high school I decided to interview all first-generation immigrants living in our vicinity who had been raised in our ancestral Volga German village of Yágodnaya Polyána. This mission led to dozens of visits, usually in the company of Grandpa Scheuerman or my heritage-minded uncle and aunt, Ray and Evelyn Scheuerman Reich. Evelyn was our indefatigable family genealogist and provided essential service in what anthropologists term the special role of "folk broker." In this way I was able to start recording a series of oral histories about life in peculiar dimensions inhabited by Old Country field spirits and dire wolves, *Braucheri* (folk doctors) and *Hexeri* (spell-casters). My special interest came to be life *an dem Khutor* ("in the country") where villagers sometimes lived for weeks on the open steppe miles away from Yágodnaya to tend fields, range livestock, and harvest their crops. These area seniors also told of hand-broadcast sowing, sharpening tools on a stone grinding-wheel, churning butter, and carding wool. I had some notion that their generation was a fundamental link between cultures ancient and modern. What they knew from self-reliant experience about raising crops and making clothes most members of my generation would only know through works by folklorists, artists, and authors.

While attending the annual convention of the American Historical Society of Germans from Russia in San Francisco in June 1977, my wife, Lois, and I met an elder and Yágodnaya native of capacious memory, eighty-two-year-old Catherine "Kedda" Luft. She had traveled from Sheboygan, Wisconsin, with her two daughters, Esther and Miriam, who shared their mother's appreciation for the history and traditions of our people. Kedda, who had immigrated to the United States in 1913 at the age of seventeen, shared much with us about Volga German folk music. She recorded many plaintive renderings of *An ihre Wolga, auf der Steppe* (Along the Volga, on the Steppe) and other songs of mystical and sacral character reaching far back in time. She described the Yagaders as "a gentle, contented people of the soil," and because her brothers had died in childhood she joined her father from the age of eight to work in the family's allotted fields. Kedda reminisced about riding in the wagon with her father some dozen miles beyond the surrounding stands of birch, oak, and hazelnut to reach their land:

> *I can see the other farms stretching out black and fruitful, or in other times, green with*
> *growing fruits, and in harvest, gold with waving, ripening grain. As far as I know, the land*
> *in our region was all black, usable earth. When the rain came at the right time, the seeds*
> *sprouted and one could just see them grow. . . . For us there were large stretches of potatoes*

and sunflowers. But we could not think of the size; the weeds had to be hoed out, otherwise the young plants would have been overgrown! . . . We often had backaches until it was done. We lived almost the whole summer in tents out in the field.

After the hoeing came rye, wheat, barley, and the oat harvest. Some of these were mowed with the scythe [Reff]; others with the hand-sickle, and then tied in sheaves. These sheaves, except the ones of wheat, were driven in wagons to the threshing place to be threshed with a flail. The wheat kernels were driven out by horses or horses and wagons. So it went from one job to the other. . . . I must admit it was a lot of hard work, but one could be happy over the harvest blessing and forget the work soon after.[27]

I sometimes joined my grandfather for early Sunday morning German language services at Trinity Lutheran Church in Endicott. I remember visiting once afterward with his old Volga-born neighbor, Conrad Blumenschein, about the unusually dry summer we were experiencing. A stately, stout fellow who always dressed for church in a double-breasted blue suit, Mr. Blumenschein had emigrated in 1913 so like Kedda had been old enough to experience the Old World seasonal farming cycle. Elders like them and my grandfather were living tributes to agrarianism. That Sunday I heard Mr. Blumenschein explain in characteristic Hessian brogue how in the Old Country they feared times of drought and the dreaded *Hohenrauch* ("High Smoke"). This withering south wind sometimes arose mysteriously from the Caspian and could reduce a ripening grain crop to tiny, shrunken kernels in just a few hours. "The Russian peasants would fall on the ground and pray for rain," Conrad said. I asked if it did any good. "*Ach,*" he smiled kindly, "*das Gebet kann nur hölfen.*" ("Well, prayer can only help.")

On a visit to see Mr. Blumenschein at his tidy St. John home in May 1980, I asked about his farming recollections as a young man in the Volga colonies. He vividly described harvest operations with scenes of the distinctive pole-frame "ladder wagons" laden with grain reminiscent of an Ilya Repin landscape and familiar *eh-eh-oókh-nyem* ("heave-heave-ho") workers' dirge:

Harvest began the last of June and early July. Sometimes folks ran out of bread by then so cut several bundles to dry and get 40 to 100 pounds of rye to get by. In July rye was pulled up to aerate in rows, two to three weeks of drying, then hauled home to a threshing yard on the outskirts of town. Four to six men flailed the bundles, one side at a time. Then the bundles were cut open. The ground had been trampled hard by the horses and watered down. Others lift the bundles with forks while others flail and stack it for the horses and cattle. Those who didn't have a granary—die ambar, often flailed bundles on the ice in the yard during winter. Fanning mills were then used after flailing, a pile of grain put in to clean it and wheat runs out on canvas then shoveled into hundred-pound sacks.

. . . In early August wheat harvest began and was done differently. We put tents up where there were no buildings an dem Khutor [in the country]. . . . Cut [the grain] with a sickle and bundled, then bundles shocked and gathered to be spread into a big circle, about 100 to 200 bundles. Then [it was] trampled out with horses and wagons, left alone in the middle, then with a pole go all around the ring and turn it over to shake the wheat [kernels] out. Then repeat with the horses. The women shake the stalks and rake it out on a pile. The chaff and wheat are piled into the center of the ring, 200 to 300 bushels. Then it is run through a

fanning mill and deposited again on a big canvas or bagged. Sometimes wheat was taken to the Volga River and loaded by hand onto boats which were stranded in the river because of low summer flow, but that's where the buyers were.[28]

Diminutive, cheery friends Mary Morasch and Mollie Bafus told of Volga flax and hemp harvests and the laborious process of transforming the dried plants into beautiful silvery-brown thread, yarn, and fabric. These spring-sown crops were pulled out by the roots, tied into small bundles, and first broken down by either dew or soak retting. After drying workers then used a wooden "breaker" to crush the outer, brittle layer of decayed stalks for separation with knives from the strands of soft inner bast that extend into the roots. After this peeling process (scutching) the threads were pulled through combs of thin, sharp prongs (hackling) to clean, split, and straighten the fibers. The long, hair-like threads were then spun and woven into three grades of fabric that was patiently boiled and sun-bleached to made into linen tablecloths and bedspreads, heavier work clothes, and coarse material for tents and sacks. In the 1890s Northwest farmers began experimenting with flax cultivation using plants and techniques introduced from Russia, Belgium, and Holland. Russian Riga and White Blossom Dutch were the most widely cultivated American varieties with vast acreages raised along Puget Sound and in the Willamette Valley. Substantial quantities were exported to Ireland and Scotland.[29]

While wistful at memories of egalitarianism communalism that lingered into modern times, our immigrant elders did not paint an Old Country pastoral idyll. They had willingly left and were grateful to have done so. After more than a hundred years of intensive agriculture, crop yields on the Volga by the late nineteenth century had diminished along with farmer landholdings since the colonists tended to have large families. Complicated property "revisions" took place in accordance with periodic censuses with land assigned on the basis of the number of males in each household. The vast majority of villagers were self-sufficient small farmers who maintained substantial *Hinnerhof* (backyard) gardens, berry patches, and fruit trees. Only in larger cities like Saratov and Pokrovsk did some families engage in milling and textile trades. Military conscription began in the late 1800s with the revocation of Empress Catherine's century-old liberal settlement provisions, and often meant miserable service in remote parts of the Russian empire. For family members who remained in Russia, the new century brought cataclysm. Following the 1917 Russian Revolution, the country was plunged into a three-year civil war from which the Bolsheviks under Lenin and Stalin gained power.

The new communist order warred against private landowners and organized religion. The pietistic Volga German farmers fell prey to some of the era's worst depredations that were part of the government's wider campaign to submit the masses to its authoritarian control. The policy led to mass murder, deportations east of the Urals, and what historian Robert Conquest termed the 1926-1937 "Harvest of Sorrow" terror-famine. As many as 14.5 million people died as grain supplies in the countryside were ruthlessly confiscated to provision urban areas. Some twenty million small farms existed across Russia in 1929 that had long supplied the nation and markets throughout Europe. The brutal collectivization campaign consolidated those holdings by 1934 into 240,000 *kolkhozes*, many managed by inexperienced or disreputable persons, that produced a fraction of earlier harvests.

Like many others in our area of the Palouse, the Morasch, Bafus, and Blumenschein families received letters from relatives in Russia during the 1930s with dire appeals for help. Against the backdrop of abundant Northwest grain yields and despite the Great Depression, generous donors throughout the

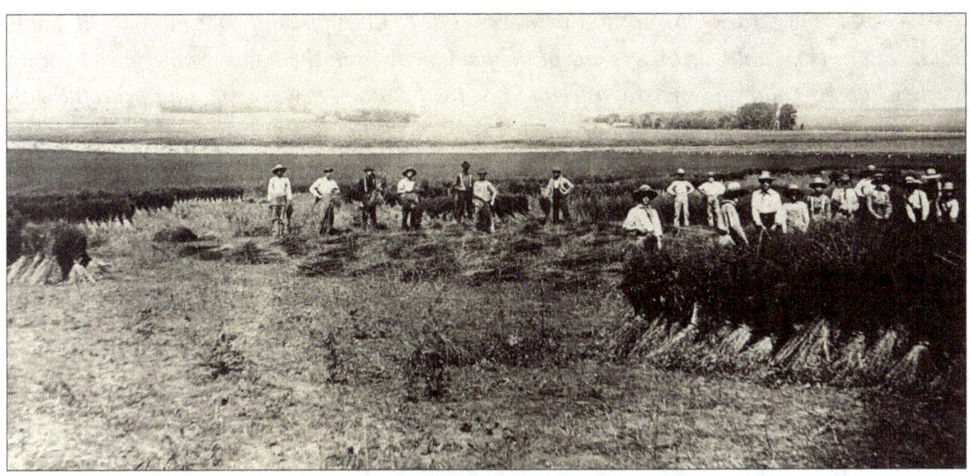

Puget Sound Flax Harvest (c. 1900)
Columbia Heritage Collection

region rallied to send funds. Although often confiscated by the authorities, in some cases what made it through meant the difference between life and death. Mary Morasch's great-niece, my wife, Lois, found a cache of these letters in 1990 that had been stored away in a basement corner for nearly six decades. A young girl pleads for help so she can buy fabric for her first communion dress; others report on the confinement of family members in "the house with no windows," and "hard life . . . here on the tundra."

The yellowed pages of flowing script attest to catastrophe that many would not survive: "To begin with, we send you greetings in the name of the Holy Spirit. I will let you know that we are without parents. Where they are is unknown to us [and] only to God. We think they received the same treatment as others so is very bitter for us children. Will you not take mercy on us because we are lost children and will not be long on this earth. . . . Please send us help because we are orphans." Family communications from the Volga and Ukraine ended abruptly at the end of the decade. The quest for knowledge of happenings in the old homeland collapsed into what researcher Conquest called "the Memory Hole." Details regarding the survivors' fate—briefly recounted here in a later essay (see pages 111-113), would not come to light until long lost relatives were finally reunited in the 1990s.[30]

"Reapers at Noontide"

As I explored more fully these stories of harvests past and present, it also dawned on me that my grandfather's treasured "Lautenschlager and Poffenroth" photograph had been taken near the John Poffenroth farm within a couple miles of our home place north of Endicott. Grandpa was there because his older sister, our short, indomitable Aunt Mae Poffenroth Geier, had married John. Beloved and sometimes feared Aunt Mae was a no-nonsense paragon of self-reliance. She forever cooked on a woodstove until her passing in 1972, and sang and played the familiar Volga German hymn *Gott ist die Liebe* (God is Love) on an oak pipe-organ in her small living room. Had she decided in her seventies to butcher a hog or drive a tractor we would not have interfered. She and John had raised their family just over the hill from where the 1911 harvest picture had been taken. In a 1963 memoir, Aunt Mae provided insight regarding women's essential and substantial roles at harvest time, even during pregnancy:

We raised big gardens. An early garden was close to the house and late garden with pota-toes, watermelon, and cucumbers out in the field away from the house. Sometimes I got up early while my children were still asleep to hoe the potatoes. It was quiet and peaceful with the fresh dew on the wheatfields and garden smell. Nothing I liked better with only the sun coming up over the hills and blue sky. . . . When my twins were born we were harvesting our winter wheat. I was cooking for ten men and just before they were born the men went out to another place to harvest. Between then and when they came back to cut our spring wheat I had my twins and was back on the job cooking again and had to help to do the chores and heavy work.

. . . For at least three summers I also cooked in harvest in the cookhouse for Conrad Her-gert's crew. From ten to fifteen men were on hand. I baked bread for all the men every other day and cooked on a big old wood stove. One morning I counted the baked things and besides six loaves of bread, I baked twenty-four biscuits, twenty-four cupcakes, and three pies. If you ever wanted to smell something good it was coffee made in a big coffee pot on the stove. I got up at 3:30 in the morning to light the stove to have it hot by 4:00 AM to fry bacon and eggs, sometimes pancakes on Sunday morning.[31]

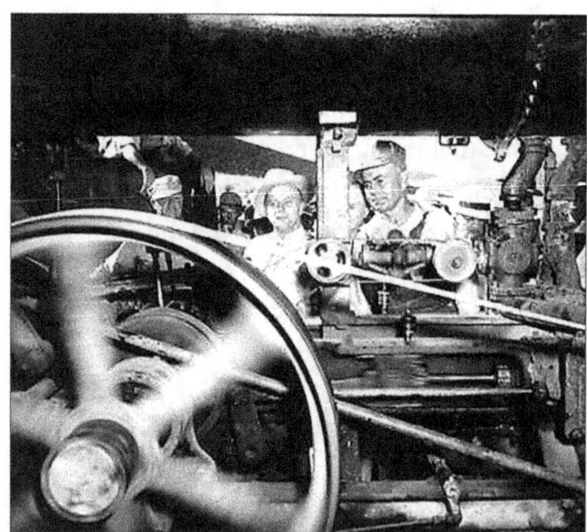
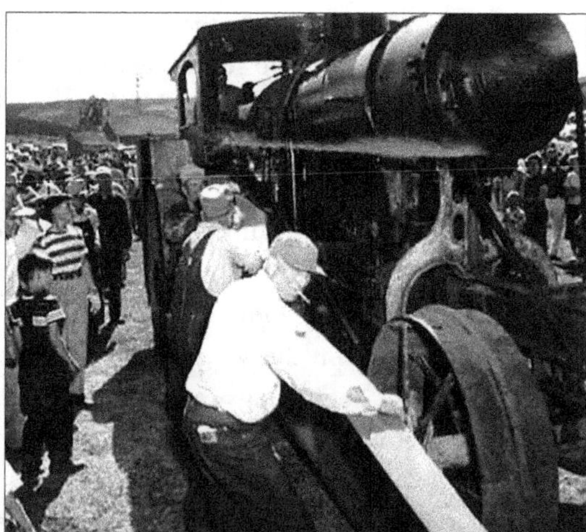

R. R. Hutchison, *Busch Threshing Bee* (Edgar Bergen & Chris Busch at left), 1953
R. R. Hutchison Studio Photograph Collection; Manuscripts, Archives & Special Collections
Holland-Terrell Libraries, Washington State University, Pullman

Among our area's most colorful and continuous vintage harvest experiences were the September threshing bees first organized by Chris Busch at his farm near Colton, Washington, in 1947. Busch had known steam-powered threshing from his Palouse Country youth. As the massive old machines and harvest wagons became obsolete, he began a collection that grew to thirty-some in various states of repair. Word spread throughout the region about the remarkable assemblage and with help from a dedicated network of other enthusiasts including neighbor Bill Druffel and C. R. Miller, retired WSC professor of agricultural engineering, Busch organized the Western Steam

Fiends Club in 1951. Members soon joined from six states and three Canadian provinces. The event kicked off with a grand banquet sponsored by local church and school volunteers followed by a parade led by Busch's favorite 1912 Minneapolis engine, 1917 McCormick reaper-binder, and Case separator. A spectacle of old-time harvest festivities followed that drew as many 4,500 onlookers, and the tradition is carried on today at the annual Palouse Empire Threshing Bee held on Labor Day weekend at the county fairgrounds near Colfax. Other Northwest threshing bees have been regularly held in recent years at Pomeroy, Davenport, Ellensburg, Riverside, Toledo, and in Oregon at Dufur and Madras.[32]

Our mother was an avid reader and may have been among the few 1960s Book-of-the-Month farm wives in the vicinity. She had also been a serious stamp collector in her youth and our small family library contained her substantial *Adventurer Album* filled with colorful commemoratives from the 1930s and '40s. Pages featured virtually every nation on earth and served as a fascinating introduction to world geography and cultures. She had grown up in the small northern Idaho mining community of Wallace where many of her neighbors were first-generation immigrants from Europe whose correspondence with relatives abroad started her collection. Unlike many of the American stamps we used that featured famous persons, older ones from Weimar Germany, Hungary, Ukraine, and elsewhere were emblazoned with scythe-wielding workers and other agrarian images.

One notable grouping was Switzerland's 1945 "*Pax hominibus*" series of five stamps that celebrated the end of World War II with a Latin theme based on Luke 2:14 translated as "Peace for Men of Goodwill." The images were a horn of plenty with grain and grapes, a spade overturning earth, and more customary olive branch, dove of peace, and door key. I learned later these were designed by Swiss artist-engraver Aldo Patoochi (1907-1986) who was known for many scenes of farm life. Patoochi designed his images to be superimposed over pictures taken by Swiss landscape photographer Hans Steiner (1907-1962). The horn of plenty rose before a prodigious field of ripened grain.

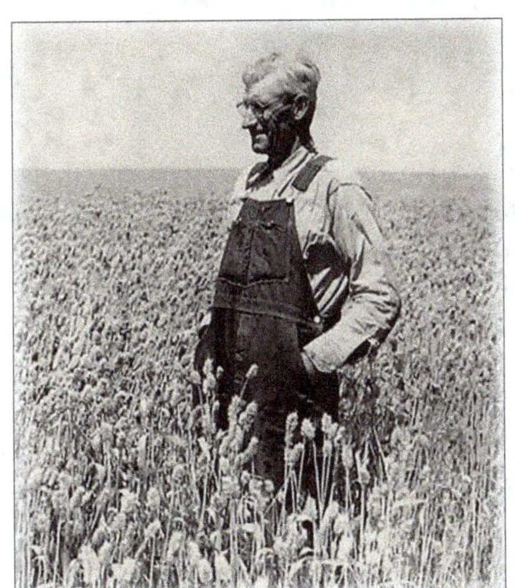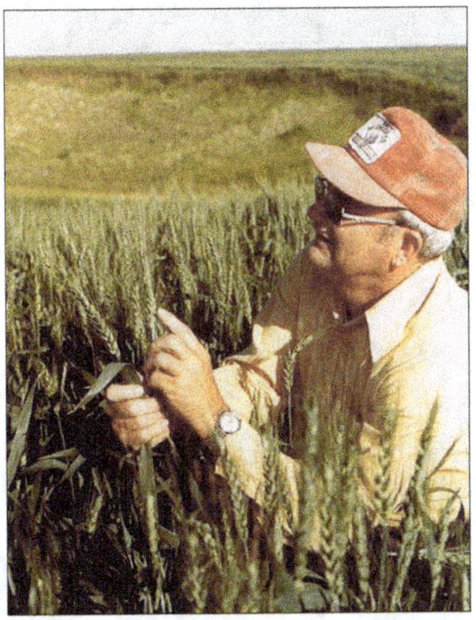

Left: Art Sunwold in his field of club wheat near Fairfield, Washington (c. 1950)
Right: Don Scheuerman counting bearded wheat head meshes to estimate yield (1980)
Columbia Heritage Collection Photographs

Mom's love of reading provided us with early access to beautifully illustrated hardbound selections by John Steinbeck, Carl Van Doren, and other leading authors as well as historical classics like Catherine the Great's *Memoirs*. That volume, richly bound and embossed in bright orange faux leather, introduced me to Romanov intrigues and life when the empress invited our ancestors to colonize the Volga region. Grandpa called her "die Kaisarina Katarina" and said she so missed hearing her native German language that she came down to the St. Petersburg docks to mingle with the newcomers. I came to regard that prospect skeptically but learned long afterward that she sometimes did so. Thanks to Mr. Yenny, our formidable school principal and eighth grade teacher, more thorough study of Grandpa's poets came our way with agrarian relevance. We read every word of Longfellow's *Evangeline* aloud in class, and my hexametric memories of "Acadie" remained evermore vivid not only because the heroine's evocative if peculiar name was the same as my maternal grandmother's middle name. The epic made recurrent reference to words familiar to our rural experience; in just the opening lines we met "goodly acres," "harvest heat," and "reapers at noontide."

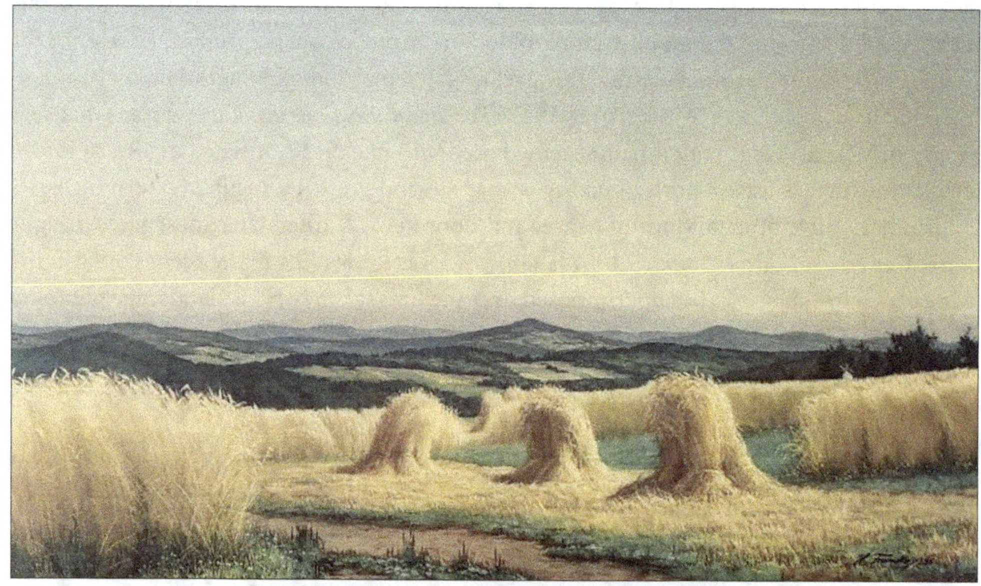

Hans Franke, Baden Harvest Landscape (1935)
Oil on panel, 20 ¼ x 34 ¾ inches
Columbia Heritage Collection

Apart from a large mirror and family pictures on our living room walls, our country home had little in terms of framed decor. But lack of popular country scenes by Millet or Bruegel did not limit the colorful and meaningful existence of a threatened agrarian lifeway. We were immersed in it from every direction. The "Northwest Drylands" grain fields popularized a generation later by Richland photographer John Clement on calendars and canvas surrounded our place. I enjoyed periodic visits to the home of many elderly relatives and local storytellers. Clara Schmick Litzenberger not only shared tales passed down about our family's Old Country life but also had wide-ranging interests in music, art and the German language. A large harvesttime painting in a sculpted gray and gold frame hung in her living room. Except for some scattered woodlands in the distance the scene could have been fields surrounding nearby Steptoe Butte. A solitary Baden scyther strolls through a prodigious stand of grain in a scene of mystical grandeur. The artist was Freiburg landscapist Hans Franke (1892-1975), who painted numerous altarpieces and views of contemporary German rural life. In his book *De Sieben Sphären* (The Seven Spheres, 1933), Franke wrote that art's highest expression was fidelity to the divine order seen in nature rather than material values and public acclaim. My boyhood interest in these topics apparently evidenced a kindred spirit. Upon Clara's passing in 1979, I learned that she had willed the piece to me.[33]

Hallowed Space

Among Dad's few books was a 1930s English translation of the German Ohio Lutheran Synod's *Gebets-Shatz*, or *Treasure of Prayers*—probably a confirmation gift. A "Harvest Festival Prayer" reminded listeners of forces beyond mortal control known to farming folk:

> *O Give thanks to the Lord; for He is good;*
> *because His mercy and truth endure forever . . .*
> *O, how we took we feared the destruction of the precious grain in the fields!*
> *O, how we took thought and troubled ourselves,*
> *lest the bread which God has yet given us . . . might be snatched away.*
> *Thou hast given us the early and the later rain in due season,*
> *and has faithfully and annually protected our harvests. . . .*

Confirmands in my day were customarily presented a similar small volume containing a "For Fields and Crops" prayer that likely harkens back to the medieval Rogation Days of Ascension Week:

> *. . . Teach me, dear Lord, to know that Thou dost supply, in due season, daily bread for us and all mankind. Give us the needed diligence and necessary skill in the sowing and gathering of our harvests. Protect our fields from hail, fire, and floods, and let the earth yield its increase. Make us a thankful people as we enjoy working amid growing things, and open our eyes to behold the beauty of Thy creation.*

Edwin W. Molander (1901-1983) based the 1949 construction of hometown Endicott's substantial Trinity Lutheran Church on Old World Northern European ecclesiastical design. The Spokane architect had received prominent regional commissions for Northwest churches and public buildings that reflected a distinctive blend of traditional and modern features. These were generally characterized by exposed rough-hewn timbers, natural stone, and decorative carvings. A regnant color scheme of Prussian blue, turquoise, and umber with gold filigree and trim was used throughout the vaulted sanctuary. An attached cloister porch featured a substantial frieze of carved wooden panels depicting the life of Christ in symbols and Latin monogram.

One respectfully approached the church's main arched entry as if entering hallowed space to uplift to the inner spirit. Members sought to honor the example of Tabernacle decoration by Bezalel ("'The Lord has . . . filled him with the Spirit of God, with skill, with intelligence, with knowledge, and with all craftsmanship, to devise artistic designs . . .'" [Exodus 35:30-31]) and applied their skills in woodworking, metal fabrication, and calligraphy. Although Trinity's sanctuary did not contain figurative stained-glass windows, other churches in the area featured spectacular examples of such art. In later years our daughter attended a Spokane fellowship, Christ the Redeemer, which met in one of the city's finest examples of English Gothic Revival architecture and completed in 1911. John K. Dow (1862-1961), one of principal architects of the Empire State Building, designed the beautiful structure which contains stained-glass masterworks like *The Sower* and *Ruth the Gleaner*.

At Trinity in Endicott, Pastor Fred Schnaible (schnai'·blee) presented weekly lectionary reading in both German and English until the 1970s. They often featured associations of such ancient Jewish harvest festivals as the Firstfruits barley sheaves "wave offering" and Feast of Harvest Ingathering. He safeguarded our church's remarkable library of rare books assembled by his predecessors. A massive volume by Christian Hebraist Johannen Lund (1638-1686), *Die Alten Jüdischen Heiligthümer, Gottesdienste und Gewohnheitenbound* (*The Old Jewish Shrines, Worship and Customs*), described Old Testament Levitical offerings. Printed in 1701, the book was bound in vellum and lavishly illustrated with German engraver Johann W. Michaelis's woodcuts of biblical scenes that included ancient pastoral and agrarian traditions. Pastor Schnaible also learned of local farmer and church member Walter Scholtz's special skill as a calligrapher. He prevailed upon him to inscribe countless confirmation certificates and other church documents in his distinct Old World script of red and black with gold ink embellishments that have become treasured works of art in their own right.

Worship services at Trinity seasonally featured hymns of thanksgiving and impressive choir cantatas like "All Praise the God of Harvest" (". . . with head and heart and voice. All praise the God of harvest; creation all rejoice") and "The God of Harvest Praise" (". . . hands, hearts, and voices raise with sweet accord; from field to store the grain bearing you sheaves again") from an old Scottish Moravian tune. We heard sermon texts from the Book of Ruth about barley gleanings and her Kinsman Redeemer, Boaz. Tattered brown hymnals contained considerable music of classical origin including John Galloway's "Lord of the Harvest, Thee We Hail," based on a Franz Haydn tune from his 1798 oratorio *Creation*. Agrarian imagery has also inspired modern hymnodists like Albert

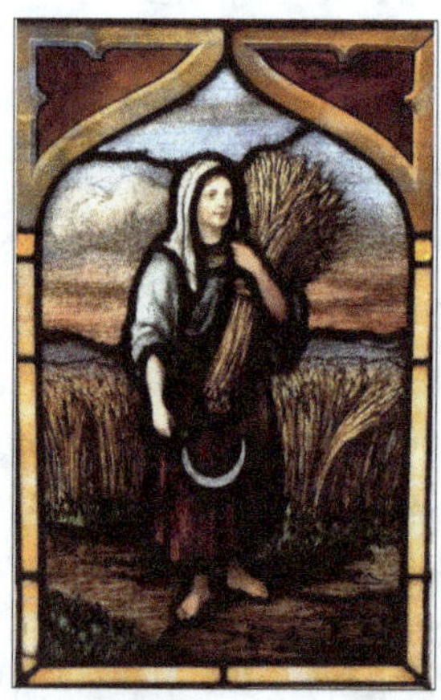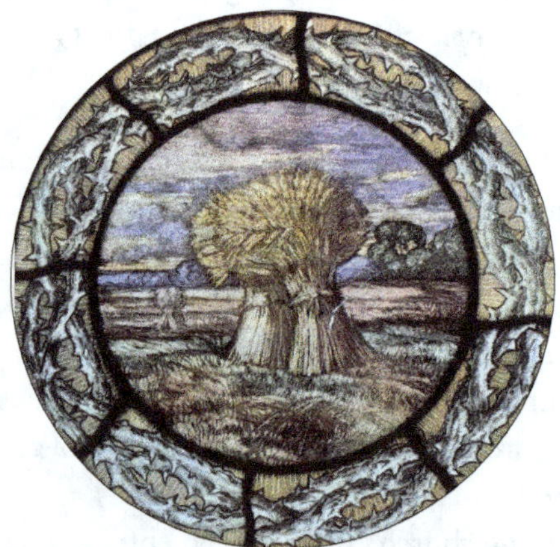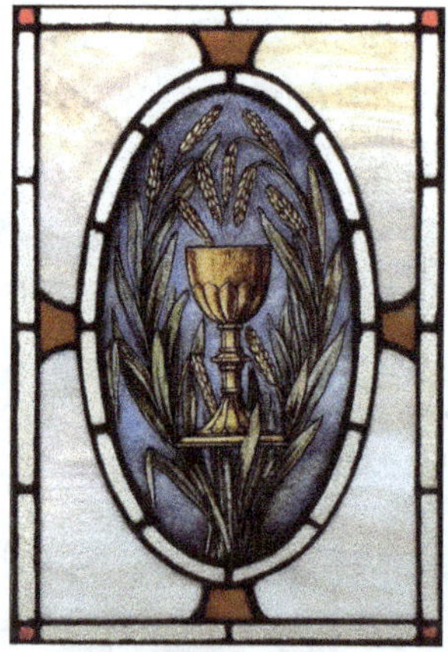

Left to Right: ***Ruth the Gleaner, Sheaves and Crown of Thorns, Christ the Bread and Wine* (c. 1910)**
Stained Glass Ensemble, Christ the Redeemer Church; Spokane, Washington
Columbia Heritage Collection Photographs

Bayly (1901-1984) and Omer Westendorf (1916-1997) whose verse have been associated with communion and thanksgiving:

"Gift of Finest Wheat" (O. Westendorf)

You satisfy the hungry with gift of finest wheat.
Came give to us, O saving Lord, the bread of life to eat.
You give yourself to us, O Lord; then selfless let us be,
To serve each other in your name in truth and charity.

"Praise and Thanksgiving" (A. F. Bayly)

God bless the labor, we bring to serve you,
That with our neighbor we may be feed.
Sowing or tilling, we would work with you,
Harvesting, milling for daily bread.[34]

Architect Molander's majestic panels also included broad window base panels displaying carved grain sheaves as if homage to pre-literate medieval times. In those days clerics valued visual expressions of biblical history and spiritual truths in cathedrals and chapels throughout Europe. At Sunday worship we heard Psalms on "abundant wheat throughout the land" (72:16) and the prophetic words on nations giving up war and beating "swords into ploughshares and their spears into scythes" (Isaiah 2:6). Most all congregants committed the weekly-sung offertory to memory: "Gather a harvest from the seeds that were sown, that we may be fed with the bread of life. . . ." Jesus' familiar parables told of sowing, bushel baskets, and sickles—these in the fourth chapter of Mark's Gospel alone, and John's soaring prose related the allegorical Parable of the Grain on sacrifice and resurrection (". . . unless a grain of wheat falls into the earth and dies, it remains alone; but if it dies, it bears much fruit." —John 12:24).

Every Sunday morning service concluded with Pastor Schnaible's benediction in the familiar words and cadence of the ancient Aaronic Blessing: "The Lord bless thee, and keep thee. The Lord make his face shine upon thee. . . ." These lines first appear in the Old Testament book of Numbers (6:22-26) as Israel prepares to depart Mt. Sinai for the Jordan River and are repeated as the opening verse of Psalm 67. Both the original blessing and psalm have special significance for the harvest

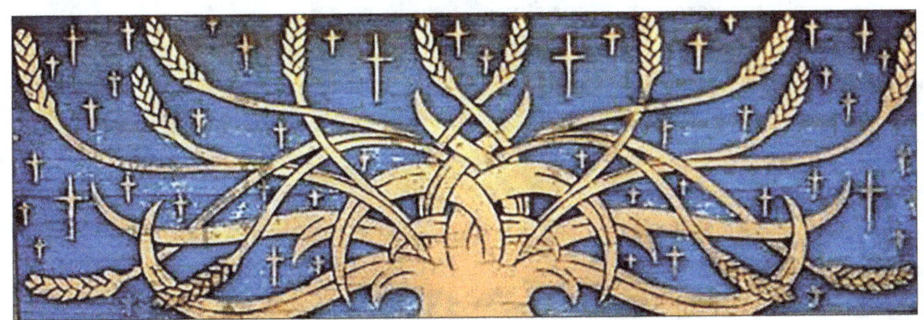

Edwin Molander, Grain Sheaf Wall Panel (1949)
Trinity Lutheran Church, Endicott, Washington
Columbia Heritage Collection Photograph

season. Numerous verses described the Promised Land as a place of goodness with grain and fruits (e. g., Deuteronomy 11:14) that were ceremonially offered in Temple worship. Reference in Psalm 67:6 to harvest throughout the earth is significant for the universality of divine blessing to "all nations," "the peoples," and "the ends of the earth." The psalm in Hebrew contains seven lines of seven words each and was commonly recited during the forty-nine-day "Counting of the Omer (Sheaf)" between Passover and Shavuot (Feast of Weeks/Christian Pentecost) when grain was harvested. Psalm 67 inspired Martin Luther to compose the popular hymn "May God Bestow on Us His Grace" (1523), and our hymnal contained a service proper with "Festival of Harvest" prayers and readings.[35]

Twentieth-century experience throughout rural America continued to impart the cultural significance of harvest and agrarian lifeways through the mutually reinforcing influences of family, school, church, and locality. For generations these forces that radiate from life on the land formed the foundation of moral values for individual meaning and community solidarity. They invigorated religious life, promoted personal health, strengthened family ties, and fostered cooperative endeavor in ways long been marked by ritual and creative expression. Contemporary mass media increasingly features spectacle and fashion to entertain those who grow increasingly distant from physical endeavor and agrarian roots that nourished body, mind, and soul since time immemorial. But convenience and economic benefit need not supplant regard for earth care. Appreciation of agrarian art, literature, and music offers enriching prospect for purposeful reconnection whether as creator or consumer, farmer or gardener.

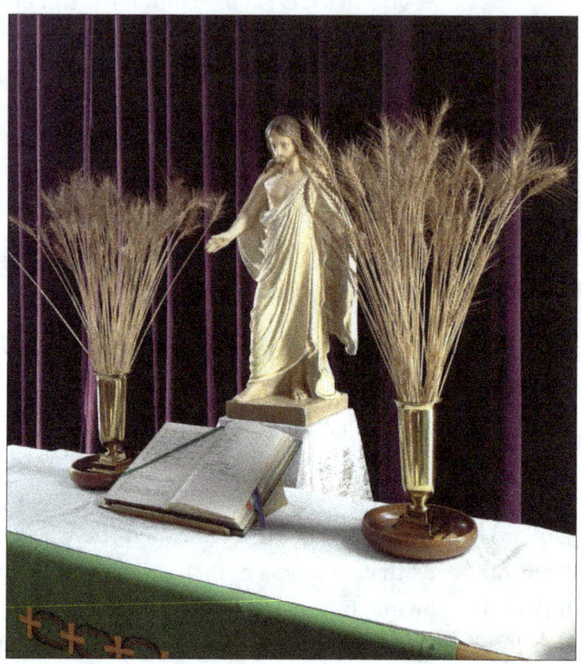

**Myrna Morasch, Altar Grain Stalks Arrangement (2014)
Trinity Lutheran Church; Endicott, Washington
Columbia Heritage Photograph**

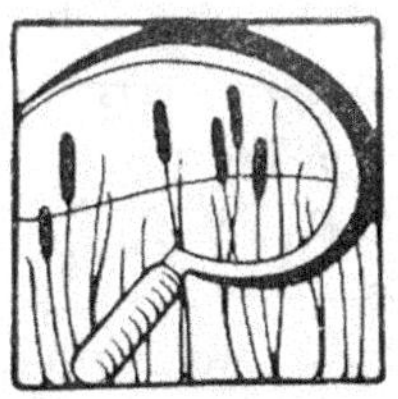

II

May Grain Abound
WESTERN HARVEST EXPRESSIONS

OUR HIGH SCHOOL ENGLISH TEACHER IN ENDICOTT, Louise Braun, was a person of capacious mind with expectations that students should read and appreciate Shakespeare and Robert Frost with the same enthusiasm shown for local sporting events. A native of tiny Viola in the Idaho-Washington Palouse prairie borderlands of scattered pine, Mrs. Braun guided our uncharted literary journeys across time and place with the peculiar incentive—highly controversial among faculty and parents, that once a week we could spend class time reading *Farm Journal, Field & Stream*, or any other periodical of our own choosing. "*Reading* is the main thing," she would say to encourage expansion of young minds. In response to adolescent complaint that many poems in our anthology of world literature seemed incomprehensible, Mrs. Braun said that poetry was commonly composed for spirited oral delivery. Soon afterward, local farmer Leonard Jones arrived as a guest speaker who had undergone a stunning transformation into Mr. Jones, country bard. "'Two roads diverged in a yellow wood,'" he thundered in the first dramatic recitation on stage that many of us had ever heard. His delivery of Robert Frost's "The Road Less Traveled" was made even more memorable in the knowledge that he was one of us, and unashamedly relished the written and spoken word as artform.

Mrs. Braun soon had us wondering about just what the poem's "long scythe" was "whispering to the ground" (from "Mowing"). Could "The Need of Being Versed in Country Things" mean that mature thinking could question some assumed claims of science and religion? To my mind, our most formidable read was *Beowulf* through which we battled for many days to make sense of alliterative Old English expressions that related ancient Scandinavian lore. This was a time before *Lord of the Rings* had ushered in a resurgence of interest in classical fantasy literature. Our teacher's reminder that many of us were descendants from these tribal peoples provided modest encouragement to continue our study of this most ancient English epic. She related that Beowulf may have been composed at the court of King Alfred the Great as early as the late ninth century, and reflected various ethnic influences of the North Sea Empire uniting the peoples of England, Denmark, and the Scandinavian Peninsula. Mrs. Braun guided our literary discovery that its monsters personify calamitous forces in human affairs, and that the poem's opening lines opened intriguing agrarian reference:

> *Lo! the Spear-Danes' glory through splendid achievements*
> *The folk-kings' former fame we have heard of,*

How princes displayed then their prowess-in-battle.
Oft Scyld the Scefing from scathers in numbers
From many a people their mead-benches tore.

We learned that Old English "Scyld the Scefing" may be translated "Shield Sheafson" or "Scyld of the Sheaf." This knowledge fostered a glimpse into mystical realms also described by J. R. R. Tolkien in his legendarium of Middle-Earth. The namesakes of the eponymous Scyldings' (Sköldings) Danish royal house founder in *Beowulf* are associated with the crucial roles as historical protector of the people (Shield) and as agri-cultural hero (Sheaf). In the mythic past he had arrived as a foundling from the west who washed up in a boat on the shore of Jutland's Old Anglia. (Elsewhere the location is given as Scandia, a place in ancient times associated with an island near the southern Scandinavian peninsula.) The golden child's head rested on a pillow of grain stalks, and he would grow into a strong and wise ruler who was the ancestor of heroic Beowulf and the Anglo-Saxon kings. Like a medieval Triptolemus, Scyld taught his adopted people the first principles of farming and husbandry. (Tolkien further explored aspects of this legendary figure who is also mentioned in lesser-known works of Old English literature like William Malmesbury's *Gesta Regum Anglorum* ["Deeds of the English Kings," 1125] and the eighth-century *Historia Langobardorum* ["History of the Lombards"]. His imaginative rendition drawn from these several accounts is beautifully expressed in the poem "King Sheave," featured in his unfinished story "The Lost Road" [1936] that is described in the modern agrarian literature section of this series.)

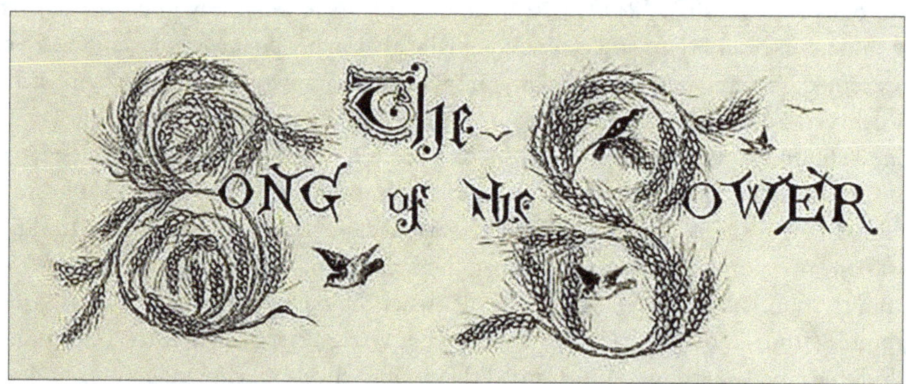

William Cullen Bryant, *The Song of the Sower* Title Page (1871)
Columbia Heritage Collection

Mrs. Braun's artful tillage of minds also included selections from *The Canterbury Tales* and several classic nineteenth-century works with rural reference by William Wordsworth (1770-1850), William Cullen Bryant (1794-1878), Walt Whitman (1819-1892) and other Romantics. I recall the challenge of appreciating Bryant's death ode "Thanatopsis" (c. 1811) while taking some pride that our region—"Where rolls the Oregon," so far removed from the author's Early American Massachusetts, merited mention in such a notable work. (The reference was revisited in his 1832 poem, "The Prairies.") But introduction to the Romantics' notions about nature, life, and death prepared the way to subsequent exploration of Bryant's agrarian-inspired verse:

Honor waits, o'er all the Earth,   *The art that calls her harvests forth,*
Through endless generations,   *And feeds the expectant nations.*

—"Ode for an Agricultural Celebration" (1832)

Fling wide the grain, we given the fields *The harvest that o'erflows the vale,*
The ears that nod in summer's gale,   *And swells, an amber sea. . . .*
The shining stems that summer yields,

—"The Song of the Sower" (1859)

Bryant's 1871 edition of his lyrical ten-stanza *The Song of the Sower* in book form was profusely illustrated by leading artists and wood engravers including Winslow Homer (1836-1910), English-born Harry Fenn (1845-1911), and Irishman William John Hennessey (1839-1917). Having been raised on a farm near Cummington, Massachusetts, Bryant grew up taking part in his parents' homestead harvests of grain, fruit, and vegetables. He became an accomplished horticulturalist and drew upon his experience and knowledge for evocative expressions used for such Sower illustration titles as "The Song of Him Who Binds the Grain" and "Consecrated Bread."

The devout Unitarian visionary also wrote hymns and became a committed abolitionist and conservationist. He expressed these views in poems like "The Planting of the Apple-Tree" in which he decrees oppression of mankind and nature. In "The Rivulet" Bryant wonders if unrestrained human activity will damage water and land, and Mrs. Braun led us to pose similar questions about our place and day. We learned that many of the most notable Romantics like Bryant and Whitman were classicists

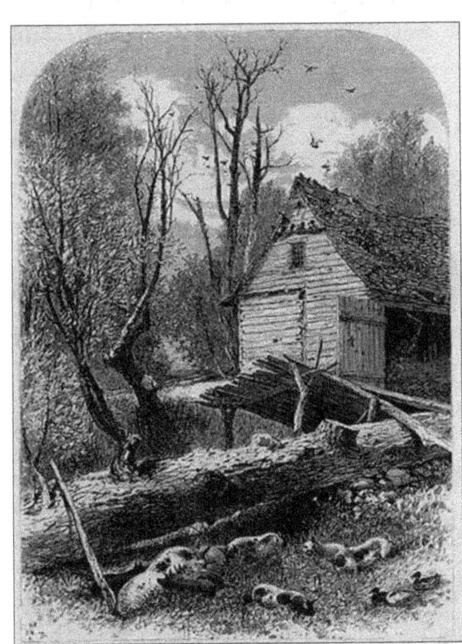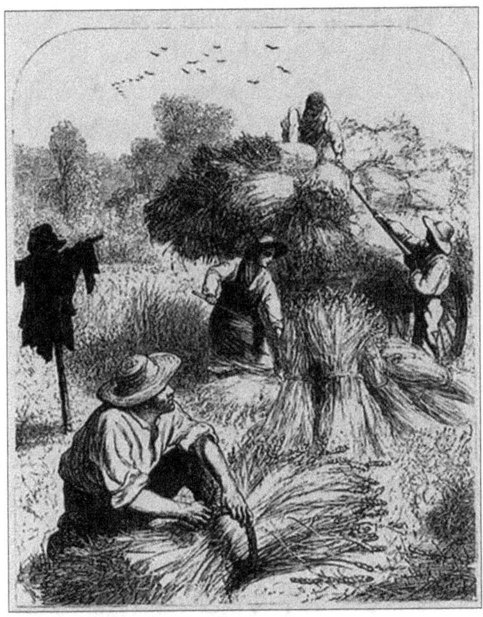

Left: Harry Fenn (artist) and John Karst (engraver), "From the Distant Grange"
Right: William John Hennessey (artist) and William J. Linton (engraver)
"The Song of Him Who Binds the Grain"
William Cullen Bryant, *The Song of the Sower* (1871)

who venerated ancient tales and ruins in allusions to the rise and fall cycles of civilization as natural and renewable as grain sprouted from seed. (Among other prodigious literary projects, Bryant completed a blank-version translation of Homer's *Iliad* and *The Odyssey*.) Hennessey's Sower illustration "The Song of Him Who Binds the Grain" begins Bryant's couplet about "those who load the wain," and "From the Distant Grange" (barn) in Fenn's engraving is heard "The clatter of the thresher's flail. . . ."

Long afterward recognition dawned that Fenn's fallen tree and deteriorating threshing barn were nuanced accompaniments to Bryant's grand if melancholy lesson on the human condition. Walt Whitman's 1871 edition of *Leaves of Grass* featured his four-line poem "As I Watched the Ploughman Plowing" that starkly expressed similar sentiment in concluding line metaphors: "Life, life is the tillage, and Death is the harvest according."[1] In such ways many authors and artists shaped lived and reflective experience into meaningful, transmissible impressions that would awaken for me years later in the art of Canadian-American painter Robert Atkinson Fox and seventeenth-century Dutch printmakers, as well as in Russian Village Prose and novels by Willa Cather and Hamlin Garland.

The Arts and Future Farmers

Art and elementary classroom teacher Arden Johnson was Mrs. Braun's and Leonard's kindred spirit who joined them in organizing community theatre and penned a weekly column on a variety of topics for the local newspaper. A pedagogical force of nature, Mrs. Johnson first came to our school as an art education consultant to rural districts in the region. In the 1970s she joined the Endicott faculty for a career that spanned three decades as an art and elementary classroom teacher. When I returned to the district as a teacher and administrator in the late 1980s, I collaborated with Mrs. Johnson, Mrs. Braun, and Leonard on a series on "Journeys" interdisciplinary curriculum units for which Arden contributed the art lessons. We often met in her exceedingly cluttered classroom that was wall-papered with student creations she had conserved over the years and displayed haphazardly with colorful cutouts from *The New Yorker*. She adored the whimsical country scenes like Ilonka Karasz's *Harvest Landscape* (1967) and *Blue Vase with Grain Stalks* (1982) by Joseph Farris that were featured on colorful *New Yorker* covers.

Through discussions and ever-changing exhibits Arden drew the attention of both students and adults to rural beauty that many of us had taken for granted. She placed numerous solicitations in the daily school bulletin for grain stalks, vegetables, flowers, and other objects to be transformed into countless sketched, painted and potted treasures that adorned hallways and homes. "Fieldtrips" took on a literal significance as she arranged busload excursions to area farms so students could draw crops, livestock, fencerows, and buildings. Arden mentored younger educators throughout the country and presented at numerous conferences and school and university audiences. Her passion for art commanded attention and many of us safeguarded the detailed lesson plans she generously shared with others: "Whatever you select to draw, place it on your desk and look CAREFULLY at it! I know this sounds obvious, but this step is often overlooked. A tree branch or grain stalk is not a straight line. A berry is not just a round ball. A good drawing takes a lot of looking before you ever put your pencil on paper. Arrange your grain or stems by placing them across some rocks, or putting them in a clear glass jar. . . ." Painting sessions back at school led naturally to discussions and viewings of similar subjects painted by van Gogh, Thomas Hart Benton, Mary Cassatt, and other important artists.

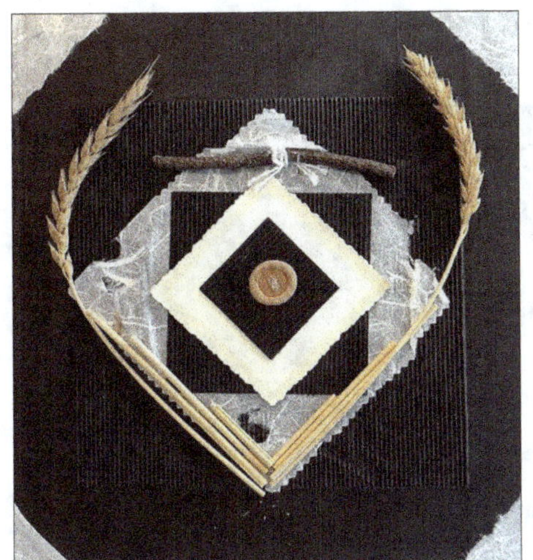 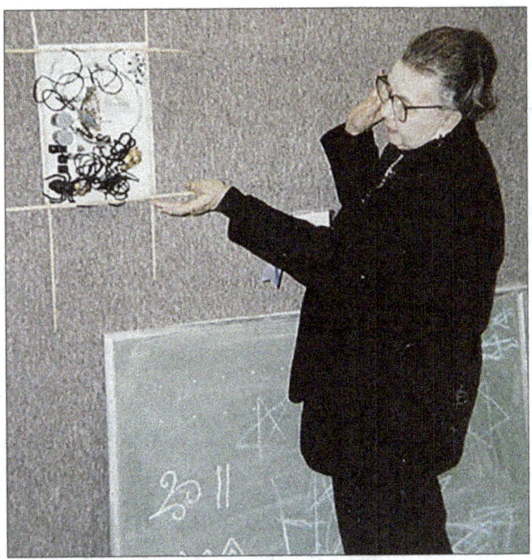

Student "Journeys" Scrapbook Cover and Art Lesson with Arden Johnson
Columbia Heritage Collection Photographs

Amidst ponderings of sheaves and sowers in English and art, our high school Future Farmers of America advisor Dan Birdsell told us of the Greek grain goddess Demeter and about the classical symbolism of ancient farm tools and crops mentioned in the organization's opening ritual that we memorized. Some forty-five of our sixty-three-member high school study body in 1969 (70%) had parents who farmed and the vast majority of their sons donned the FFA's blue and yellow jackets. (Girls were finally allowed to join that year. In 2000 when our son graduated from the district a generation later, the number of high schoolers with farmer parents had fallen to 37%.) As youths we may not have always understood the deeper meanings behind the emblems, but we came to know the difference between a sickle and scythe, spring from fall grain varieties, and that a bushel of wheat weighed about sixty pounds. Mr. Birdsell also arranged to have us periodically attend Grange meetings in the nearby hamlet of Winona for extracurricular practice in parliamentary procedure, memorized ritual presentation, and public speaking. The local membership, that admitted both men and women, was famous for enormous county fair mosaics exquisitely crafted from colorful grains—golden wheat and barley, green lentils, red beans, yellow peas.

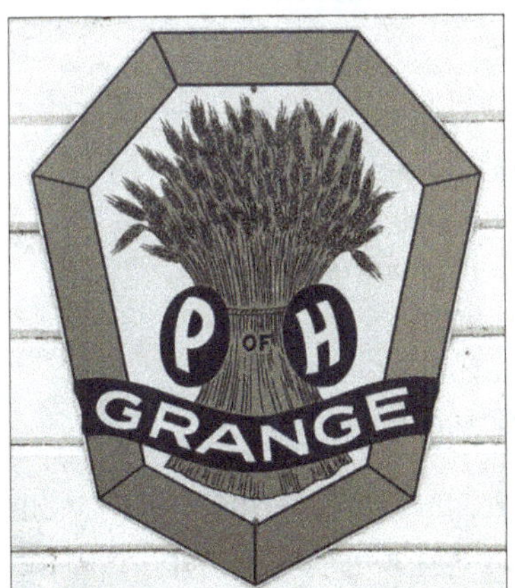

Washtucna Grange Sign
[*Patrons of Husbandry*] (2020)

My substantially forgettable speech on world hunger somehow made it to competition at the 1967 National FFA Convention in Kansas City. One of the opening lines of that talk has stayed with me: "Vietnam and moonshots may get the headlines, but hunger remains the world's number one problem" (Unfortunately, by simply naming another nation in crisis the line remains apt.) Research for the speech had led me to Rachel Carson's 1962 classic *Silent Spring* and the dire predictions on overpopulation and famine risk by nineteenth century British economist Thomas Malthus. The sources I consulted heralded the prospect of sufficient world food supply through agricultural improvement.

The Grangers I knew as a youth met monthly for socializing, to promote agrarian interests in state and national politics, and to pledge themselves to honorable goals pledged in Grange rituals for higher degrees like Harvester (male) and Gleaner (female): "Brothers and Sisters, as Harvesters and Gleaners, reap for the mind as well as for the body. Natural history is replete with both the wonderful and beautiful, and its study enables us the better to carry out the principles we inculcate of Faith, Hope, and Charity. Cultivate an observing mind. It is delightful to acquire knowledge, and much more so to diffuse it." Clutching a small sheaf of grain and flowers, the initiates and others then raised the roof with songs like the Patrons of Husbandry unofficial anthem, Knowles Shaw and George Minor's familiar hymn "Bringing in the Sheaves" ("Waiting for the harvest, and the time of reaping. . . ."). The organization's official songbook, *The Patron* (1925), contained numerous selections commemorating the significance of field labors:

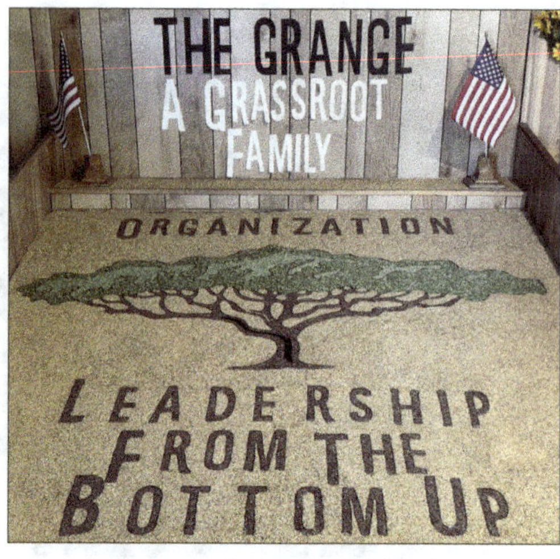

Pomona Grange Seed Mosaic
Palouse Empire Fair, Colfax, Washington
(2023)

"Harvest Song"

Grain that was in verdure waving,
Weareth now a hue of gold,
And the yellow heads are bending,
With the fruitage they hold.
That the ripened fruit be gathered,
Speed the sickle to and fro;
For the countless hosts of kernels,
Snowy loaves ere long will show.

Soon from out the noisy thresher,
There shall golden streams be pour'd,
That the farmer's heart will gladden,
And shall bring his just reward.
Smiles the land today with plenty,
Plenty for the needy throng;
Let all classes and conditions,
Join to swell the harvest song.

"The Gleaner"

When the earth is crowned with fatness,
And the yellow harvest yields
To the sickle of the reaper,
Toiling in the sunny fields;
Mark the glad, contented gleaner,
Gather one by one her store—
Ev'ry act of cheerful labor
Makes her richer than before.

Golden treasures, thickly scattered,
Strew the world's surface o'er;
Man is but a humble gleaner,
Finding knowledge, seeking more.
Step by step he plods his way,
One by one his blessings rise;
He who binds his store together,
He alone is truly wise.[2]

Such tunes were not part of our high school music repertoire, but choirmaster David Wells introduced us to a range of popular and sacred selections that included songs and composers with rural associations. Raised on a farm in southern Idaho, Mr. Wells came with degrees in music, vocational agriculture, and theology. Though most credentialing programs today discourage teacher preparation in such certification areas deemed "unrelated," he drew from these perspectives to greatly enrich our learning. We had known the '60s blues protest tunes of Arlo Guthrie, but through Mr. Wells's guidance also came to appreciate his father Woody Guthrie's Columbia River ballads. We heartily sang the folk icon's memorable "In wheat fields waving and dust clouds rolling" line of "This Land is Your Land" and performed other similar works. (Years later at late night song-fests in Russia I was regularly impressed when listening to my hosts loudly regale in memorized English with all stanzas.) Guthrie's 1940s *Talking Columbia* carried the refrain "Uncle Sam needs wheat," and *Washington Talkin' Blues* told how the life-giving waters of Columbia River irrigation had delivered those who exchanged the "black dust bowl" drylands for our region's productive wheat fields and orchards.

Enduring Traditions

Some of our community religious traditions contrasted with certain enduring aspects of local agrarian folklore. We gleaned aspects of this from Norwegian farmer elders on our mother's Sunwold side of the family who lived in the northeast Palouse "Upper Country" environs of Fairfield and Waverly. I could grasp the significance of planting field potatoes on Good Friday, but among many members of our grandparents' generation the seeding of grain crops took place during a waxing moon, and harvesting commenced when it waned. In some cases women, who fed the harvest crews five times a day, were not allowed out-of-doors during the combine's first pass around a field of grain. Still in my day harvest concluded with spirited shouting and the ceremonial threshing of the straw fedoras worn by many of our fathers.

Delving into the medieval European folk traditions occasioned by this study provides some explanation for why such traditions persisted through the years of my adolescence amidst harvest truck radios blaring rock music and news of Vietnam. Among those connecting influences were the pensive, high-pitched lyrics of country-rock musician Neil Young. The album title *Harvest* boldly appeared on his 1972 album cover in swirling calligraphy emblazoned against an orange sun as a stark symbol of the recording's pastoral summer vibe. Like its kindred successor *Harvest Moon* (1992), the Canadian prairie native's music harkens to love of place, disgust at its loss, and contentment through family and friends. Young's commitment was not ephemeral. In the 1980s he joined Willie Nelson and John Mellencamp in organizing Farm Aid to help distressed farmers for an annual concert series that has continued for over thirty years.

Through verse by an accomplished local poet, Harry Helm (1906-1987), claimed as kin through some vague ancestral connection, we also knew poetic expressions about the beauty of our own local landscapes. Harry's grandparents, John and Mary (Kleweno) Helm, had been among the first group of Volga German immigrants to settle along the Palouse River not far from our home in the 1880s. In a bucolic setting along the river that came to be known as the "Palouse Colony," some half-dozen immigrant families established an Old World peasant commune. They used farming methods suggesting medieval origins—long, narrow *Langstreifen* fields (akin to English furlongs) in three-crop rotations (*Dreifelderwirtschaft*), and *Almenden* commons for grazing and gardens. Many colony families were associated with the *Brüderschaft* (Brotherhood), a pietistic lay movement begun by Moravian missionaries in Russia that had thrived among Volga German Protestants in the 1800s.

The industrious Brethren, a term that included both men and women, also evidenced care for creation as a tenet of the faith.

These hardy Palouse pioneers harvested grain with sickles and scythes and threshed by "hoof-treading" the stalks with horses led round and round on a packed-earth *Denn*. Other Volga Germans led by J. Friedrich Rosenoff came by train and wagon from Nebraska in 1882 and settled in the Walla Walla Valley and Ritzville-Odessa area where many also became farmers. In tribute to their culture and contributions to Northwest agriculture, the work of contemporary Northwest artists who share this heritage like Jeffrey Hill and Arlon Rosenoff have been influenced by their cultural experience. Hill's murals and life-size bronze sculpture *Harvest Memories* (2018) are prominently displayed in Walla Walla public spaces. Thickly textured palette knife canvases like *Wheatfield Showers* (2018) by Rosenoff, a self-taught artist, are reminiscent of van Gogh and depict colorfully expressionistic combinations of farm workers, equipment, and the landscapes they inhabit.

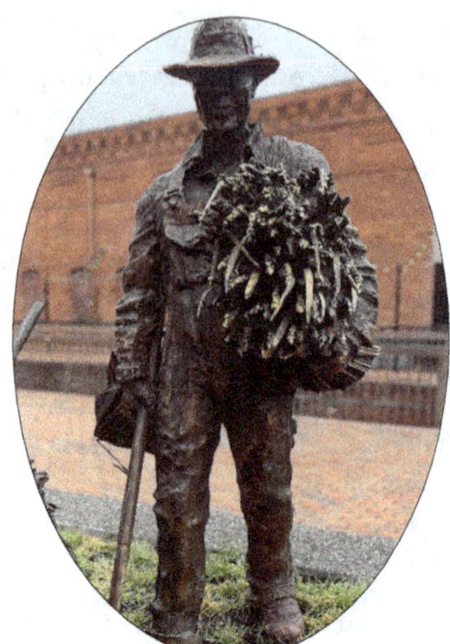

**Jeffrey Hill, *Harvest Memories*
bronze sculpture (2018)
Land Title Plaza
Walla Walla, Washington
Courtesy Susan Chlarson**

Harry Helm grew up hearing similar stories about his people's old ways, and his reflective eye wove heritage and horizon in "Endicott Wheat Field" (1962) and other poems that curated local landscapes:

Grandpa said:
The grass was like Europe's grass,
Soft and waving like a sea.
It hissed and whispered like a friend
In well-known German words to me.
The hills were like German hills,
Green plumed against a feckless sky.

And I went riding bunchgrass trails,
Where the prairie chicken fly.
Clear waters tumbled through the trees
In every golden, sun-swept vale.
While flowers tipped their hats to me,
As they touched my prancing pinto's tail.[3]

The art of Robert R. Helm (1943-2008), great-grandson of venerable John and Mary, also reflected family aesthetics. Helm's mysterious, exquisitely painted and collaged arrangements of landscape, rocks, and architectural fragments were crafted on cherry, birch, and pine. These masterpieces reflect his relationship with heritage and imagination. Though Helm's style is sometimes associated with Surrealism and Luminism, its distinct physicality makes it more akin to that of medieval artisans using unique combinations of format, composition, and color that explore meaning and memory. Characterized variously as "icons of stillness" and "neo-trecento renditions of rural America," Helm's dreamy, rustic creations like *Untitled House* (n.d., PLATE 9) led to global recognition with his works exhibited in Paris and Berlin, and in the permanent collections of the Smithsonian and New York Metropolitan Museum of Art in New York.

Helm's works evoke timeless themes of natural beauty and isolation, and he drew inspiration from frequent travels along the Washington-Idaho rural borderlands between the Coeur d'Alene Valley and the Palouse. Stories of his pathfinder ancestors were often retold in my visits with elders who had resided in the colony. Oil on panel works like Helm's *Spring Thaw* (1985) and *September Burn* (1991) are notable for the placid depictions of hills, as well as for the artist's meticulously built hardwood frames. "The Palouse has nurtured and reinforced his formal, psychological and metaphysical vision of life and its meaning," writes Marti Mayo, director of the University of Houston's Blaffer Gallery, which hosted a one-man exhibition of Helm's art in 1994. Northwest author William Kittredge offers further commentary on how place shaped Helm's art, just as it has power to influence others who take time to know a landscape relationally:

> *Go back to a place, Helm knows, and purely physical memories of who you were when you were there before may begin to echo in your body. You may remember exactly how things first felt and something of how it felt to be the person you used to be.*
>
> *Go back enough times and your sense of yourself in that place may begin to stack up before you in layers. It's a way to recall your story of yourself through the years of change and to relearn the reasons for your work and the consequences. It is a way to keep reinventing your knowledge of who you are and how you are trying to make your work matter in the world.*[4]

The Helms passed on appreciation for the creative writings of lesser-known Northwest authors like Cyrus Gilbert and Robert Sund (1929-2001) whose observations on rural life are too often confined to the shelves of family libraries and local used bookstores. Gilbert worked in the early 1900s as a teacher and pastor in Burien when the Puget Sound community was still flanked by small farms along the Duwamish Valley. In "Summer Song" and "Seed-Time and Harvest" (1948), he invoked classical allusions to sing of Ceres, "abundant offerings of grains," and a thresher's "insatiable maw . . . munching down sheaves and spilling the cereal gold" to yield "Heaven's bounty . . . in the miracle of growth."

Sund's fifty-four evocative "harvest poems" were inspired by summertime experiences in 1963 on a farm and tending the nearby Ennis grain elevator north of Walla Walla and published as *Bunch Grass* (1969). The time had been especially poignant as the Puget Sound native heard on a radio broadcast of the unexpected passing of his literary mentor, University of Washington English professor Theodore Roethke. Harvesters commonly mark life events by season and location. Sund channeled uncommon appreciation for the inhabitants and farmlands of the Inland Pacific Northwest that he expressed in crisp, calm descriptions drawn from nature and daily labor at his isolated location. He penned many of his creations with India ink in illuminated longhand and wrote of field silence and cricket song, combine clatter and crumbled dirt clods.

> *Let these poems be like bunch grass,* *flash floods, and sunlight*
> *in ground winds,* *holding together. . . ,*

Sund proposed with images that bespoke the inner peace of Asian religious masters he had long admired.[5]

Inland Empire Farmer Innovators

As elsewhere throughout rural America, representatives of harvesting equipment manufacturers like John Deere, International Harvester, and Allis-Chalmers kept abreast of notable improvements made by innovative twentieth-century Columbia Plateau farmer-inventors. Local mechanics frequently adapted machinery to the steeply rolling terrain there and in places like central California's hilly Stockton district. In 1932 James Love and Horace Hume of Garfield patented a flexible "floating" cutter-bar to harvest dry peas and lentils with a gas-powered pull-type combine. It operated efficiently on field contours and greatly reduced shattering incurred by the customary mowing and pitching of the plants. Two years later Love and Hume patented a tined header pickup-reel that gathered stalks and downed grain with improved efficiency.

Until the 1940s tractor-pulled combines throughout the hilly Palouse and in many other grain districts required manually operated rack-and-pinion leveling connected to a gearbox and left wheel housing to operate safely. In 1942 Raymond Hanson of Palouse conceived of an automatic leveling system comprised of a free-flowing mercury switch in a glass frame connected electrically to a solenoid and cylinder that could raise and lower the wheel. Hanson sold the first commercial models in 1945 which became the genesis of his Palouse and later Spokane-based equipment fabrication company. No harvest would commence on our place until local Hanson serviceman Jim Leonard first climbed atop our pull-combine to remove the glass panel and check the flow of the silver fluid and electrical connections. Always smartly dressed in khaki pants and shirt, Jim was an electrical wizard who served as our sixth-grade teacher during the school year and introduced us to the wonders of circuits along with tales of Ulysses.

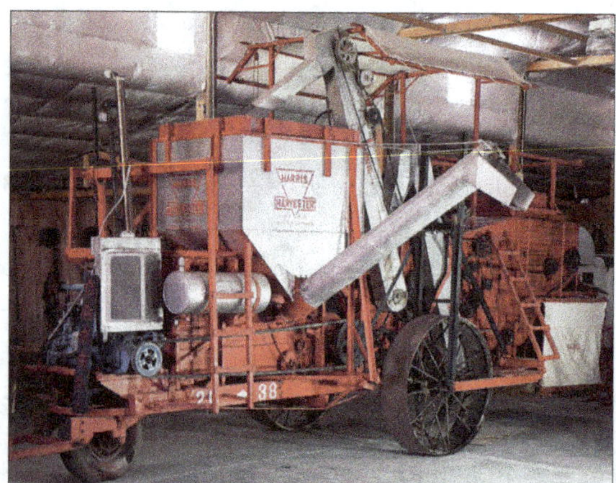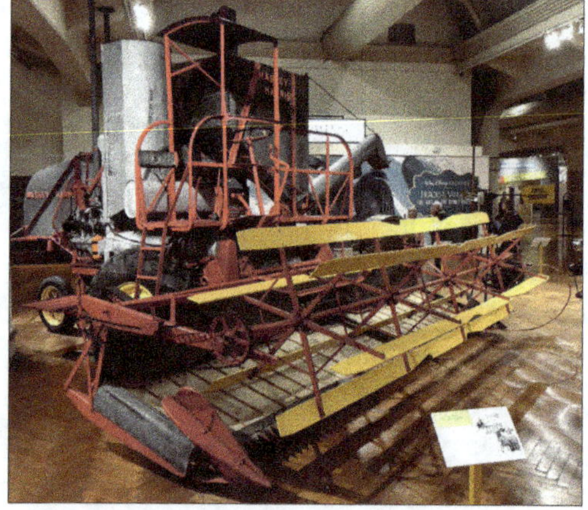

Left: Harris Model 26-36 Dual-Leveling Pull-Combine (1932)
Eastern Washington Agricultural Museum, Pomeroy
Right: Massey-Harris Model 20 Self-Propelled Combine Prototype (1938); Henry Ford Museum, Detroit

"Killer Hill" was an exceedingly steep slope not far behind our house that earned its name as a mythic destination for wintertime sledding with the Leonard family and others. Sometimes when we were very young and rode the combine with Dad, he would stop and order us off when approaching such places. Once while slowly driving along the upper brow of Killer Hill, the combine hit unusually soft ground and slid down so precariously that Dad barely managed to stop the tractor before the combine would have tipped over. He raced to town to retrieve a crew who brought a McGregor Company truck-mounted derrick and spent the rest of the day righting the near catastrophe. Through it all the Hanson leveler never failed.

In the late 1940s John Deere Spokane regional manager Clyde Moody teamed with Everett Kroll and Elwood Widman of Moscow, Idaho's Everett Will Tractor Company, Dave Neal of Garfield, and Endicott's Conrad Hergert, owner of one of the state's first John Deere dealerships, to prepare specifications for a prototype self-propelled and self-leveling "hillside" grain combine. Holt Manufacturing Company of Stockton, California, had manufactured several hundred self-propelled threshing machines from 1917 to 1921 but did not continue production as farmers preferred the reliability and maneuverability of combines pulled by draft animals and tractors. Brothers Curtis, George, and Ernest Baldwin of Nickerson, Kansas introduced their first trademark silver and red Gleaner combine in 1912 that was built for level land operation and mounted on a Fordson tractor so not a fully self-propelled machine. Production ended in 1928 when Henry Ford discontinued production of the Fordson Model F. The first commercially successful self-propelled combines were the level-land models Massey Harris 21 (1940), International Harvester 123 (1942), John Deere 55 (1947), and J. I. Case 9 (1948).

The area team assembled by Clyde Moody had remarkable talent for visualizing complicated solutions to structural and mechanical challenges ranging from steel bracing for heavy equipment to establishing standards for combustion engines. In the early 1950s Moody arranged for John Deere to bring some of the group to company headquarters in Moline, Illinois, headquarters to help design a hydraulic self-leveling mechanism using an activated lever and valve device. During the same period Frank Farber of Moscow's Idaho Machine Company and local Helbling Brothers International Harvester dealership (successor to McCormick International) to fabricate a similar system for IH combines that could level in four directions. Such cooperative efforts inaugurated a new era of harvest machinery for the Palouse with 1953 field testing and the 1954-1955 release of John Deere's 55H and International's McCormick 141H. In 1957 Kroll joined John Deere's Arrow Machinery in Colfax and received a patent for an improved leveling device that used a water-antifreeze mixture to push against a diaphragm-activated microswitch. Dealerships debuted John Deere's larger 95H combine

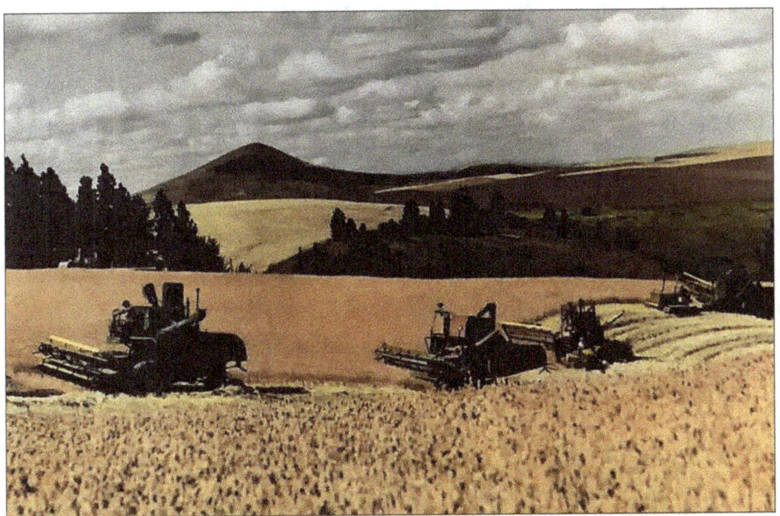

"Harvest on the Edwin Curtis Farm near Colfax, Washington" (1954)
Prototype John Deere 55H followed by Case Self-Propelled and John Deere Pull-Combines
Bill Walter Photograph, Dennis Solbrack Collection

in 1958 as did International's popular IH-151, followed by the IH-403 in 1962 with all these models widely used throughout the decade.[6]

Improvements in leveling technology did not eliminate the dangers of combine operation in the steeper districts of the central Palouse Hills where drivers were at risk given the machines' relatively high center of gravity and narrow stance. Most folks in our neighborhood had acquaintances who had been killed in tragic tip-overs that took place on the unruly terrain before the advent of wide-body combines in the 1970s. When starting out as a schoolteacher, I would return to help during harvest and periodically drive the John Deere 95H co-owned by Dad and his lifelong farmer friend Don Lust. One sweltering afternoon while cutting our home place I noticed the flashing red out-of-level warning light as I maneuvered higher and higher along a notoriously steep swale with each passing swath. The light often appeared on such ground so was not particularly worrisome but the lengthening slide during each pass while approaching the top was palpable. The normal drill was to gently brake the downhill front wheel while steering the back pair slightly uphill and hope for the best. Leaving even a narrow skip would have broken my elders' code of field conduct. Perched in the grain truck at some distance, Don had been observing the situation while Dad was off to town with the previous load. When dumping wheat from the combine into the truck after another round, Don asked if he could finish up, and I obliged. With the graceful hand and foot motion of an experienced operator, Don deftly executed the dance with his six-ton slow-motion partner and scarcely missed a single stalk. I marveled at such skill, and never forgot that he kept the incident to himself. Folks like him who spend their years on the land have countless other examples of the abilities and experience needed to meet the challenges of country life.

Local farm equipment dealerships and shops operated by Kroll, Love and Hume, Art and Joe Helbling, and other latter-day McCormicks and Deeres circulated colorful promotional calendars of fine art that featured harvest scenes by Jean Millet, Maxfield Parrish, and other notable artists. Technical equipment schematics have also long represented an important genre of documentary art like the exquisitely rendered scythes and sickles seen in Diderot and Le Breton's beautifully illustrated 1760s *Encylopédie*. Some examples of this genre represent important intersections of agriculture with American history as in George Washington's 1792 design of Mt. Vernon's threshing barn and briefs from the fateful Manny-McCormick Reaper Trial of 1855 that first acquainted country attorney Abraham Lincoln with Pennsylvania's Edward Stanton. A century later the self-propelled combine enabled a single worker to increase a day's harvest a hundred-fold.

John Deere's completely redesigned 6602 hillside combine in 1971 featured a wider header, hydro-static drive, and a significantly wider stance for safer operation on steep slopes and slick straw. But on sloping terrain the accumulation of unthreshed stalks on the downhill side of a combine's interior could still limit the complex machinery's capacity to function optimally. This tendency also contributed to aggravating internal slugging that could damage equipment and consume precious operating time. One of the most notable late twentieth-century innovations in harvesting technology was the advent of the "axial-flow" rotary system that replaced the threshing cylinder and concave base with a substantially longer rotor situated parallel to the main frame. This substantial change allowed for a wider distribution of kernels onto augers below the rotor and greater expulsion of straw and chaff and out the back. In the 1970s and '80s, Spangle native Edward Hengen served as a senior project engineer at John Deere's sprawling harvester works in Davenport, Iowa, to devise an axial-flow threshing drum and higher capacity header-auger feeding system. Similar equipment was patented by New Holland and International Harvester in the 1970s and introduced in the popular high-capacity NH TR70 in 1975 and the IH 1440 and 1460 in 1977. International's larger 1470 hillside model equipped with Hanson mercury switches came one year later. Gleaner introduced its first rotary combine in 1979 with the N6 powered by an Allis Chalmers engine.

Local lore long held that the axial-flow concept had been originally formulated by Endicott inventor-mechanic Arthur Edstrom, though the concept had been worked on independently in workshops across the country for decades. The Edstroms' high-school daughter helped our mother with harvesttime cooking chores in the 1950s, and I remember Dad's favorable reference to her father's new-fangled ideas when passing by their home in town. Kindly Mr. Edstrom's revolutionary cement mixer-sized, two-wheeled contraption was usually parked near a lilac bush in their side yard, and he later mounted a full-size prototype on a Case combine chassis. That model also featured his novel air-flow header mechanism that reduced shattering the brittle ripe heads of grain. The header blew the stalks into the combine instead of striking them with the conventional wooden slats. Like many

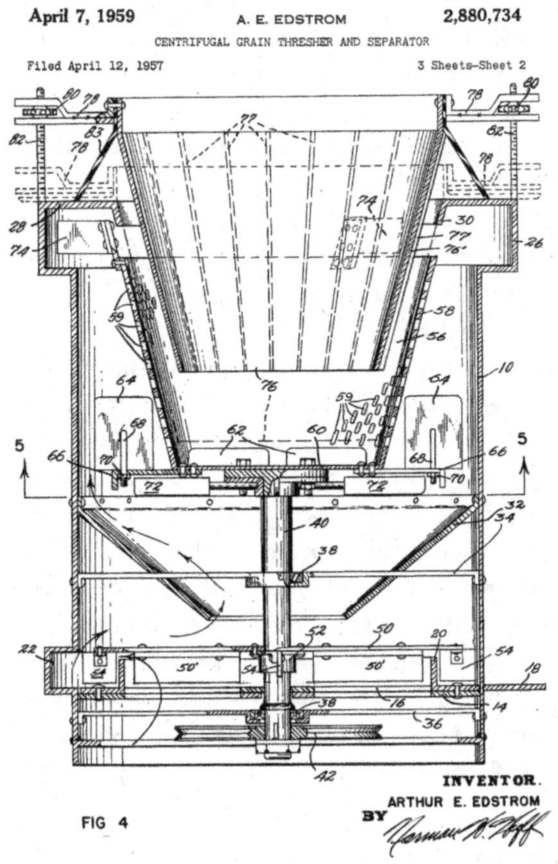

Norman Huff, *Edstrom Centrifugal Grain Thresher and Separator* (1957)
United States Patent Office Schematic Drawing
Columbia Heritage Collection

inventors, however, Mr. Edstrom never managed to bring his visionary concept into commercial production. His technical designs remain, however, as important documentary art akin to detailed drawings of earlier innovative threshing and tillage equipment.

Other significant developments in harvesting technology took place in the Columbia Basin. Although Grand Coulee Dam was completed in June 1942, the outbreak of war the previous December tabled plans for the massive Columbia Basin Irrigation Project envisioned to bring water to one million arid acres in central Washington. After the troops returned home in 1945, the Bureau of Reclamation organized a demonstration project in Franklin County to rekindle federal support for the ambitious undertaking. Bureau officials teamed with Pasco-area advocates to construct a seven-mile canal system to divert water from the Columbia River and in 1948 the first irrigated flow reached a field of grain in historic 5,200-acre Block One north of town. The success of the demonstration renewed government efforts to commence work on a grand scale to bring water pumped from the Lake Roosevelt backwaters of the dam into the Banks Lake Reservoir and extensive canal network.

Lotteries were held for the newly productive farmland with preference given to veterans. Among others who were selected and moved to a newly opened quarter-section near Pasco in 1951 were Oklahoma transplants Ben and Alma Grant. A highly decorated Army Air Corps bomber pilot, Ben Grant had long shown a knack for mechanical prowess and soon applied his skills to the self-propelled 1946 level-land Massey Harris 26 moved to Washington. Grant's ambition commenced a legendary fifty-five-year custom harvesting career during which he owned a total of 125 Massey combines, deployed as many as fourteen in a season, and patented seven major modifications to improve threshing. Massey and other manufacturers had been challenged with fabricating machines with a sufficiently adjustable cylinder and concave to thresh a variety of seeds ranging in size from beans and wheat to radish and grass without cracking. Grant solved the problem by installing an air-inflated cylinder and rear beater rubber panel.

In the 1960s Grant confronted the problem of inadequate threshing the dense loads of high yielding semi-dwarf wheats developed by agronomists at Washington State University. Under the sunshine and controlled irrigation conditions of the Columbia Basin, yields of Oroville Vogel's soft white Gaines wheat set world records at an astonishing 160 bushels per acre. But the abundance also presented challenges to conventional harvesting that Grant was determined to solve. He devised an ingenious third shaker shoe sieve—what Massey Harris soon featured as the factory-standard "Cascade Shoe," to more thoroughly thresh and move the heavy cuttings. Grant's 1970 hydraulic concave lowering system addressed a related problem that greatly vexed combine operators when grain bunched up around the cylinder and required manual cleaning. Devoted to the wellbeing of his young drivers and embracing the "spending money makes money" ethic, Grant was also among the region's first farmers to custom build comfortable air-conditioned cabs equipped with bucket seats and two-way radios. For Grant's success as a farmer, visionary, and philanthropist, Massey management credited him with significant contributions to the brand's success across the country.[7]

FarmHouse and University Influences

In the fall of 1969, I began studies in history and literature at Washington State University in the eastern Palouse city of Pullman near the Washington-Idaho border. Although less than an hour from home and in a familiar place from numerous previous visits to campus, university experience fostered appreciation for the region in the larger context of national and world affairs. Numerous professors and scientists at the school valued scholarly study of agriculture and country life. I joined FarmHouse Fraternity which traces its origins to a group of University of Missouri agriculture majors led by D. Howard Doane (1883-1984) who had attended YMCA meetings in 1905. Doane would go on to national prominence as founder of the nation's first farm management and research firm, Doane Agricultural Services. He served at the USDA under William J. Spillman (1863-1931) who had established the grain breeding program at the Pullman school in the 1895 and organized the USDA's Office of Farm Management a decade later. Under Secretary of Agriculture Henry Wallace, Spillman would become a leading architect of the New Deal's Agricultural Adjustment Administration (1933) that stabilized crop prices by reducing surpluses. Appointed by President Truman in 1947 to the Hoover Commission to reorganize the USDA, Doane went on to play a leading role in post-war struggles with the oil industry to transition from fossil fuels to renewable synthetics from grains and other crops for ethanol, plastics, and rubber.

In spite of the name and heritage, FarmHouse members in my day pursued studies in numerous fields, but many were from farm families and all aspired to a fellowship represented by the organization's green and gold shield that prominently features sickle and sheaf. The pledge manual informed us that the colors signified "the high calling of agriculture" upon which all of society's institutions rest. The harvest motif was to remind us of the Apostle Paul's words regarding commitment, responsibility, and reward: ". . . [W]hatever one sows, that will he also reap," and "[L]et us not grow weary of doing good, for in due season we will reap" (Galatians 6:7,9). We held an annual Harvest Ball and became acquainted with prominent FarmHouse alumni including Nobel Prize winner George Beadle (1903-1989), and a prime mover in founding and sustaining our WSU chapter, National Medal of Science winner Orville Vogel (1907-1991). As the university's senior agronomist, Dr. Vogel had long been respectfully known by many of our farmer-fathers. Both graduates of the University of Nebraska in the 1920s, these two individuals made significant contributions through plant genetics to work of Nobel laureate Norman Borlaug and the 1960s Green Revolution. When Borlaug visited Dr. Vogel in Pullman he lodged with us at FarmHouse.

Studies in Pullman also led to fortuitous encounters with emeritus professor and longtime history department chair Dr. Herman Deutsch. Although in his eighties at the time, Dr. Deutsch remained active in department affairs and was the grand old man in state and regional history organizations. At some point I risked voicing complaint that the history of the American West seemed to be given short shrift with only one or two courses devoted to the subject amidst an array of offerings ranging from ancient Greece to modern Europe. Dr. Deutsch took me aside at a gathering of faculty and history majors, and later wrote to me, emphasizing the primacy of regional history and agrarian studies as the basis of understanding larger trends in national and world history. His encouragement opened a world of scholarly possibilities related to my nascent interests and led to an encounter with the scholarship of Palouse Country native Donald Meinig, Maxwell Professor of Geography at Syracuse University and author of *The Great Columbia Plain: A Historical Geography, 1805-1910* (1968).

Professor Meinig's prodigious study of relationships among Inland Northwest landscapes, regional agriculture, and grain trading networks remains endlessly fascinating, and serves as a valued vade mecum for ready reference. The substantial tome is filled with the author's signature maps and descriptions of transportation networks, field rural communities, and crop production. The book would be a precursor to his magisterial three-volume "Shaping of America" series by Yale University Press (1986-2004). Hoping for some prospect of academic kinship through shared nativity and mutual interests, I wrote to Dr. Meinig with questions about his research methods. He generously responded to inaugurate periodic correspondence over the years that has deeply enriched my perspectives on geography and the humanities. In letters, lectures, and publications he has made the case that exploring the work of ordinary folks enhances our ability to find meaning in the commonplace and reflect beyond sentimentality to abiding themes of responsibility and place.[8]

These values are evident in the story and images in *National Geographic's* "North with the Wheat Cutters" (August 1972) by Noel Grove, and in a film based on the story, *Amber Waves* (1980). The article, illustrated with photographs by staff photographer James Sugar, profiled the Max Louder custom cutting brigade that had spent several months each year for decades following the grain harvest from Texas to Montana. Outfitted with a phalanx of Massey-Ferguson combines and diverse crew of experienced hands and young newcomers, the Louder army began its demanding continental run of twelve-plus-hour workdays that stretched from June to October. Grove credited his knowledge of "combines, trucks, and farm people" with his commission for the story—first of several for *National Geographic*. At the time Grove's article appeared, Louder was in the process of relocating the headquarters of his operation from the Midwest to Pasco, Washington, in the heart of the Columbia Basin's irrigated farm country. Louder adjusted his operation to follow the western harvest cycle from Arizona to the Pacific Northwest established a world record in 1990 for the most grain harvested in an hour (54 tons) while threshing near Irrigon, Oregon.

I returned to WSU in 1978 for graduate studies in history and learned more about the tension between advocates of conventional farming methods using synthetic inputs and those promoting sustainable organic approaches. An influential proponent of the latter was soil scientist Robert Papendick who had come to Pullman in 1965 to conduct research for the USDA's Agricultural Research Service branch and teach at the university. During the 1970s he conducted research and published extensively on sustainable approaches to grain production, minimum tillage, and multiple cropping systems. In 1979 Papendick was tapped by Secretary of Agriculture Bob Bergland chair the USDA Study Team on Organic Farming which issued its influential *Report and Recommendations on Organic Farming* (1980). Although criticized by some conventional farming advocates and commodity organizations, it did succeed in establishing the USDA Office of Organic Resources Coordinator. Several of the report's recommendations were implemented including creation of the Sustainable Agriculture Coalition in 1988 and subsequent Sustainable Agriculture, Research, and Education (SARE) grant program. Along with conservation, economic development, and numerous other initiatives, SARE funding also supported National Institute of Food and Agriculture initiatives to promote farm-themed arts education in public and private schools.[9]

When asked to speculate on the future of American farming for a Smithsonian Bicentennial Symposium on Agriculture in 1975, longtime USDA chief economist Don Paarlberg (1911-2006) cited the advent of minimum tillage systems along with introduction of biodegradable farm chemicals and legitimate public interest in the environmental improvement of privately-owned crop and grazing lands. Paarlberg, who had been raised on a farm in Indiana and served as founding director of US-

AID's Food for Peace program during the Eisenhower Administration, also urged his listeners to appreciate the "poetry of agriculture" while considering recent technical developments. "The wonders of life, growth, and death, the cycles of the seasons, the marvels of continuous creation—these will still be with us. In fact, it might be said that these are the most enduring things about agriculture. . . ."[10]

Depression Era Agrarian Art

While at WSU I also encountered Northwest Works Progress Administration (WPA) projects of the 1930s and associated oral history collections. As elsewhere across the country, WPA initiatives included ambitious efforts to employ painters and sculptors to create works for public buildings and writers and to compose detailed state travelogues. In 1936, University of Idaho Art Professor Mary Kirkwood (1904-1995) painted murals for the Agricultural Science Building and directed the work of several students whose Palouse farm frescoes decorated the concrete walls of the YMCA building through WPA support. (The panels were later relocated to the University Commons building.) Kirkwood, a graduate of the University of Oregon in Eugene who later studied are in Stockholm and Paris, introduced Modernist concepts during her forty-year career (1930-1970) at the Moscow school. Her painting *Palouse Harvest* (1956, PLATE 10) shows a self-propelled combine in the center foreground cutting across a hill of tawny grain. The machine is dwarfed by immense swells of gamboge yellow and Verona green.

Area WPA authors investigated historic sites as part of the popular American Guide Series and conducted extensive "Pioneer Project" oral histories. Many Northwest libraries and museums still have tattered copies of the three-volume *Told by the Pioneers: Tales of Frontier Life* (1937-1938). Project director Charles Ernst lauded the series as a resource to present "colorful and courageous background" the cultural heritage of future generations. In colorful vernacular elders tell of harvesting with cradle scythes and primitive mechanical reapers and relate other lesser-known aspects of pioneer self-reliance.

A *Told by the Pioneers* account recorded by Augusta Eastland from John Nelson family elders, who traversed the Oregon Trail in 1843, describes Yakima Valley frontier threshing labors:

> *Primitive methods were used in threshing the wheat grown in the early days. A threshing ground was prepared by removing some of the dirt from a tract of ground, making it slope toward the center. This threshing ground was tamped down until it was quite hard; then the wheat was piled on it and the boys and girls of the family rode their horses over it until the grain was threshed out. . . . The wheat was then taken to the government mill at Ft. Simcoe and ground into flour, the grinding paid in flour.*[11]

Catholic social activist Dorothy Day (1897-1980) made several trips to the Pacific Northwest from New York before World War II to lecture at churches and schools on perspectives of faith and progress through rural renewal. In January 1940, she traveled from Spokane to Pullman in the heart of the Palouse Country to address students at Washington State College. Since founding the *Catholic Worker* with Peter Maurin in 1933, Day had rigorously voiced support for the National Catholic Rural Life Conference and New Deal initiatives benefiting farmers as well as urban workers. She participated in the summer "Harvest" retreats of The Grail, a countryside community of faith founded in the 1920s for small groups of lay women. They lived in self-reliant farm settings closer to nature for a sustain-

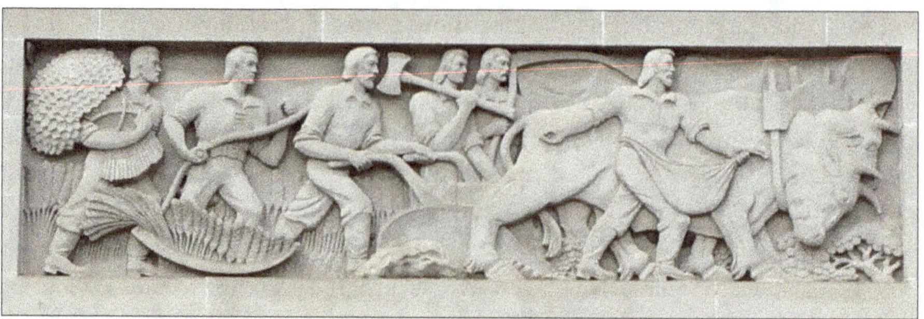

Men at Work **Stone Bas-relief (detail) and Brushed Steel Grain Stalk Baluster (1941)**
Adams County Courthouse, Ritzville; Columbia Heritage Collection

able "sacred agriculture," and to grow spiritually through fidelity to the liturgical seasons. Grail communities featured agricultural classes, arts and crafts centers to develop skills in ceramics, woodcarving, and calligraphy, and to undertake social ministry in area cities. In these ways the fulfilled spiritual life embraced a natural union of ecclesiastical worship, intellectual pursuits, and manual labor.

Day found "tremendous interest" in the work from students and others during her time in Spokane and Pullman and continued on by train to similar meetings in Portland and Seattle. She viewed the vast grain lands of the inland Northwest and wrote of wheat farms covering several thousand acres that made Grail and *Catholic Worker* farms pale in comparison. But she also noted problems inherent in managing larger operations when low commodity prices meet dryland yields and prop-

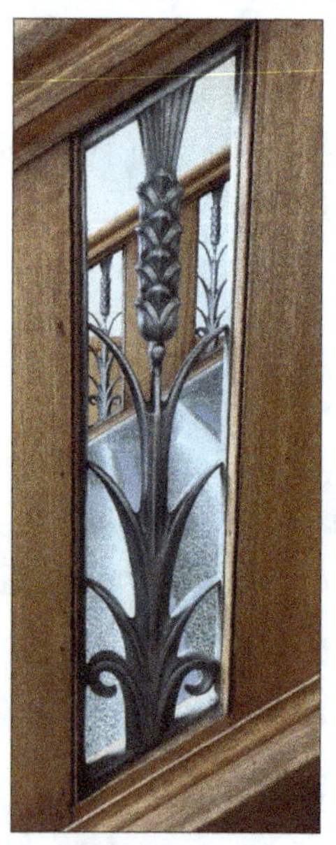

erty taxes. "I met one woman," she wrote during the trip, "whose husband was ruined by the collapse in the price of wheat . . . ," she wrote to a friend during her trip, and followed by warning against land greed and asking if he had read Tolstoy's essay, "How Much Land Does a Man Need?" Despite economic dislocations, Day marveled at the region's scenic rural vistas while expressing concern about the plight of the region's itinerant farm workers.[12]

Portland artist brothers Arthur (1891-1971) and Albert (1894-1971) Runquist, who both attended the Art Students League in New York in the early Thirties, returned to the Northwest where they shared a studio and painted scenes laden with social commentary on the experiences of minorities and laborers. Arthur, who began working for the Federal Arts Program in 1935, was once severely beaten for his socialist leanings. He painted numerous landscapes including the richly colored *Early Oregon* (1941, PLATE 11) mural as a state Federal Arts Program commission for Pendleton High School on which he was assisted by the brothers' "self-described sister" and fellow activist Martina Gangle (1906-1994). The immense painting includes a substantial harvest scene that shows unsmiling field hands resting amidst the stubble in the foreground of a passing threshing machine while other workers stack grain sacks on a truck. A red elevator rises in the distance against a range of barren hills and the pensive pose of the central figure casts a mood of resilience amidst despair upon the idyllic landscape. The harvest scene, now framed in three panels with other sections of the mural, were salvaged during renovation at the school for exhibition at the Umatilla County Historical Society Museum in Pendleton.

Although still in the throes of the Great Depression, most Northwest farmers had long since made the transition to mechanized farming. Only one large farm along the lower Columbia River route taken by Day still used animal power to pull the combine behemoths. George Wagenblast of Dufur, Oregon, harvested rugged slopes near the mouth of the Deschutes and could not bear to part with his beloved team of twenty-seven mules. But times were changing and in 1941 they would make their last appearance before being sold for wartime service by the U.S. Army for about $45 a head. (He had paid $175 apiece in 1929). Comparing New Deal era and twenty-first-century themes in public art and critical discourse indicates a modern trend away from intellectual consideration of the land and its toiling masses. Of over 45,000 entries in the most recent edition of the authoritative thirty-four-volume *Grove Encyclopedia of Art* (2011), for example, no subject headings are included for agrarian, agriculture, rural, or rustic. To be sure, countless numbers of regional artists and authors continue to create important interpretive works, and their enduring appeal is evident in the listings of agency websites like Saatchi Art and Mutual Art that feature hundreds of contemporary harvest-themed works. But their inclusion in gallery exhibitions has diminished in recent years. Israeli historian Yuval Harari, author of *Homo Deus: A Brief History of Tomorrow* (2015), suggests that the pace of technological innovation is increasingly associated with the volume of personalized digital postings that threaten shared values that have long knit cultural identities and connected peoples to landscapes. Such preoccupation, Harari asserts, not only increasingly distances people from employable skills, but risks humanity's wellbeing by neglecting regard for land care and sustaining social values.

My graduate studies at Washington State University led to a memorable meeting with history professor Clifford Trafzer in Graham House, an older wooden structure of Craftsman design, where some faculty members kept offices. An enormous portrait of the Snake River-Palouse Chief Cleveland Ka-

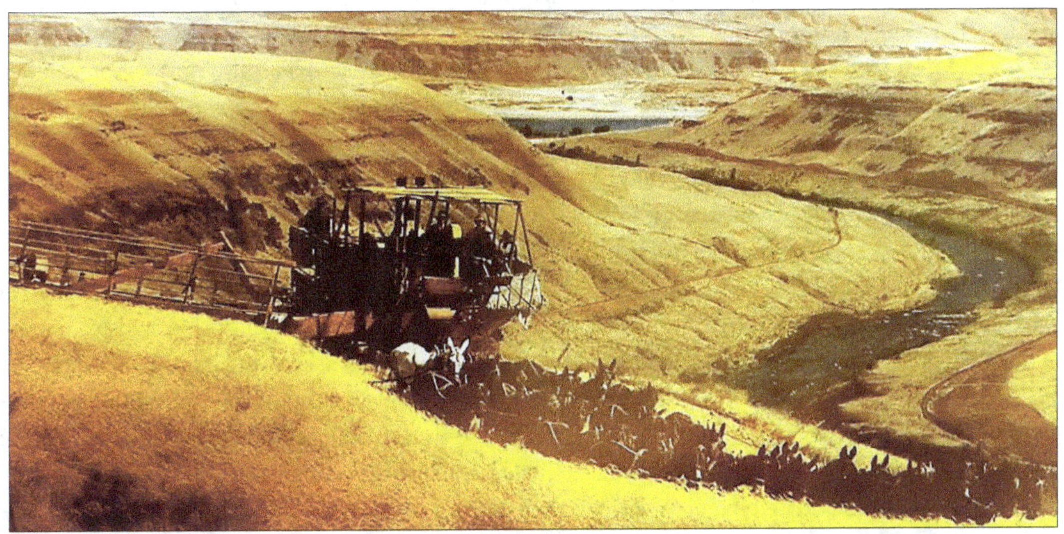

Wagenblast Harvest Outfit on the Kuck Ranch near the Deschutes River, Oregon (1932)
Columbia Heritage Collection

miakin hung on a stairway wall though it did not bear any obvious identification. I recognized the revered leader from old photographs safeguarded by family members on the Colville Indian Reservation in North Central Washington where I had been conducting oral histories of Columbia Plateau tribal elders. Cliff and I took the painting down for closer inspection and found that it had been painted by longtime art professor Worth Griffin (1893-1981). We soon learned that Griffin and Clyfford Still (1904-1980), a 1935 WSC graduate and influential contributor to Abstract Expressionism, had established the 1937-1941 WSC Nespelem Art Colony. Similar initiatives had been underway at that time by Grant Wood in Stone City, Iowa; Guy Williams in Essex, Connecticut; and Wayman Adams in the Adirondacks. Among the several Nespelem project art instructors was Pomeroy native Vivian Kidwell (1908-1996), who helped Griffin and Still administer the program, and married Griffin in 1940. She painted numerous inland Northwest rural scenes in a unique style as seen in *Elevators and Fields* (c. 1953, PLATE 12) that combined realism with geometric modernism.

While Native American subjects remained the hallmark of the WSC project, some colony instructors and students also painted agrarian scenes as did many of their Midwestern contemporaries like Wood, Thomas Hart Benton, and John Steuart Curry. In his 1935 booklet essay *Revolt Against the City*, Wood had written of artistic "habitat" in the agricultural heartland as broader, proper consideration of the American Scene: "[The artist] finds it quite contrary to the prevailing Eastern impression, not a drab country inhabited by peasants, but a various, rich land abounding in painting material." While energetic, affable Griffin handled much of the summertime colony's administrative needs in addition to teaching classes, brilliant, enigmatic Still served as intellectual mentor to colleagues and students. Born in the farming community of Grandin, North Dakota, Still's family moved to Spokane but soon relocated to a small wheat farm in Alberta where the family struggled to make ends meet. Still wrote to a friend during the 1927 harvest season that country life was "a scarcely tolerable pastime" that motivated his commitment to art. "The farm will provide inspiration while I'm painting," he mused, "and nourishment while I'm not." Still later returned to Washington and graduated in 1933 with an art degree from Spokane University and then taught at WSC for the next eight years. His aesthetic inclinations, formative experiences, and periodic study with Wayman Adams at the famed Yaddo retreat near Saratoga Springs, New York, led to his association with Jackson Pollock (1912-1956), Mark Tobey (1890-1976), and philosopher-artist Robert Motherwell (1915-1991) as the founders of combustive, elegant Abstract Expressionism.[13]

Although agrarianism is not generally associated with modern art, many works by Still, Motherwell, and other Northwest abstract painters like William Ivey (1919-1992) show strong rural influence with harvest themes and colors. Still's early Expressionistic tendencies can be seen in *PH-77* (1936) that shows a pair of male gleaners in almost simian pose stooping down to gather small sheaves of grain. The figures have blood-stained arms and labor beneath a dark sky to foster a sense of exhaustive agony, perhaps reminiscent of a memory from the artist's youth. *Row of Elevators* (c. 1929) suggests greater optimism in the contrast of activity on both sides of a towering line of grain elevators, probably near Alberta. On the right is a horse-drawn bulk grain wagon approaching the open doors of the immense structure, while to the left are a truck, boxcar, and train engine spewing forth steam. But a dilapidated wooden fence to the side and discarded remains of a wooden structure in the foreground evoke a sense of hardship and the footprint of humanity upon the land.

Still, who considered landscapes the most complex artistic subject, went on to teach in the late

1930s at the federally-sponsored WPA Spokane Art Center with Z. Vanessa Helder (1904-1968), James FitzGerald (1910-1973), and Jane Dunning Baldwin (1908-1991). National director Holger Cahill lauded Spokane with similar federal initiatives in New York and Chicago as hosting the finest such schools among several hundred in the nation. Vast numbers of area residents clamored to enroll in the group's hands-on art classes and thousands attended exhibitions at the center's gallery that included paintings, drawings, and prints of regional agrarian landscapes. One of the Northwest's early Modernists, master watercolorist Vanessa Helder was raised near Lynden, Washington, and known for her tightly controlled strokes of radiating color in a medium known for fluidity and blurry lines seen in paintings like *Palouse Rhythm, Spangle Wheat Country,* and *Snow and Stubble* (c. 1939). Before relocating to Los Angeles in the 1940s, Helder was known for her interpretations of Columbia Plateau rural landscapes.

A native of Aberdeen, Washington, Robert Motherwell studied philosophy at Harvard and Columbia in the 1930s, briefly taught at the University of Oregon, and developed a friendship with Still. Both men were influenced by the ideas and art of Picasso and Miró. They eventually relocated to New York and came to disparage the American Scene art of Wyeth, Rockwell, and others as idealized representations oblivious to the harsh realities of a national experience deeply affected by wartime destruction. Their often-misunderstood abstract canvases of jagged lines, blurred shapes, and splashes of vibrant color reflected free aesthetic associations as well as non-conformity to a post-war new world order of consumerism and persistent social inequities. Association with these themes and growing urbanization also influenced Modernist artist decisions to reside in New York, Chicago, Los Angeles, and other major American cities where galleries exhibited and promoted their work. Commercial success that sometimes followed conflicted with the communal ideal of rural art colonies that had flourished in places like Yaddo and Old Lyme and in pre-war Europe.[14]

In the late 1940s, Still, Motherwell, and Mark Rothko (1903-1970), who came of age in Portland amidst radical union circles, founded Subjects of the Artists, a teaching group that became New York University's influential Studio 35. Although Motherwell considered the East home for the rest of his adult life, he frequently returned to the West Coast and worked extensively with Los Angeles master printmaker Kenneth Tyler. Together they produced a series of Motherwell's novel collages combining elements of photo-offset lithography, hand-screen, and pochoir stenciling. The pair's collaboration led to the creation of Motherwell's landmark Summer Light "Harvest" collage series (1974) of mixed media lithographs. Forming a kind of visual autobiography from studio detritus, the relentlessly experimental Motherwell arranged discarded cigarette labels in various positions on austere fields of the artist's favorite colors—yellow ochre, ultramarine blue, and white. In these and other newly creative ways, artists like Motherwell, Still, and Ivey expressed deeply personal associations with pleasures and necessities as well as primal relationships with life-sustaining earth and water.[15]

El Camino Real Agriculture

From 1973 to 1974 my wife and I lived in Monterey, California, where I studied Russian at the oceanside city's historic Presidio military school—the Defense Language Institute. We often visited nearby Carmel-by-the Sea where some half-dozen art galleries and studios showcased paintings of scenic coastal environs, Colonial Spanish historic sites, and the exquisite photography of local resident Ansel Adams (1902-1984). In the wake of a major regional and national known gallery exhibits on Western art in the mid-1970s, sleepy Carmel would be transformed into a mecca for aspiring artists as well as dealers marketing to upscale tourists. Local painter-printmaker Howard Bradford (1919-2008) produced numerous agrarian serigraphs including *Summer Wheat* (1974), *Harvest Time* (1980), and his *Summer Sky* series that are rich tones of yellow ochre, russet, and gold. Bradford, who received a Guggenheim in 1960 for creative silk-screening, had been raised on the outskirts of Los Angeles where he tended family gardens and chickens. The experience imparted a lifelong affinity with country life depicted in his many Carmel and Salinas Valley farm scenes.

Prominent American landscapists George Inness (1825-1894) and William Keith (1838-1911) visited the Monterey Peninsula in 1891 and contributed to other artists' awareness of the area's uncommon natural beauty and Alta California adobe architecture. The Carmel Art Colony emerged between 1900 and 1910 with a small group of painters including Charles Judson (1864-1946) and Jane Powers (1868-1945) who embraced the moody realism of the Barbizon School and atmospheric forms of Tonalism that had influenced Inness and Keith. By the 1920s the group had shifted toward Post-Impressionism and was sponsoring annual exhibitions, lectures, and a summer school. Oakland native Alice Geneva Glasier (Gene Kloss, 1903-1996), who studied with Judson at the University of California in Berkeley, was among those attracted to scenes of Southwestern country life and Native American culture. She also devised an innovative etching technique by painting acid directly on copperplate for sharp-edged landforms and luminous spaces seen in masterful prints like *New Mexican Harvest* (1939). Her work garnered recognition as the region's only artist named a National Academician by the National Academy of Design.

Glasier married poet Phillip Kloss in 1925 and began many years of seasonal work and exhibits at Carmel. The couple maintained residences both in Berkeley and Taos where she sketched Pueblo life and attended tribal ceremonies which she later painted in works like *Tribute to the Earth* (1972), an aquatint of singers and dancers. "The economic wealth of America was and is the harvesting of crops," wrote Phillip Kloss about his wife's 1943 print *The Wheat Field*, and noted the significance of wheat, corn, beans, potatoes, and other produce to the nation's prosperity. The image shows two fieldworkers carrying bundles of grain up a slope that glows in whiteness as if sacred space where another pair loads a wagon. The art of Kloss and such Ansel Adams's Taos Pueblo photographs (1930) as *South House, Harvest,* and *Winnowing Wheat*—the first of his published oeuvre, offer place-based meditations on Native American care for the land that confronted prevailing Euro-American dichotomies of wilderness and colonization. The aesthetics and ethnography of these artists brought to wider attention what native agriculturalists of the Americas had known for generations: the basis of human flourishing is environmental stewardship.[16]

Harvest of grains, fruits, and vegetables around Salinas provided the Great Depression backdrop for writings by Nobel laureate John Steinbeck (1902-1968) that explore everyman protagonist confrontations of fate and injustice. His life and work are memorialized at the National Steinbeck Center that we visited in Oldtown Salinas during my military time in the area. Steinbeck's works like *The Harvest Gypsies* (1936), *Of Mice and Men* (1937), and *The Grapes of Wrath* (1939), Steinbeck's characters confront economic upheaval without interventions to mitigate crop failures, low commodity prices, foreclosure, and unemployment. Dismissed by some political elites as overly sentimental,

Steinbeck stories gave human faces to tragedies of the time with allusions to organized labor, social safety nets, and agricultural policies that could moderate the plight of dispossessed families.

Of Mice and Men can be seen as a parable on care for one's fellow man and secondarily as commentary on conflict between America's Arcadian ideal and brutal reality of migrant farm worker life. Intelligent, resourceful George Milton looks after his strong but mentally incapacitated friend Lennie Small as they seek seasonal work on California farms and ranches. George periodically consoles Lenny with their shared dream of a future when they can "live off the fatta the lan,'" but the circumstances of their itinerancy and Lennie's condition conspire to threaten such hope. The two find work in a Salinas harvest field loading bags of barley onto wagons and meet other ranch hands who dream of a better life. But what may be a season of abundance for the landowner turns tragic as the two fieldworkers prove unable to overcome the evil of economic powerlessness and abuse.

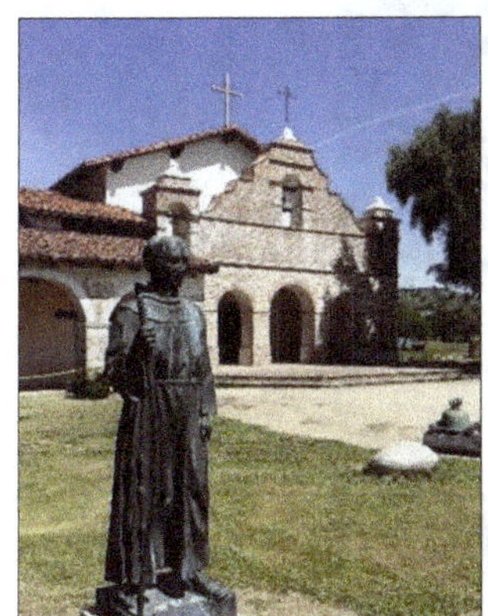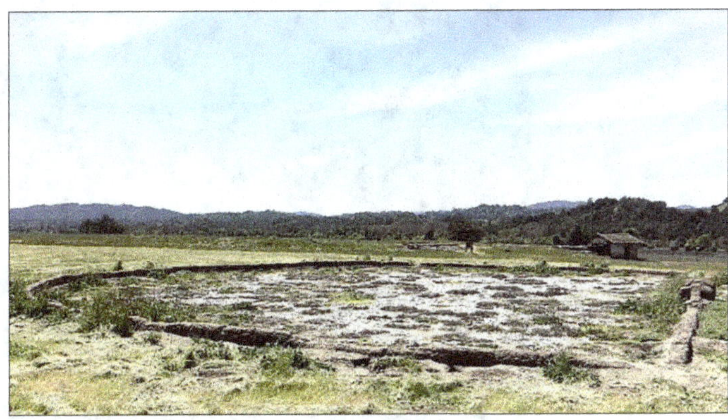

Left: Father Serra Statue and Mission San Antonio du Padua Façade
Right: Mission San Antonio Threshing Floor and Grist Mill (far right), near Jolon, California
Columbia Heritage Collection Photographs

While stationed at Monterey we explored many of the region's Spanish missions and reduction farm remnants. Concerned two centuries earlier about colonial ambitions by Russia and Great Britain, King Carlos III of Spain authorized an expedition from San Diego led by Gaspar de Portolá and Franciscan friar Junípero Serra to travel overland to Monterey to claim the region "for God and the king of Spain." They reached this scenic area in 1769 and six years afterward Monterey became the capital of Alta California. Father Serra established the Royal Presidio Chapel in 1770 and in the following year founded Mission San Carlos Borromeo as his headquarters in nearby Carmel Valley. San Carlos and San Diego were the first in a chain of twenty missions and farming outposts constructed along the 600-mile El Camino Real from San Diego to present Sonoma (San Francisco Solano). Exhibits at the Presidio of Monterey Museum and Carmel's San Carlos Borromeo showcase many treasured art objects associated with the missions' spiritual and agrarian heritage. Crosses meticulously decorated with lustrous grain straw appliqué are displayed with tapestries featuring colorful rural scenes, and deftly hand-wrought metal forms wheat stalks and grape cluster candelabras (PLATE 13).

Until the United States seized control of California during the Mexican-American War in 1846 and later secularized the missions, these centers of regional development utilized Indian workers to raise substantial wheat, barley, corn, and other crops as well as vast livestock herds. Areas of highest production, with a total of some twenty thousand tons of wheat produced from 1782 to 1832, centered around the missions San Gabriel Arcángel west of present Los Angeles, and at Santa Inéz and La Purísima Concepción to the northwest. San Antonio du Padua, founded by Father Serra in 1771 near present Jolon in central California, is the most fully restored of El Camino Real missions and features an adobe granary, water-powered grist mill, and the original stone threshing floor—among the last extant in North America. Standing on the hardened surface, one can imagine the sound of treading hooves that accompanied songs and sighs of workers using pitchforks to separate life-giving kernels of grain from windblown chaff.

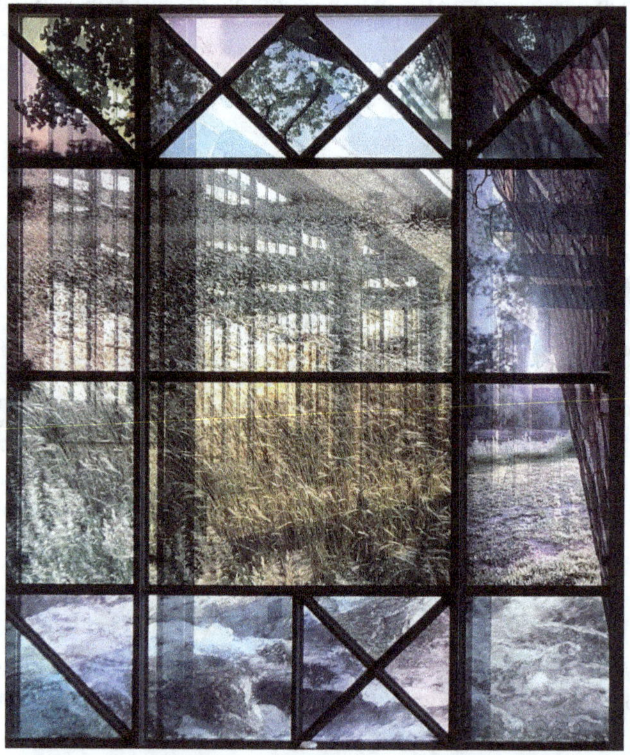

JoAnn Verburg, *Between Now and Then* (detail, 2003)
Steel-framed Photographic Glass Collage of Grain and River, 14 x 25 feet
Mill City Museum; Minneapolis, Minnesota
Columbia Heritage Collection Photograph

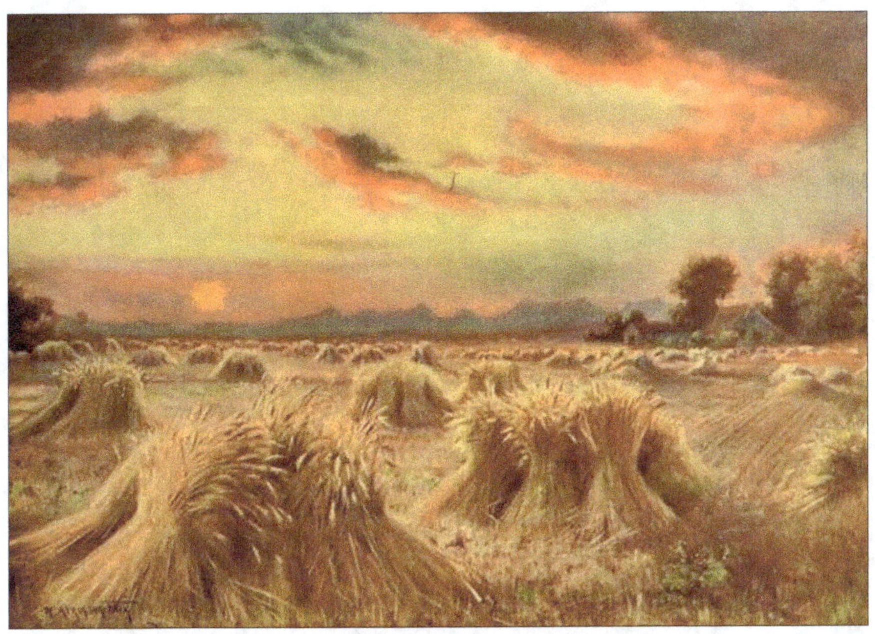

PLATE 1: Robert Atkinson Fox, *After the Harvest* (c. 1910)
Lithograph on paperboard, 4 x 6 inches
Columbia Heritage Collection

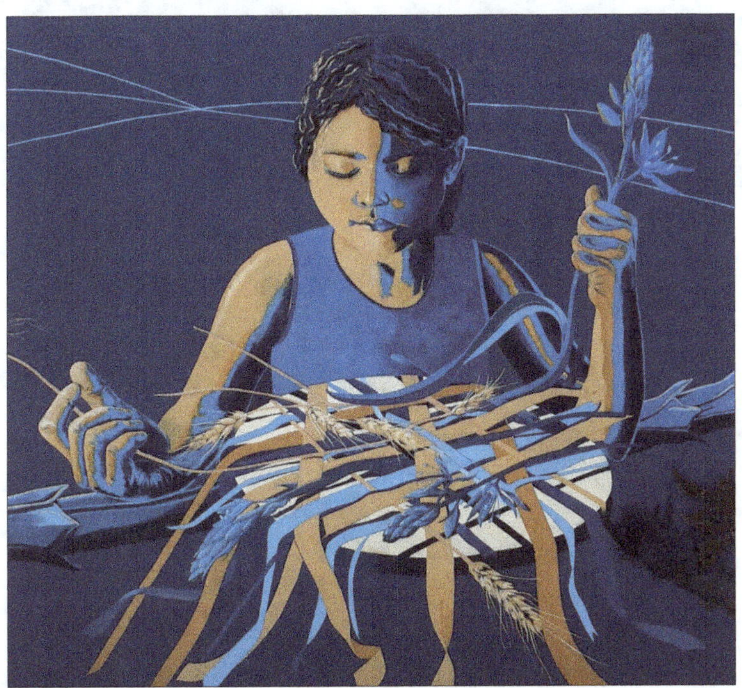

PLATE 2: Esteban Camacho Steffensen and Jessily Brinkerhoff, *Working Forward, Weaving Anew* (2017)
Oil on masonry, 38 ½ x 72 feet
Prairie Line Trail Historic Interpretation Project, Tacoma, Washington

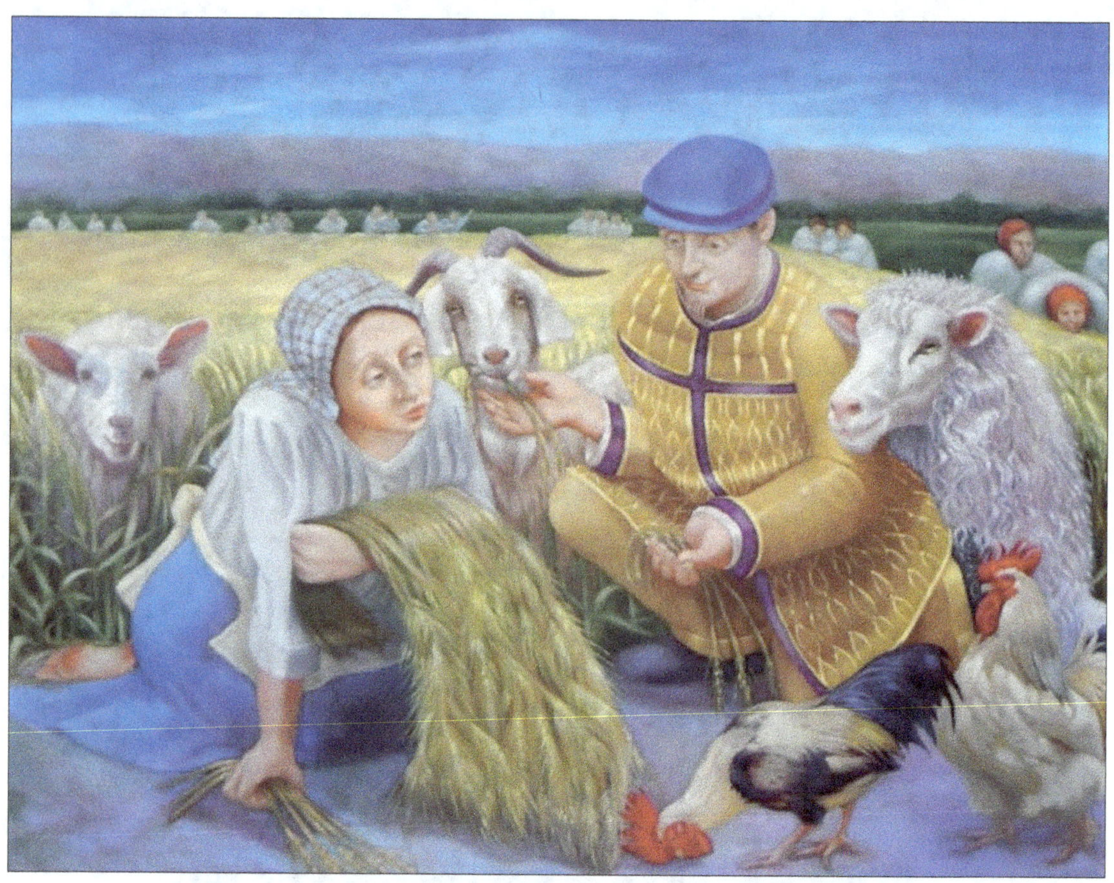

PLATE 3: Rosemarie Oehler Adcock, *Ruth and the Barley Harvest* (2014)
Oil on canvas, 40 x 60 inches
Arts for Relief in Missions Collection, Peoria, Illinois

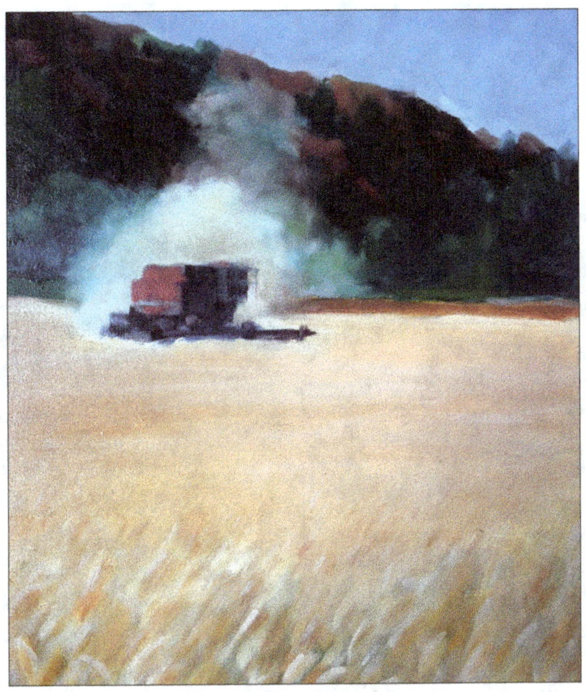

PLATE 4: Jim Gerlitz, *Palouse Colony Harvest* (2016)
Acrylic on canvas, 17 ¾ x 24 ¾ inches
Columbia Heritage Collection

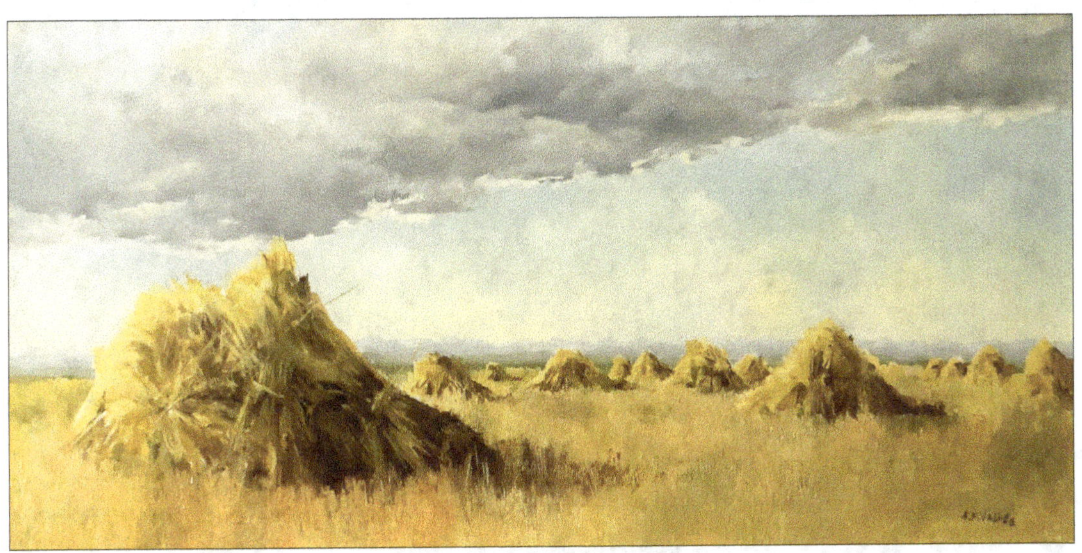

PLATE 5: Armand F. Vallée, Untitled Alberta Harvest Scene (c. 1965)
Oil on canvas, 18 x 36 inches
Courtesy of David and Michelle Bertrand O'Callaghan

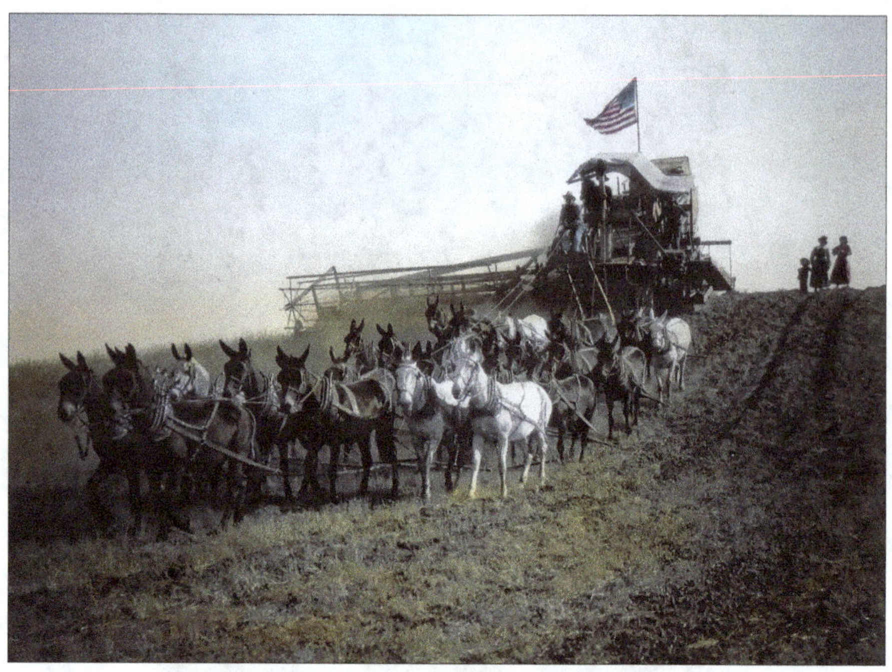

PLATE 6: Asahel Curtis, Combine-Harvester (1915)
Lantern slide print on paper, 5 x 7 inches
Columbia Heritage Collection

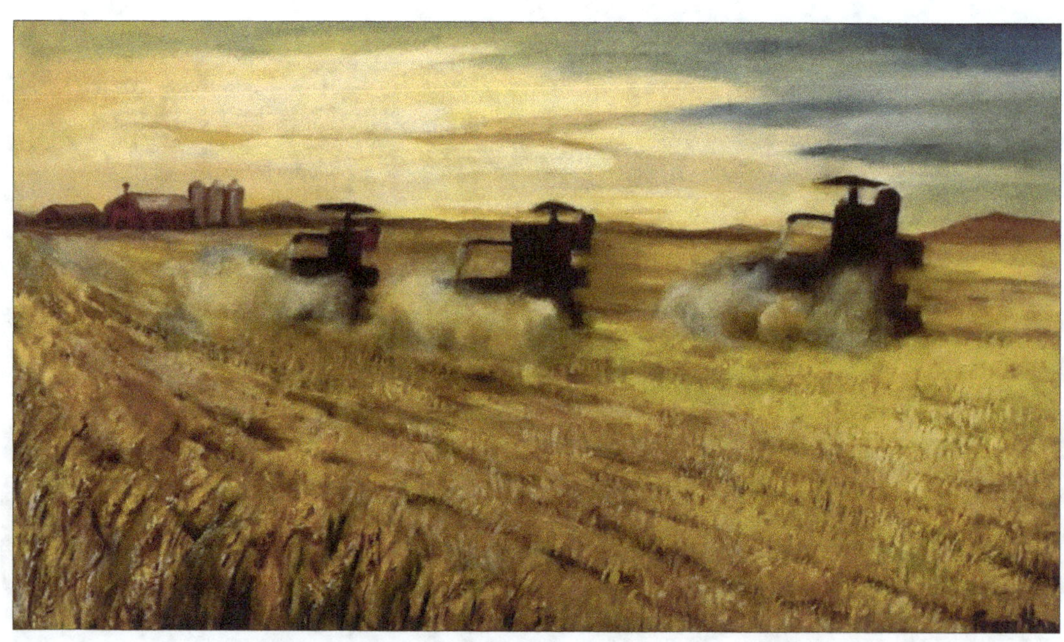

PLATE 7: Peggy Hanover, Untitled Harvest Scene (c. 1960)
Oil on canvas, 18 ½ x 32 inches
Whitman County Library; Colfax, Washington

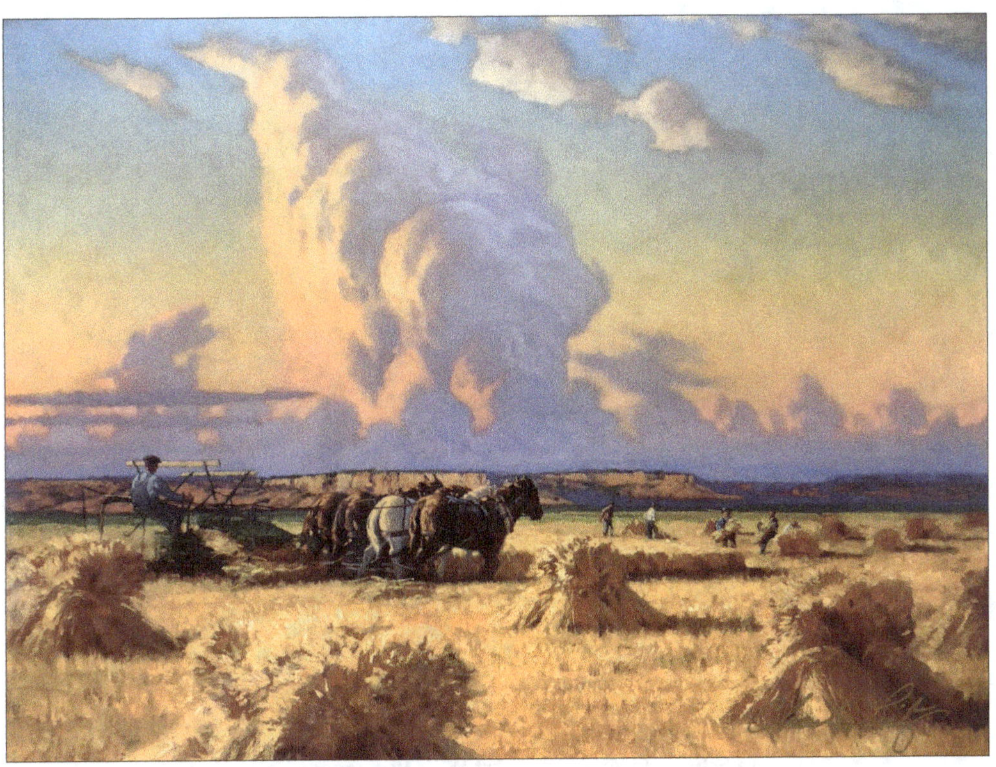

PLATE 8: Charles Fritz, *When Toil Brings Bounty* (*Big Sky Harvest*, 2010)
Oil on canvas (original), dye sublimate on aluminum panel; 8 x 10 inches
Courtesy of the Artist, Columbia Heritage Collection

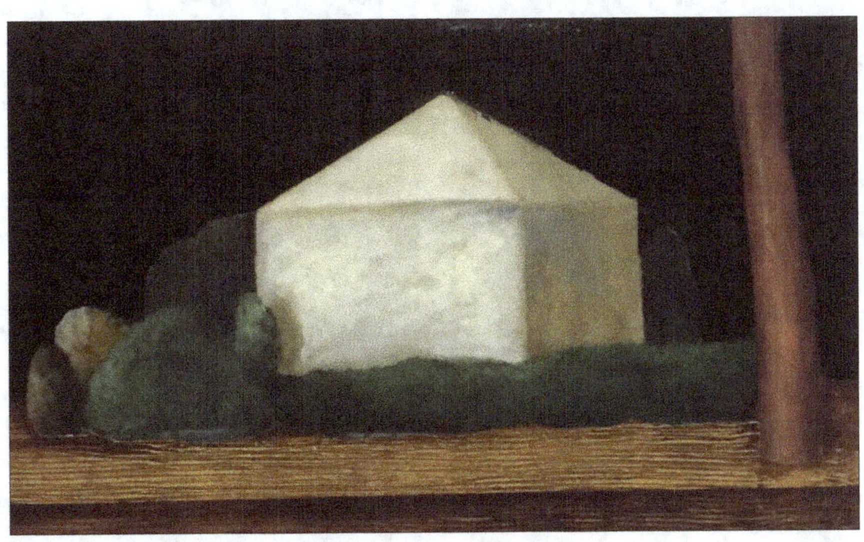

PLATE 9: Robert Helm, *Untitled House* (n.d.)
Mixed Media on Panel, 7 x 10 inches
Jordan Schnitzer Museum of Art WSU Permanent Collection
Robert Helm Memorial Fund and Partial Gift of Tamara Helm

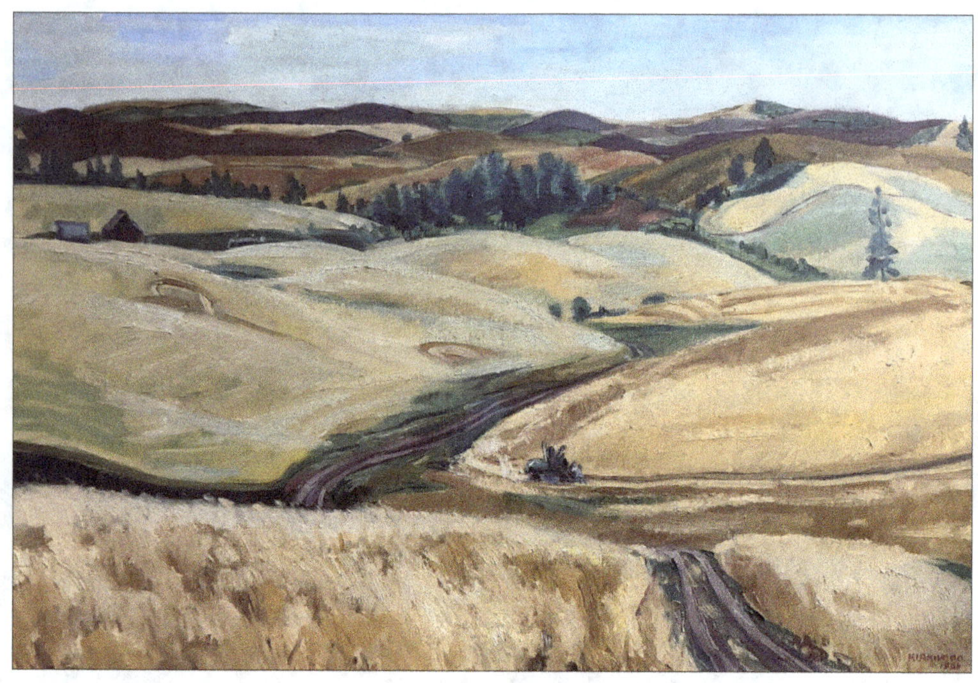

PLATE 10: Mary Kirkwood, *Palouse Harvest* (1956)
Oil on canvas, 26 ¾ x 37 ¼ inches
Jordan Schnitzer Museum of Art WSU Permanent Collection
Gift of the Eastlick Estate

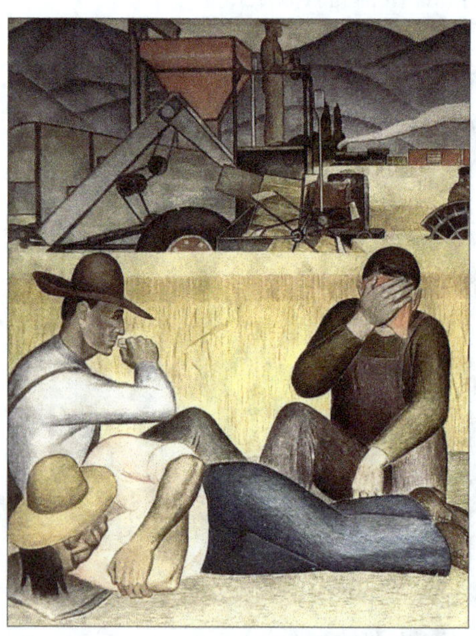 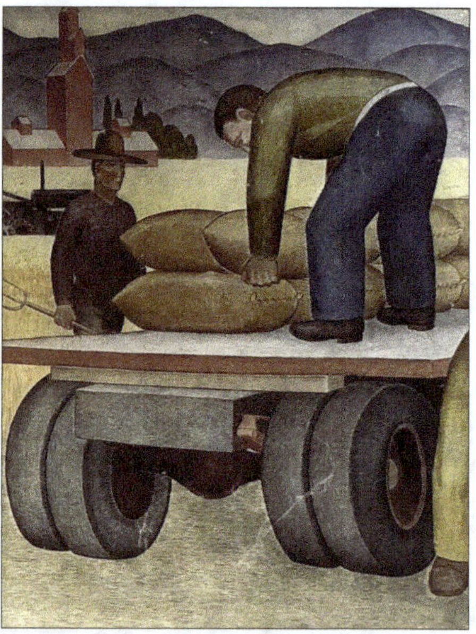

PLATE 11: Arthur Runquist, *Early Oregon* Harvest Panels (1941)
New Deal Art Project Mural, Pendleton, Oregon, High School
Relocated to the Umatilla County Historical Society Museum, Pendleton

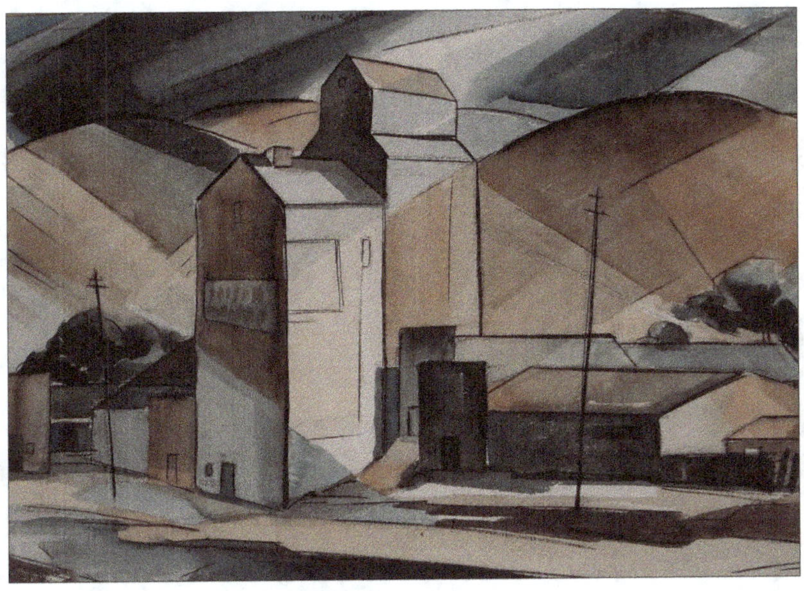

PLATE 12: Vivian Kidwell Griffin, *Elevator and Fields* (c. 1953)
Oil on hardboard, 18 x 24 inches
Gift of the Artist, Washington State University Schnitzer Museum of Art

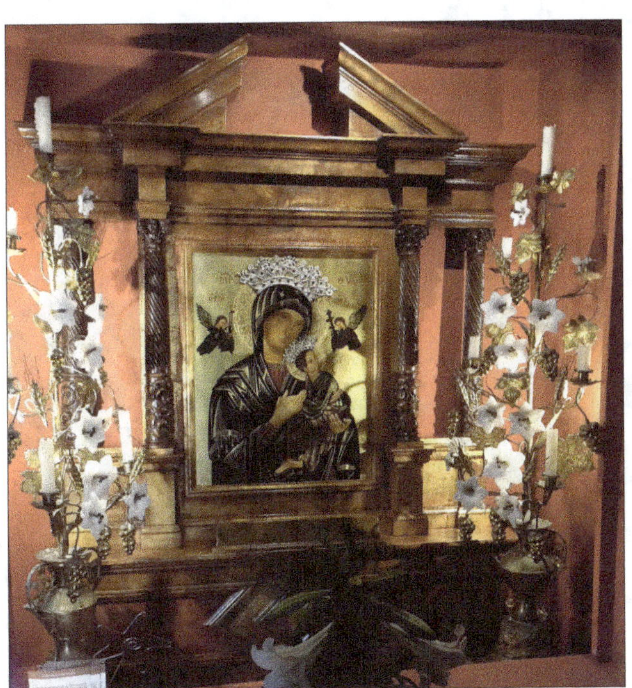

PLATE 13: Our Mother of Perpetual Help Icon
with Brass and Glass Grain and Grape Candelabras (c. 1850)
Mission San Carlos Borromeo, Carmel, California
Columbia Heritage Collection Photograph

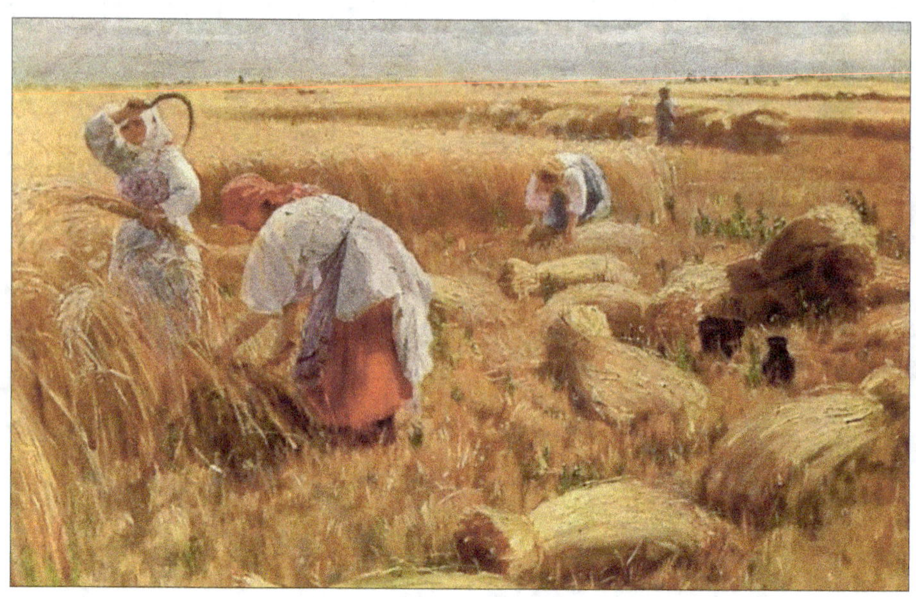

PLATE 14: Fedot Sychkov, Untitled Harvest Scene
Rishar Postcard Publishers, St. Petersburg (c. 1910)
Columbia Heritage Collection

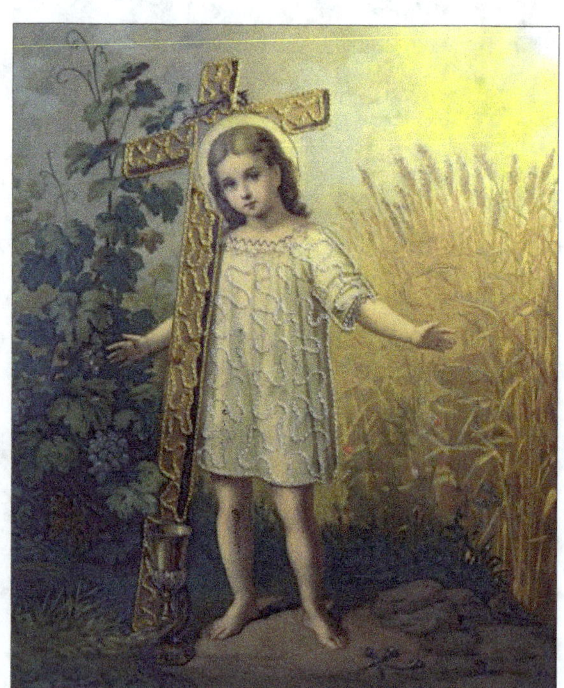

PLATE 15: Fridolin Lieben, *Jesus the Bread and the Wine* (c. 1900)
Lithograph on hardboard, 26 ½ x 34 ⅞ inches
Vogelsberger Heimatmuseum; Schotten, Germany

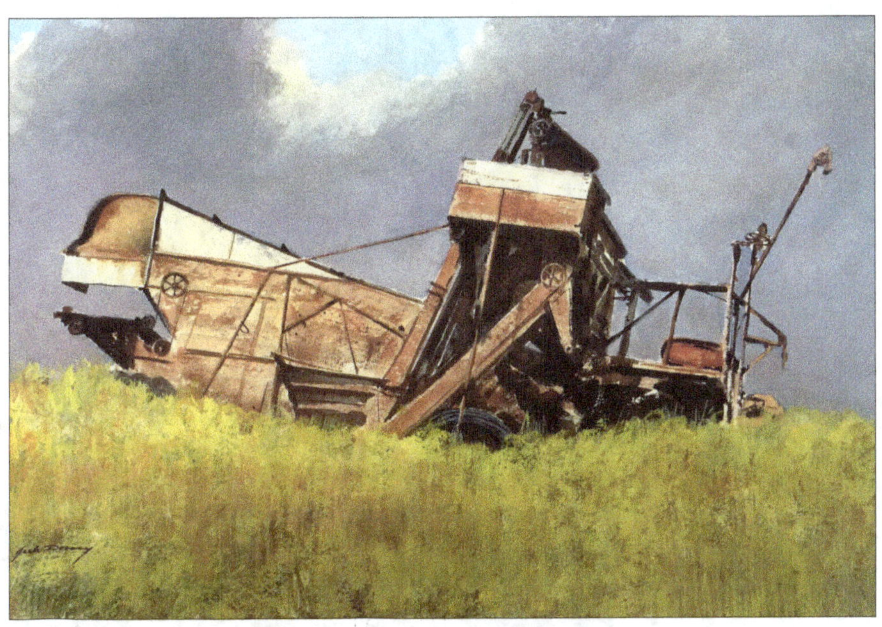

PLATE 16: Jack Dorsey, *Harvester at Rest* (2002)
Watercolor on paper, 16 x 22 inches
Columbia Heritage Collection

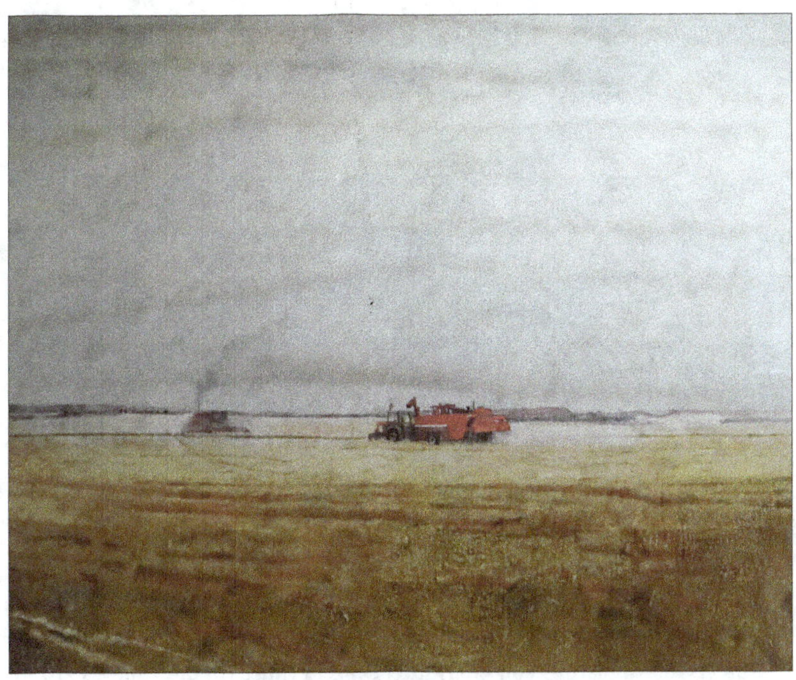

PLATE 17: George Stillman, *Summer Wheat Harvest* (1983)
Oil on canvas, 48 x 48 inches
Washington State Arts Commission, Washington State Investment Board, Olympia

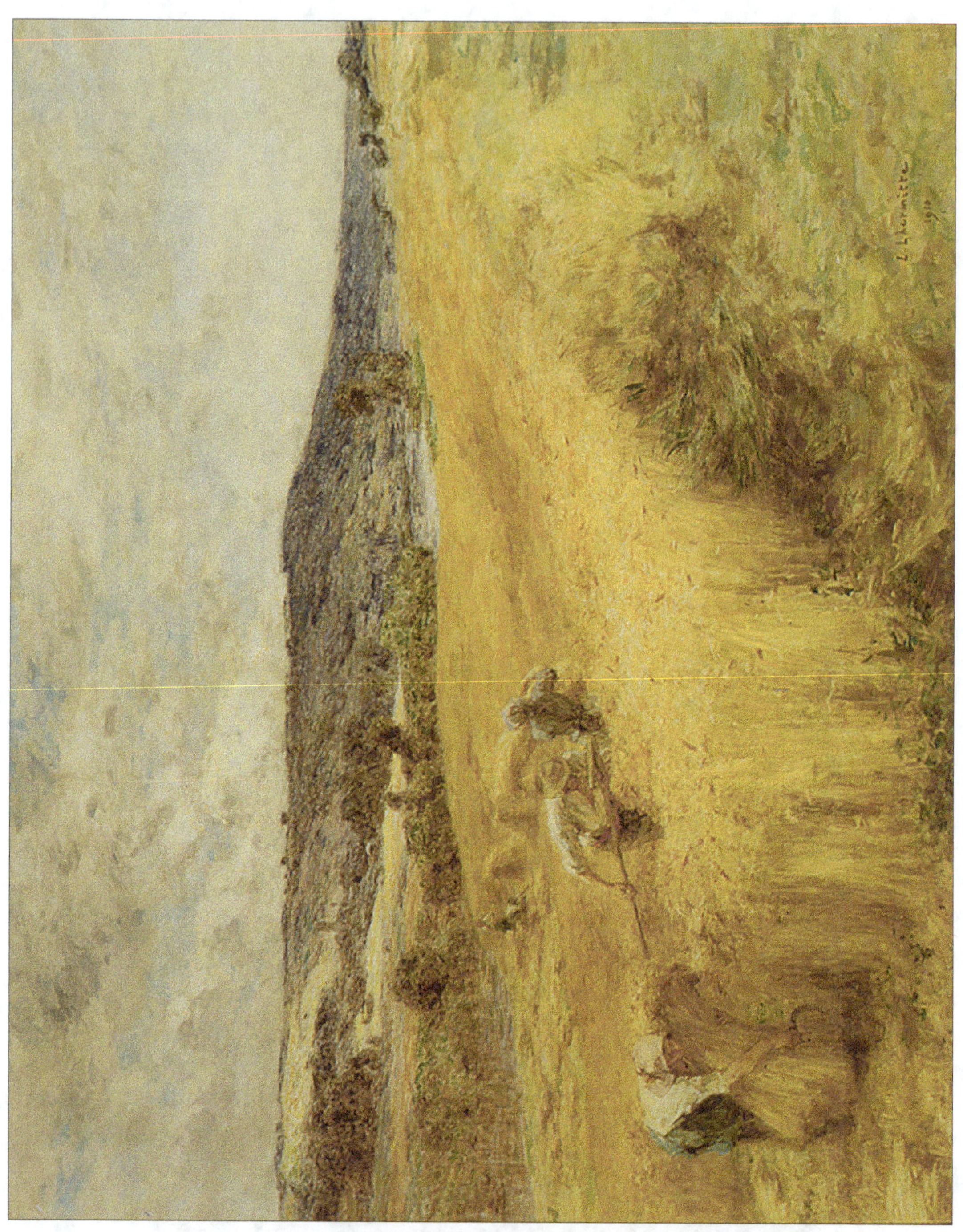

PLATE 18: Augustin-Léon Lhermitte, *Dans la vallée* (*In the Valley*, 1910)
Oil on canvas, 48 x 61 inches
Gift of Charles and Emma Frye, 1952.105, Frye Art Museum

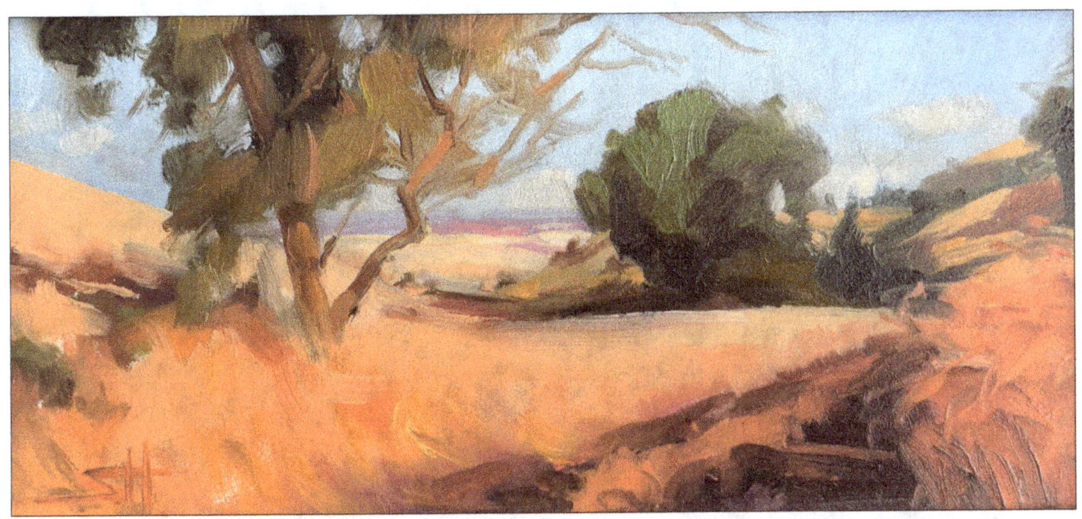

PLATE 19: Steve Henderson, *Harvest Time* (2018)
Oil on board, 4 x 8 inches
Columbia Heritage Collection

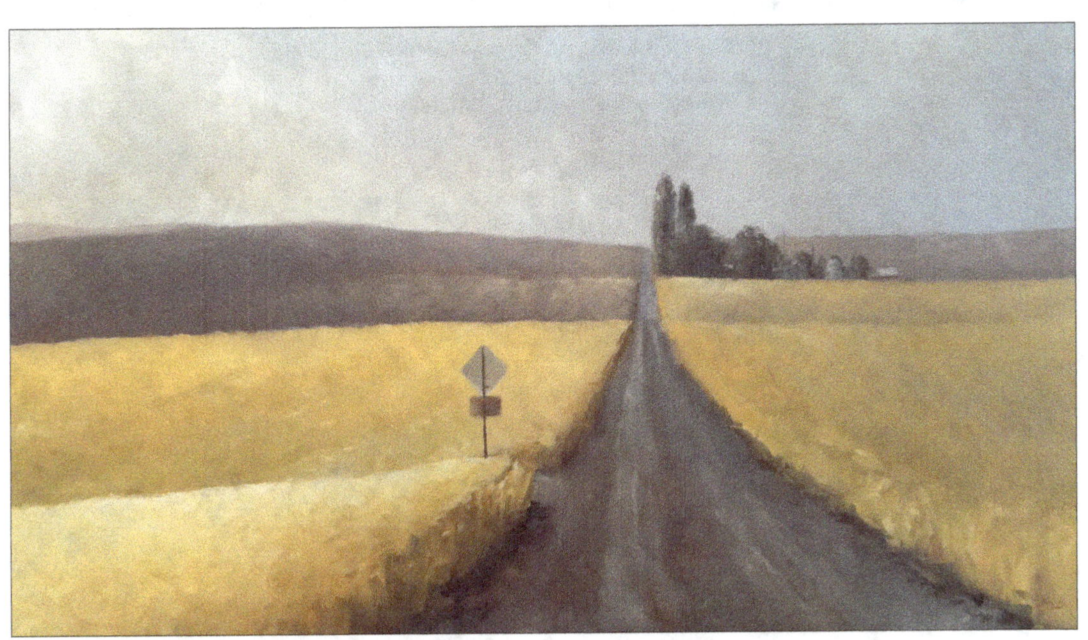

PLATE 20: Kathleen Hooks, *Dryland Wheat* (2002)
Oil on canvas, 14 x 30 inches
Franklin County Museum; Pasco, Washington

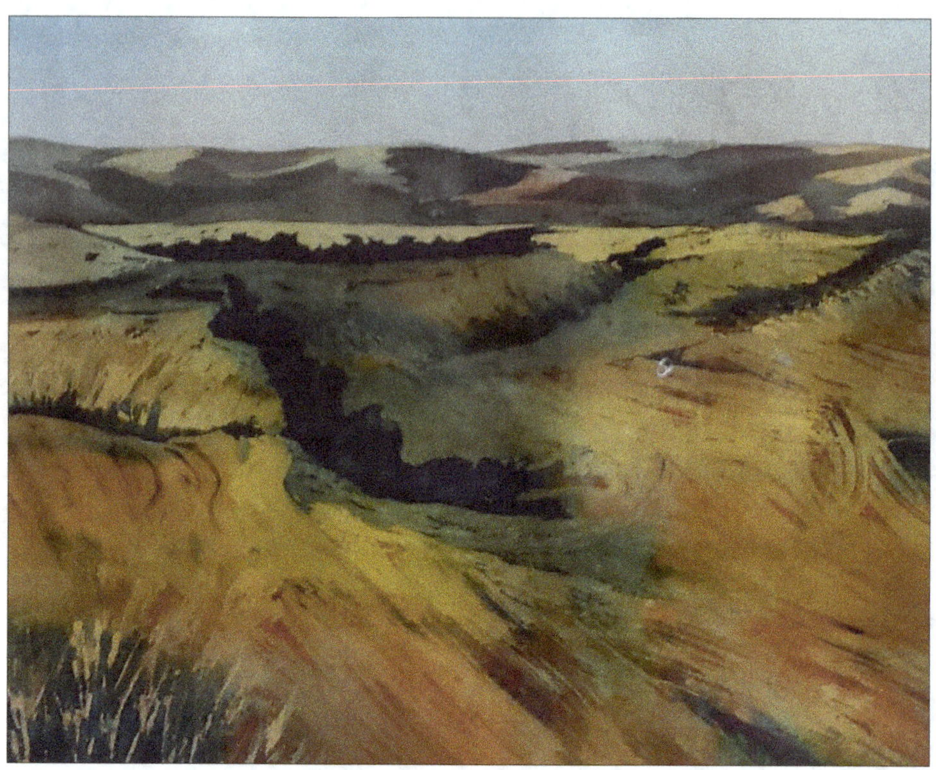

PLATE 21: Maja Shaw, *Palouse Harvest* **(2017)**
Watercolor on paper, 13 x 16 inches
Columbia Heritage Collection

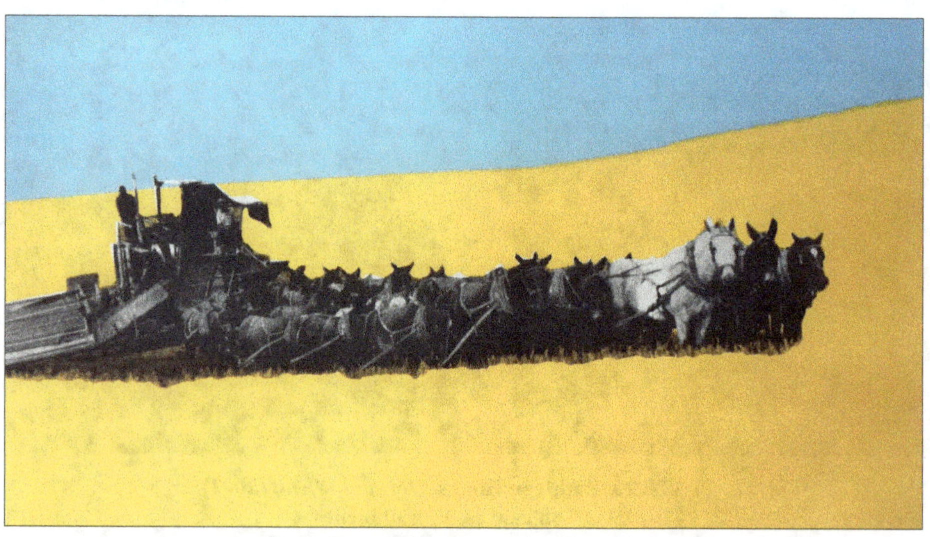

PLATE 22: John Rankin, *Home Place Harvest* **(1995)**
Silkscreen serigraph on paper, 20 x 32 inches
Courtesy of the Artist, Columbia Heritage Collection

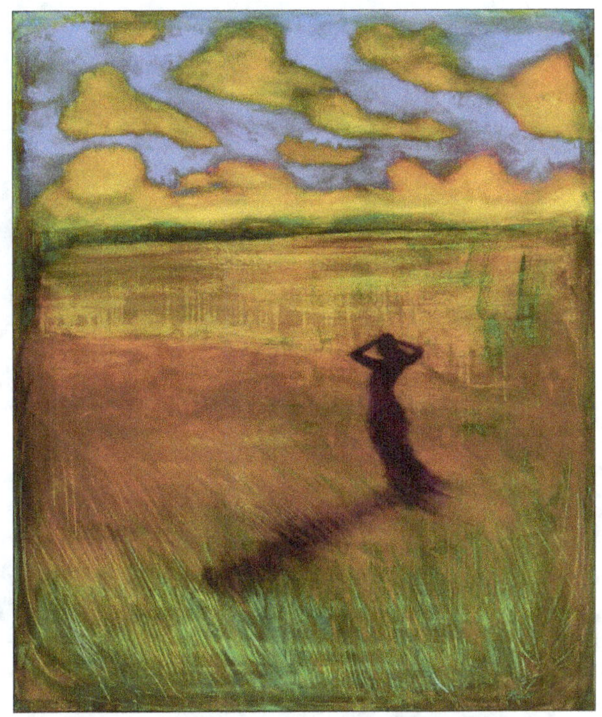

PLATE 23: Ricco DiStefano, *Winds of Harvest* (2012)
Acrylic and charcoal on canvas, 72 x 60 inches
Collection of the Artist

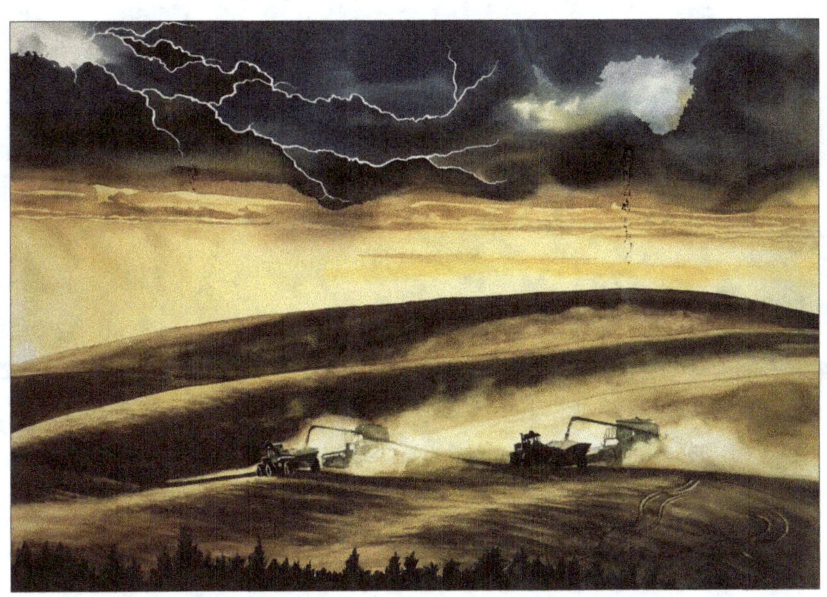

PLATE 24: Andy Sewell, *Doubletime Before the Storm* (2021)
Watercolor on paper, 13 x 18 inches
Courtesy of the Artist, Columbia Heritage Collection

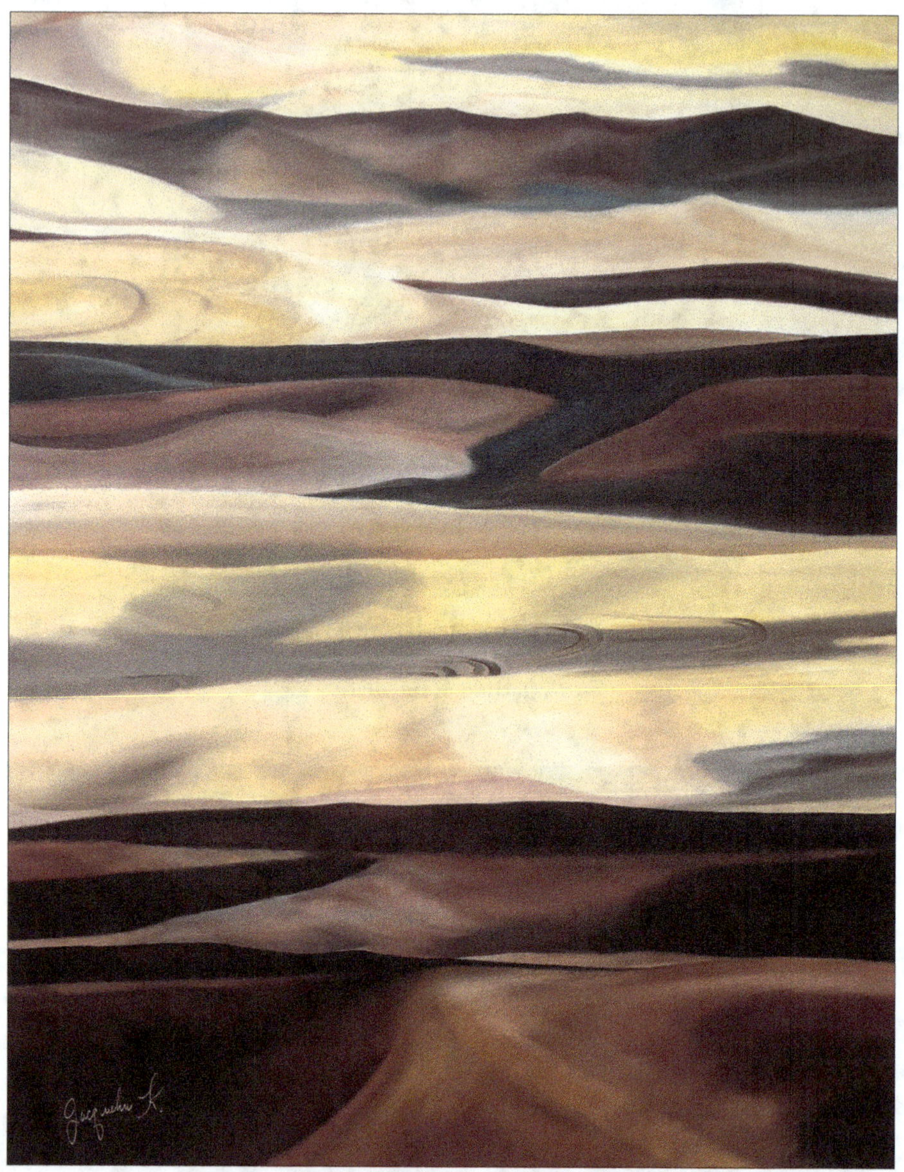

PLATE 25: Jacqueline Daisley, *Palouse Hills—End of Harvest* (2018)
Oil on canvas, 42 ½ x 32 inches
Columbia Heritage Collection

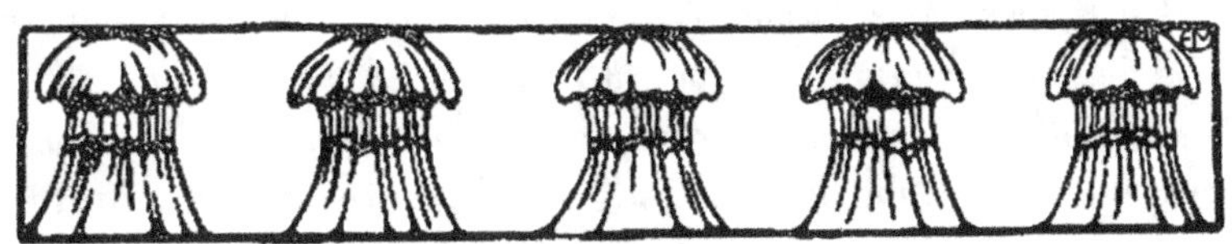

III

In the Valley
EUROPE, FRYE, AND RURAL AESTHETICS

IN THE MID-1970S I EMBARKED ON A THIRTY-FIVE-YEAR HISTORY and English teaching career that began in Cashmere, Washington, located near the geographic center of the state on the eastern slopes of the Cascades. East of the Columbia is the region's Big Bend district, a vast plateau named for the broad irregular arc formed by the Columbia before it heads northward into Canada. Since the earliest days of nineteenth-century settlement, Big Bend country like the Palouse has been synonymous with dryland wheat production. The school mascot in one of its chief rural trading centers, Waterville, is still the Shockers, which invariably requires some explanation even when visitors see a logo of gathered grain sheaves instead of electrical cables. Waterville native Arthur J. Burks (1898-1974) became a prolific author and nationally renowned for his adventure and sports stories. But one of his first novels, *Here Are My People* (1934) is a work of historical fiction that awakened me to the enduring significance of regional agrarian literature.

Reading Burks's book was reintroduction to an engaging cast of young characters drawn from turn of the twentieth-century experience who seemed to be the very elders I had known a lifetime later. Through scarcely veiled early Waterville area characters like Len and Becky Toler and Cephas and Lila Jane Ogle, readers encountered what one reviewer termed "history in their own vernacular" with a "rugged honesty" vividly described through family squabbles and harvest toil. I found Burks's telling especially compelling since many incidental characters were the pioneering Germans from Russia neighbors of his youth. This marked the first time I had ever seen familiar surnames in print in a work of fiction:

> *When the grain was ripe Charlie Slusser—who had bought a McCormick header—was employed by some of his neighbors to cut their grain. The neighbors exchanged work, figured wages in such fashion that they could pay everyone and still pay their own share when the harvesters moved from ranch to ranch. Len Toler's timber culture was heavy with a fine stand of grain. So was his west eighty [acres]. His oats were superb. He was happy in his possessions. The Lebsacks and Golls had the best yield per acre, in prospect, of anybody in the neighborhood. The Russian-Germans, however, banded together to exchange work, keeping themselves apart from all others. This was just as well, tho, as they worked longer each day in the fields than the others, who would not have kept the Lebsack-Goll hours had they ex-*

changed work with them. Most farmers considered they had done a good day's work when they started at four in the morning and got to bed by nine. The Lebsacks and Golls started an hour earlier and quit an hour later. Their farm animals were usually skinny from overwork, but they could manage to stand up under the grind. If they couldn't, well, the Russian-Germans had saved enough money, even in bad years—because of excellent management—to buy extra horses and mules.

Here Are My People concludes with the following year's harvest season which relates more details of the era's transition from equipment pulled and powered by horses to internal combustion engines. With the approach of World War I, many of the Big Bend's stolid German-speaking residents voice sympathy for Germany's cause. Hot summer winds shatter standing grain as if to presage greater political and economic conflicts abroad and at home. But the book's narrator concludes by taking stock of timeless benefits known to farm folk in spite of recurrent life complications:

I never thought of becoming anything but a farmer. I loved the farm. It was part of me. My people had all been farmers, still were—and I should be one, too. There was no nobler work for a man to do. It was a sort of glory to turn the soil with a plow and smell the sweet odor of it; to plant the wheat and hear it trickle through the machinery of a seeder, behind sturdy horses or mules, to harrow it in—and then watch the first green shoots start breaking through the soil. Day after day, then it would rise higher and higher until it had reached its full growth, when it began to turn yellow—the Blue Stem wheat and the Turkey Red, the oats and the barley.[1]

While in Cashmere in the 1970s I was among many teachers who greatly benefited from a remarkable support staff that included our reliable custodian, Jack Knowles. He kept our ancient

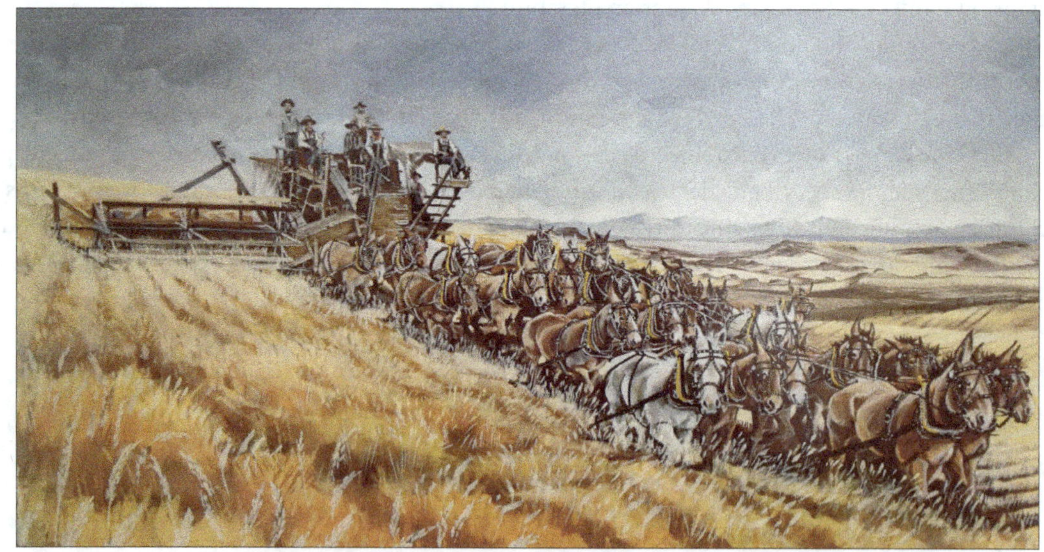

Sherryl Evans, *26-Mule Hitch with Holt Combine* (c. 1925/1978)
Lithograph, 10 x 20 inches
Courtesy of the Artist, Columbia Heritage Collection

school spotless and also served as a kind of faculty confessor-encourager. Jack had enlisted in the Marines in World War II and was among those who witnessed the raising of the American flag on Mt. Surbachi after the heroic assault on Iwo Jima. One day I sought him out regarding some classroom need and for the first time entered his closet sanctum of unfinished lathe and rafters off the main hallway. I discovered that Jack's unofficial responsibilities also included being keeper of the districts' substantial collection of castaway art. Crammed under workbenches and mounted along a narrow passageway few had ever seen were remarkable framed canvases and prints gathering decades of dust. Gilbert Stuart's ubiquitous print of George Washington peered across to Caproni plaster bas-reliefs of *Columbus Discovering America* and other famous scenes from American history. Jack pulled out several from his trove for me to admire including Adolf Dehn's *Minnesota in August*. A richly colored 1939 lithograph of a painting Dehn had completed the previous year, the picture of grain shocks and shade trees seemed to radiate harvest heat from its thin gold frame. I had never heard of the artist but the piece summoned a peculiar affinity. The view bore uncanny resemblance to the arrangement of buildings and fields of my wife Lois's family farm near our Endicott hometown.

When our old brick edifice was later condemned for safety reasons and scheduled for demolition, the neglected pieces of Jack's gallery were stacked unceremoniously on the maple flooring of the gymnasium for public auction. Each faculty member was invited to first take one as a souvenir of school service, probably a breach of some administrative code. In any event, Jack knew which one would be my choice and I soon found myself in possession of Dehn's mystical handiwork. It has long remained a family treasure, and further prodded a latent interest in agrarian art.

About that time I also became acquainted with the writing of Annie Dillard, whose essays, poetry, and prose have been compared favorably to works by Gerard Manley Hopkins and George Eliot. Her 1975 Pulitzer-winning nonfiction book *Pilgrim at Tinker Creek* describes the natural world around her home in Roanoke, Virginia, and soon after its publication she moved to an island in Puget Sound near Bellingham, Washington. Her story "Total Eclipse" (1982), often included in anthologies of notable twentieth-century "best essays," caught my attention because I had witnessed the same natural phenomenon not far from the swath of eastern Washington shadowland she and her husband visited that morning of Monday, February 26, 1979.

Dillard's peculiar retelling relates numerous happenings the morning of the solar eclipse that seem to annoyingly detract from the anticipated event. She describes banal hotel television programming, millet on the carpet beneath a canary cage, and bald men who sit silently in the dimly lit lobby. The contrast between cultural preoccupation with humdrum amidst impending cosmic grandeur begs the reader to call self and others to attention. During the climactic moments of umbra when time seems eerily suspended, Dillard conjures a slow-motion harvest scene from humanity's prehistoric past as if seen in bronze-plated silent film noir. We see wild barley and einkorn wheat growing at our feet on the flanks of the Zagros Mountains overlooking the Euphrates Valley, and remember cutting it with stone sickles. Then a terrorizing blackness threatened to overcome the sun until rays of light emerged to renew hope for life. The jarring shift in time and place impresses one with the singular significance of our relationship to natural phenomenon in a day of increasingly disconnected, virtual reality and superficial daily diversion.[2]

Keston and British Country Life

Among most rewarding experiences while composing this study has been to learn about the influence of agrarian literature and fine art on longtime acquaintances from all walks of life. While at Washington State University in the 1970s, I attended a campus ministry conference in Dallas, Texas, and met Peter Deyneka, Jr., of Wheaton, Illinois. He shared my academic interests and mentioned that his wife, Anita Marson Deyneka, had been raised on a small farm near Cashmere and then directed a Russian studies institute affiliated with Wheaton College. Peter and Anita encouraged me to learn Russian, a sobering prospect that I postponed for many months. This encounter began a lifelong friendship with the couple and in 1976 my wife and I traveled to London for graduate studies in Eastern European history at Keston College funded through a Lutheran World Federation scholarship. Once again, Palouse Country hometown associations may have played a crucial role. The Geneva-based Federation's president, Rev. Carl Mau, was an Endicott classmate of my father's who had gone on to become an influential Cold War era bridgebuilder promoting East-West religious and political dialogue.[3] Keston's founding director, Michael Bourdeaux (later knighted "Sir Michael" for his scholarly contributions to religious freedom), and the center's invigorating mélange of experts and visiting scholars introduced new cultural horizons including country life influences in the art and literature of England, Scotland, and Russia.

Keston is located near the midpoint of the 140-mile medieval Winchester-Canterbury Pilgrim's Way in the Croydon district south of London. We stayed in the upstairs loft of a nineteenth-century school designed as a Victorian country house that overlooked Keston Common not far from stony remnants of the old road that brought to mind verse from *The Canterbury Tales*. The normally ver-

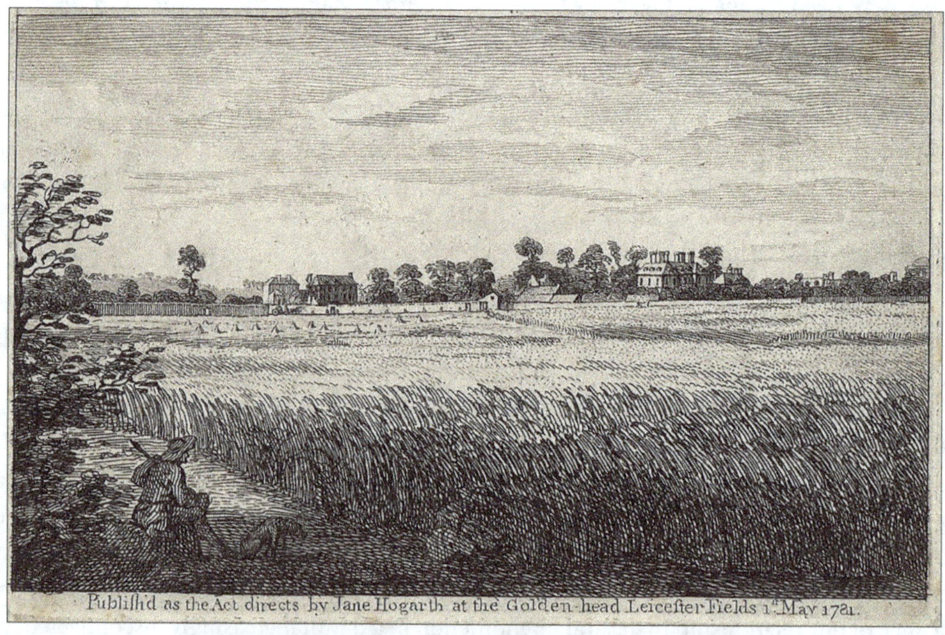

William Hogarth, *View of Ranby's House* (c. 1750/1781)
Etching on paper, 4 x 6 inches
The British Museum, Public Domain

dant setting was suffering from England's driest summer on record that parched the gently rolling grain fields of nearby Surrey. We toured the incomparable art collections of London's Victoria and Albert Museum and journeyed west to the grounds and Royal Apartments of Windsor Castle. En route to Windsor in the southwest London borough of Richmond are the Kew Royal Botanic Gardens which we learned had been a royal farm estate until George III organized previous haphazard exotic plantings there with guidance from Sir Joseph Banks. Although George III's father and grandfather were not known for aesthetic appreciation, George III was an avid collector of art including works by influential social critic and printmaker William Hogarth (1697-1764). Best known for satirical depictions that brought notoriety and income, Hogarth also did engravings for personal pleasure including a London area harvest scene. By 1749 Hogarth had acquired sufficient means to purchase a country home, present Hogarth House, on Chiswick Common across the Thames from Kew. Titled *View of [John] Ranby's House* (1781), the scene is one of Hogarth's few landscapes. It shows a horizontal row of houses across the middle of the picture with a seated figure in the left foreground who gazes contentedly upon a prodigious field of grain that seems to pulse with Hogarth's shadings. Shocks appear in the distance where shops today line fashionable Burlington Lane.

Beyond Chiswick and Heathrow is Shaw Homestead Farm at Windsor, where we saw the fields, granaries, and flour mill in which Prince Phillip was said to have special interest. German travel writer Paul Hentzer (1558-1623), who visited England in 1598 as part of his three-year Grand Tour, recorded in his *Itinerarium* one of the earliest accounts of harvest festivities at Windsor: "As we were returning to our inn, we happened to meet some country people CELEBRATING THEIR HARVEST HOME. Their last load of corn they crown with flowers, having besides an image richly dressed, by which, perhaps, they would signify Ceres. This they keep moving about, while men and women, men and maid servants, riding through the streets in the cart, shout as loud as they can till they arrive at the barn." Windsor Castle's monumental art collection contains numerous representations of grains and harvest including an enormous Gobelins "Summer" tapestry (1774) that measures approximately 9 x 12 feet and depicts Ceres crowned with grain spikes and clutching a sheaf. Intricately carved wooden panels (c. 1700) with wheat stalks and flowers flank Windsor's Garter Throne Room, and the King's Bed Chamber contains a spectacular oak and marble chest-of-drawers (1774) with intricate grain sheaf and flower marquetry that was originally built for Louis XVI at Versailles. Although we did know it at the time, we were in close proximity to some of the earliest works of art depicting rural life for its own sake. The Royal Library's *Man Ploughing with Two Oxen, and other Farm Workers* (c. 1503) by Leonardo da Vinci (1452-1519) has probably been in the Royal Collection since the late seventeenth century. The drawing in reddish-brown chalk is a magnificent example of anatomical detail and includes seven figures tending to various field operations including hoeing, raking, and scything.

Croydon was the site of one of the region's oldest buildings—the sixteenth-century Croydon Manor Palace which served for centuries as the summer residence of the Archbishops of Canterbury and regularly visited by Henry III and Queen Elizabeth I. The place has a distinguished association with agrarian aesthetics as the site of the first known appearance of an English stage play featuring harvest events—*Summer's Last Will and Testament* by Anglican playwright Thomas Nashe (c. 1567-1601). Although published in 1600, the courtly "masque" (pageant) was probably first performed at the palace as early as 1592. Set in ancient Rome, it features "wheaten crowned" Summer personified as a dying man making his final testament before the other seasons as well as Harvest and Christmas.

Harvest tells of summer labors while accompanied by sickle-bearing reapers who sing of labor and end-of-season celebration. The discourse, presented against the backdrop of plague then ravishing London, contrasts the period's shift from a more broadly shared cultural experience to emergent social disparities and economic individualism.[4]

The stirring British folk music of labor and celebration we heard that summer on weekend visits to Keston's rustic Fox Inn pub and other London area venues made lasting impressions. The 1970s British electric folk-rock scene was creating a sensation by recasting tales of country life with newly composed melodies played to a mix of modern and period instruments. Among the most influential and enduring was Steeleye Span, formed in 1969 by Maddy Prior and Tim Hart of St. Albans with three other friends. The group, which is larger and still active five decades later after various personnel changes, became a staple of the British folk revival scene. They produced a series of acclaimed albums combining traditional tales a century old or more set to spirited and haunting tunes along with tracks of reels and jigs featuring guitar, fiddle, flute, oboe, and drums.

That summer of '76, when we sang "America the Beautiful" on the Fourth of July Bicentennial on Keston Common, the airwaves frequently carried the mournful lovers title song from Steeleye's album *All Around My Hat*. The tune had just reached number five on the national singles chart. Although the group's songs about hard times and hornpipes brought to mind some of the music my father favored, subsequent compositions by Prior derived from harvest folklore has resonated through the years. Her song "Marigold/Harvest Home" (1980) recalls the time when fieldworkers would gather with the landlord's family for a lavish dinner marking the end of the exhausting harvest season—a long-awaited day of jubilation, gratitude, and opportunity to settle debts.

> *The rake has reaped, the blade has mown*   *The quiet of a heart at rest*
> *Nights draw in to call the harvest home.*   *In peace abounded.*

The song concludes with lines from Henry Alford's familiar nineteenth-century harvest hymn, "Come, Ye Thankful People, Come" ("All be safely gathered in / 'Ere the winter storms begin"). Prior's "Fabled Hare" song cycle on the album Year (1993) is based on the ancient mythical significance of the grain spirit and reminded me of the jack rabbits I had seen flee from their sanctuary in remote parts of our rolling cropland.

> *I have outrun dogs and foxes*   *I've survived your persecution*
> *And I've dodged the tractor wheels*   *and your ever changing field.*[5]

The Keston environs had been an enchanted nineteenth-century *locus amoenus* that blended artistic imagination with geographic reality for artists like John Linnell (1792-1882), Samuel Palmer (1805-1881), Henry H. La Thangue (1859-1929), and plein air watercolorist William H. Allen (1863-1943) who idealized English rurality of heath and harvest. Linnell sometimes took over a year to complete opulent canvases like *The Storm in Harvest* (1872), which also suggest apocalyptic themes present in Puritan thinking of the time. But they also indicate Linnell's intimate agrarian knowledge. His long-standing popularity among prominent British art patrons earned significant commissions which enabled him to support the talents and imaginations of young Romantic visionaries like Palmer and William Blake (1757-1827). Palmer became Linnell's son-in-law in 1837, and Linnell introduced him to Blake and his artistic mysticism.

Linnell's prosperity enabled him to purchase the small farming estate of Redhill west of Keston in the summer of 1849. The original eleven-acre holding eventually grew to nearly eighty where he built a home and regularly oversaw operations in the fields of wheat and barley. Linnell's demesne provided rare opportunity for a sketchbook artist to closely follow the seasonal activities of a working farm. He came to know the feel of freshly turned, soft fallowed ground beneath his boots in autumn, and summertime sounds of scythers and winnowing machine. His Redhill harvest paintings like *The Storm* may well have metaphorical aspect related to the uncertain influences of metropolitan modernity. But hurried field workers gathering to escape threatening clouds could often been seen in the late weeks of harvest.

Pastors and pamphleteers of the time often invoked lines from Psalm 65—Charles Spurgeon's "harvest hymn for the ages." The verse inspired Samuel Palmer so listeners and readers might discern divine "manifold witness" through the noble toil of agrarian life and creation's beauty in the English Canaan. The artist inscribed the King James text on the frame of his glowing sepia and gum varnish landscape of visionary intensity and fecundity, *The Valley Thick with Grain* (1825). Harvesters seem to toil silently as if in holy, idyllic service of divinity's ultimate creative purpose. Palmer completed other notable drawings and paintings of spiritual insight during this period including *Naomi Before Bethlehem* (1826), *Meditation upon the Wonderful Providence of God* (1828), and *Ruth Returned from Gleaning* (1829).

Palmer resided in the village of Shoreham east of Keston during these years and make frequent forays in the surrounding Kentish countryside to find suitable settings for such works. Palmer prayer-fully approached each composition to impart radiant details in the position of graceful figures, fertile hillsides, tranquil sheep, and oversized fan-like heads of grain. He was a prodigious reader, and in addition to the Bible studied works by Virgil, John Bunyan, and German Lutheran mystic Jakob Böhm. Palmer was also influenced by Presbyterian clergyman John Flavel's *Husbandry Spiritualized* (1671) which contained a line that might represent the raison d'être for his kindred "Shoreham Ancients" artist spirits: "Tis a great part of our holiness . . . to mind spiritual things, when we are conversing with the clods of the earth, and the furrows of the field."

Palmer's visionary creations inspired contributions by English artist and educator Robin Tanner (1904-1988) to a post-war revival of the Arts and Crafts movement. Tanner, who first viewed Palmer's

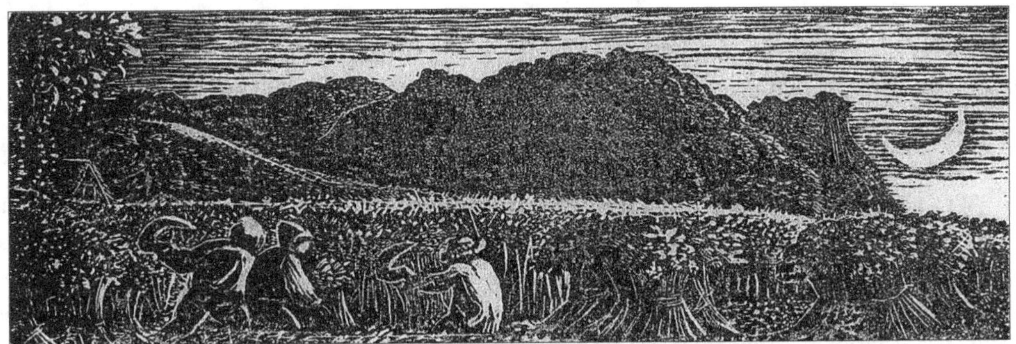

Samuel Palmer, *Harvest Under a Crescent Moon* (c. 1826)
Woodcut on paper, 4 ⅛ x 5 ¼ inches
Wikimedia Commons

works in 1926, went on to serve for three decades as Royal Inspector of Schools (1935-1964) and promoted exploratory learning and creative expression through public education. Following retirement, Tanner pursued his longtime printmaking interests in earnest and produced numerous pastoral works like *Martin's Hovel*, *Harvest Festival*, and a monthly series that featured shocks of grain, fencerows, and ancient trees. In 1970 he founded the Crafts Study Centre in Bath which relocated in 1977 to the Surrey Institute of Art and Design in Farnham, presently the University Centre for Creative Arts.

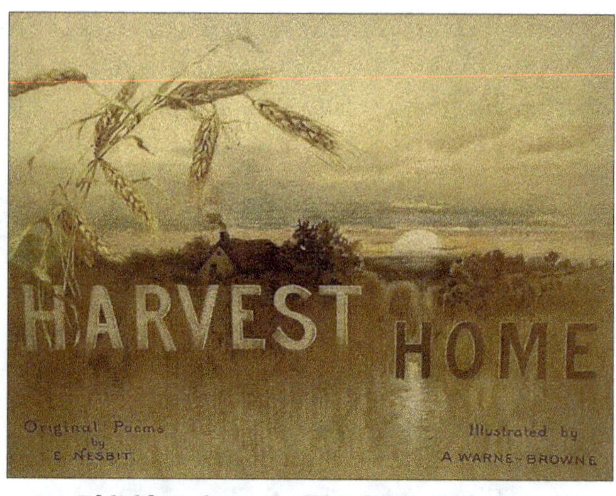

E[dith] Nesbit and A[lfred] Warne-Browne
Harvest Home **bookcover (c. 1895)**
Columbia Heritage Collection

Author Edith Nesbit (1858-1924), daughter of a Surrey agronomist, often vacationed in Dymchurch, Kent, and spent her last years at nearby St. Mary's Bay. Nesbit's collaborated with London-based painter Alfred J. Warne-Browne (1855-1915) on several books of her verse including *Fading Light* (1890) and Harvest Home (c. 1895), a celebration of cooperative labor and harvest completion. A free-thinker whose life spanned the late Victorian and Edwardian periods, Nesbit wrote romantic novels and poetry but is best known for illustrated children's stories. Her writing influenced Noel Coward, C. S. Lewis, and many other later authors for realistic depictions of adventure and English countryside summers as evidenced in a stanza from her poem, "August" (1886).

> *I want to walk through crisp gold harvest fields,*
> *Through meadows yellowed by the August heat;*
> *To loiter through the cool dim wood, that yields*
> *Such perfect flowers and quiet so complete—*
> *The happy woods, where every bud and leaf*
> *Is full of dreams as life is full of grief.*[6]

Our summer at Keston led to familiarity with the work of Croydon native Henry H. La Thangue who trained in the 1870s at London's Royal Academy and with Jean-Léon Gérôme at the prestigious Ecole des Beaux-Arts in Paris. While in France La Thangue came under the influence of the Barbizon Colony's en plein air painters of peasants and he developed an evocative style of Rustic Naturalism. In paintings like sunny if cheerless *Return of the Reapers* (1886) and *The Harvesters' Supper* (1903), La Thange captured the seasons of life in the fields showing laborers and livestock from sowing time to harvest. His rich canvases are also notable for capturing commonplace rural endeavors not usually seen in the grand works of La Thangue's more well-known contemporaries in Great Britain and France like John Linnell and Jules Breton. His workers include both young and old in Sussex and Norfolk who share a nighttime meal under a makeshift canopy and stolidly stroll down country lanes en route to another field. *The Man with the Scythe* (1896) is a rare La Thangue allegorical painting that shows a horrified mother when she realizes a young girl seated before her, likely her daughter, has died. A field worker bearing a scythe appears at a gate in the background symbolizing the arrival of the "Grim Reaper" in a day when disease claimed many children.

La Thangue established a home in the Norfolk countryside in the 1880s and read works by prominent local author and agricultural reformer H. Rider Haggard (1856-1925). His popular country life memoir, *The Farmer's Year* (1899), offered vivid descriptions of East Anglia farm operations depicted in many of La Thangue's paintings. An August vignette from Haggard's book conjured the reaper enmeshed in the cherry tree of my boyhood home, and mentioned harvest hazards similar to ones I had heard from my grandfather:

> *This morning we set the reaper to work on the glebe field of oats, which are bearing a good crop for so scaldy a piece of land, owing doubtless to the wet of early summer. Before the machine can be put in, a pathway must be mown round the field with a scythe. Then the thing starts, drawn by two horses. It is beautiful to see work, for it cuts wonderfully clean, the arms sweeping the bundles of grain from the platform in sheaves, ready for the binder. By a clever contrivance of the mechanism, . . . this act is not always done by the same arm. The limb that at one revolution delivers the bundle from the table to the ground, at the next merely bends the straw over the knives, while another dips down to the platform and clears it.*
>
> *. . . Care, however, must be taken at the corners, where the reaper turns, or it will jam. Indeed it is well for a man to round these off with a scythe. Some people yoke three horses to such machines, but I use only two, which are changed at noon. Half a day's work with a reaper behind them is quite enough for a pair of horses. . . . I see in the papers that a terrific accident has just happened with one of these machines. A pair of horses attached to it bolted, and in try to stop them their owner was thrown to the ground and so cut by the knives that he died.[7]*

Views of East Anglian fields and fens are also subjects of many paintings by contemporary landscapist John Bond, who studied art in Norwich and London. Bond's humid summer views like *While the Sun Shines*, *Late Summer Harvest*, and *August* bathe farm scenes in pale saturations of oil, watercolor, and gouache. They show diminutive scythers in bowler hats and long-sleeved white shirts working beneath atmospheric skies as they might have appeared in the time of La Thangue and Haggard.

John Bellany (1942-2013), head of the painting faculty at nearby Croydon College of Art, organized art exhibits at the school that summer of 1976. Although a native of the fishing community of Port Seton on the Firth of Forth in Scotland, Bellany was also known for his many farming scenes and stark monumental allegorical depictions of commonplace labor. As a young man, Bellany traveled widely on the continent where large paintings and altarpieces by Delacroix and Brueghel made a deep impression in his own oeuvre as did Modernists Max Beckmann and Oskar Kokoshka. Bellany's lifesize painting *The Harvester* (c. 1965) shows an expressionless man clad in black who looms between two dense stands of grain or hay. This pale, severe harvester holds an upright hayfork as if Moses parting the Red Sea. He is flanked by a family of weary field workers that includes a small boy who peeks from behind one of the men. A man at the far-left leans against a spade while a woman on the right side stands behind a sack of potatoes. The group appears downcast and resigned to an uncertain fate as crows circle high above. Death and judgment are harbingers of new life.

The Museum of English Country Life was located an hour's drive west of Keston at the University of Reading and featured nicely curated displays on the pre-industrial farm year, tillage equipment and field operations, and village life. A notable exhibit consisted of enormous wall hangings by English Modernist textile artist Michael O'Conner, who died just a few weeks after our time in Keston. O'Conner was inspired by the nineteenth-century Barbizon experience expressing traditional agrarian lifeways and had named the studio he established in Australia in the 1920s for the French art colony. By 1950 O'Conner had returned to England and was commissioned to weave six massive tapestries approximately 12 x 22 feet showing various aspects of British rural life for the London's grand Festival of Britain pavilion the following year. The lively Yorkshire hanging commemorated grain production with manor houses and smaller farm homes surrounded by trees and stooked sheaves of bright orange and yellow.

About the same distance east of Keston we found Kent's scenic Elham Valley where the 1830s Swing Riots violently marked a momentous turning point in modern agricultural, social, and economic history. Named for the motion of wooden threshing flails, the Swing Riots were the festering result of the peasantry's assault by forces of the commons' enclosure, industrialization, and unwillingness of British policymakers to effectively moderate the consequences. At the time perhaps five thousand wealthy proprietors owned about one-half of the United Kingdom's real estate. Rebellion was inevitable in a land where increasingly embittered, unemployed landless masses had little recourse to progressive legislation or unionism. The uprising had begun during the harvest of 1830 when protesters in eastern Kent set fire to recently introduced threshing machines, and the movement soon spread to grain producing districts throughout southern and central England.

Between Keston and Elham near the River Thames is the ancient village of Southfleet with its fourteenth-century parish Church of St. Nicholas. A color lithograph of the historic structure (c. 1865) by English architect, watercolorist, and printmaker James K. Colling (1816-1905) is one of the earliest examples of commercial chromolithography. The noted English artist was drawn to medieval ecclesiastical architecture after visiting Southwell Minster in Nottinghamshire and being struck by

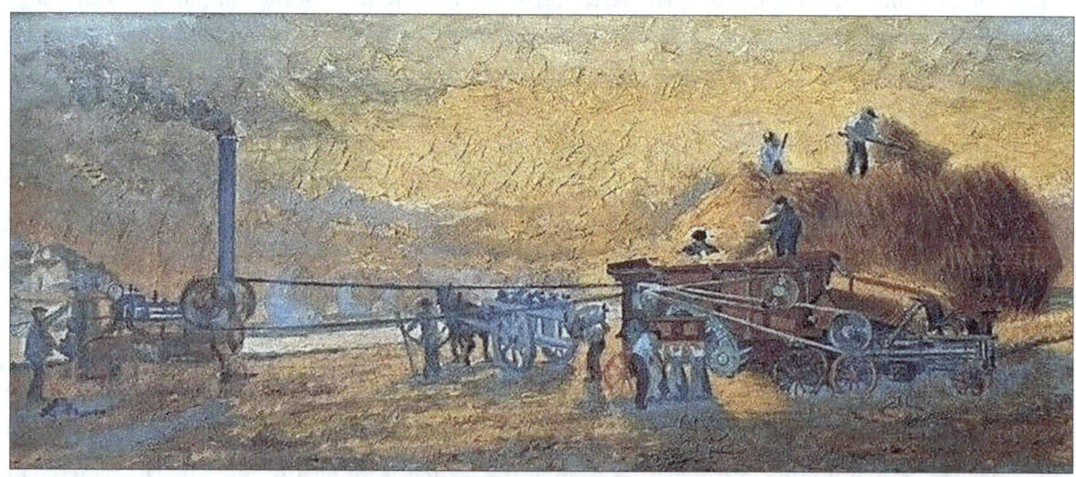

R. Gautreaux, Untitled French Harvest Scene (1929)
Oil on canvas, 35 ½ x 13 ¼ inches
Private Collection

the Early Decorative style's harmonious interior colors, elegant forms, and lighting. Colling went on to write and publish the authoritative *Gothic Ornaments: Enriched Details and Accessories of the Architecture of Great Britain* (1838) and *Art Foliage for Sculpture and Decoration* (1865). Like Ruskin, he believed representations of creatures and natural foliage like grain ("always beautiful in form") and flowers to reflect divine creation and filled his books with detailed examples of their symmetrical adaptations for architectural ornament, sculpture, and furnishings.[8]

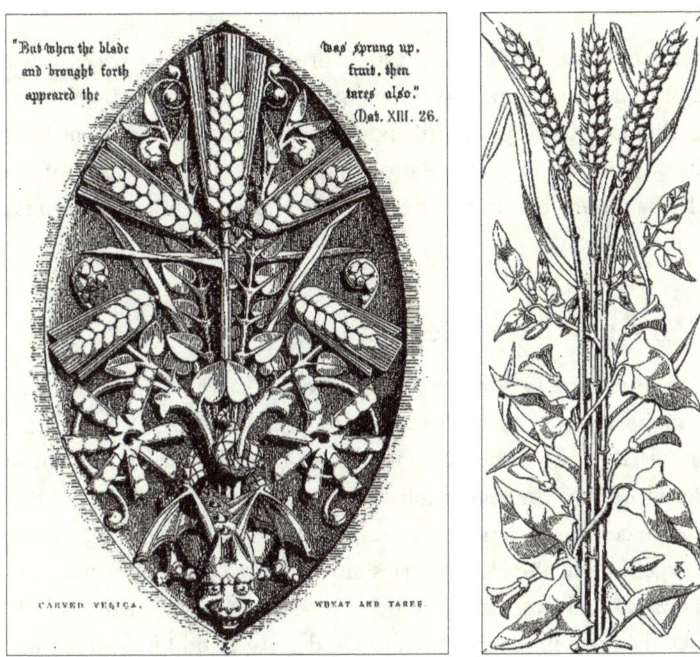

James K. Colling, *Carved Vesica—Wheat and Tares* and *Wheat and Field Bindweed*
James K. Colling, Art Foliage for Sculpture and Decoration (1865)
Columbia Heritage Collection

Chaucer himself might well have passed Southfleet's ancient church as the Reeve, Miller, Wife of Bath, and other pilgrims may have en route to Canterbury. Poet Edward Thomas (1878-1917) lived in rural Rose Acre near the village of Maidstone some dozen miles east of our Keston residence and studied medieval pilgrimages through the region. He wrote extensively about the salutary effects of country life on mental wellbeing in poems and essays like "Caryatids" that described early twentieth-century Kentish harvest scenes recalling ancient harvest ways:

On the downs we heard the reapers chanting a song in the motionless grain. All day they reaped, reaped, and never turned to behold the sun, in whose rebounding beams their faces whitened, whilst their napes and shoulders became brown as they stooped. . . . To the sun, which had been gracious to them all the year, and now was cruel, they were praying that he would still be kind; for then, after he was gone, or at least when they saw him not, in the muffled winter, they would suspend a fruited branch of his own ripening over the chimney-piece, thankfully in memoriam; or if he would not listen, they seemed to say, in half-laugh indignation, they would evoke a rain-shower that should veil his glory before evening, or trample upon this triumphs at dawn. But evermore the burden was—that the sun would lightly deal with the hills where the reapers reaped, as with the valley where he raised a beatific haze. Now charging angrily at the grain rows with sickles, now resting a minute, the reapers presently disappeared in the gulfs they hewed. The women piled the harvest in shining heaps, and after nightfall traveled home, Caryatid-like, with children upon their arms, and the wreck of sunset was scattered round them with a pomp which in human things we should call grandiose.[9]

The profound sense of solitude in Thomas's lucid prose and later poetry expresses a sense of modernist alienation that ranks him as an early existentialist. Unlike works by many overly sentimental Georgian authors, Thomas's writing remains popular for giving voice to the search for inner tranquility, and the effects of war on the natural order. He enlisted in the infantry in 1915 and was killed two years later at the Battle of Arras in northern France while the first edition of is *Poems* (1917) was being prepared for publication.

Russian Golden Age Literature and Landscapes

Our weekend jaunts to nearby London proper included visits to bookshops with numerous titles describing these regional happenings. We explored Foyles on Charing Cross, said to be the largest purveyor of new and used books in the world with miles of shelf space, but the most memorable such destination was Heron Books. The firm published fine bindings that included the thirty-seven-volume Greatest Masterpieces of Russian Literature. I had passing acquaintance with novels by Tolstoy, Dostoevsky, and Turgenev, but the Heron series also included important works by nineteenth-century authors I had only vaguely known by name—Goncharov, Leskov, Kuprin. We purchased the set to begin a self-directed course in Russia's literary Golden Age, and I was struck by the many cultural connections between the stories and recollections of Old World life on the Volga related to me back home by family and community elders. Principal subjects of many of these writers were the countryside and the peasantry to whom they assigned their own feelings and aspirations. Like most other members of the intelligentsia, Russia's leading novelists were overwhelmingly opposed to serfdom which was not abolished until 1861. Since publication anywhere in Russia required approval from tsarist censors, authors cleverly used illative dialogue and description to criticize the institution, nobility, and regime. In such ways many ordinary characters inhabiting these stories—both lowly and landed, offer philosophy on social conditions and humanity in general.

Many prominent artists and writers of late imperial Russia also had personal associations with the lower Volga region. High in the pantheon of the era's great painters, "mood landscapist" Isaac Levitan (1860-1900) depicted the seasonal grandeur of the Moscow and Crimean countrysides before traveling throughout the Volga region from 1887 to 1890. Serene *The Quiet Haven* (1890) created a sensation among art critics and contemporaries for its stunning blend of Realist, Impressionist, and Expressionist tendencies. The scene shows a village bathed in late afternoon sun that lights the white facades of two green-roofed Orthodox churches rising above the darkly forested Volga banks and the steeples' dappled reflections in the river. Levitan's paintings greatly impressed author Anton Chekhov (1860-1904), who provided summertime lodging to the artist at his country dacha, and the spirituality of his guest's paintings made a deep impression on the writer. Chekhov's semi-autobiographical short story *The Steppe* (1888) is a lyrical prose poem relating a youth's journey across Russia's vast southern plain. The scenes are evocatively expressed in terms suggested by such Levitan paintings as *Grain Stacks and Village* (c. 1883) and *Harvested Field* (1897): "Beyond the poplar stretches of wheat extended like a bright yellow carpet from the road to the top of the hills. On the hills the grain was already cut and laid up in sheaves, while at the bottom they were still cutting. . . . Six mowers were standing in a row swinging their scythes, and the scythes gleamed gaily and uttered in unison together 'Vzhee, vzhee!'"[10]

Mordvinian painter Fedot Sychkov (1870-1958) spent his peasant boyhood in the lower Volga's Penza region and later studied with Ilya Repin (1844-1930) in St. Petersburg. Sychkov's summertime homeland landscapes feature scenes of rural toil and festivities in combinations of imperial yellow and green beneath glaucous skies that seem to radiate the steppes' oppressive humidity. Notable works include untitled harvest scenes (c. 1915, PLATE 14), *Reaper* (1931) and *Harvest Festival* (1938). Novelists Nikolai Leskov (1831-1895) and Mikhail Saltykov-Shchedrin (1826-1889) served as bureaucrats in Penza, and Leskov became a leading expert on Russian agriculture. Alexander Kuprin (1870-1938), author of *The River of Life*, was a native of the area. In *Oblomov* by Ivan Goncharov (1812-1891), the indolent eponymous owner of large estates is advised by pragmatic, hardworking Andrey Stoltz, a Russian-German. The picaresque novel *Dead Souls* (1842) by Nikolai Gogol (1809-1852) is narrated by anti-hero protagonist Pavel Chichikov. He travels the land seeking landowners willing to sell the identities of deceased peasants still officially on census records and therefore of potential chattel value to schemer Chichikov. The estates are typically owned by complacent nobles and managed by cruel overseers who care little about the land or squalid conditions of the serfs. Near the end of the story, however, Chichikov hears of an enterprising and compassionate estate owner, Konstantin Kostanzhoglo—a name that also suggests foreign association. Over a sumptuous dinner of spring root and herb soup served with Hungarian wine, Kostanzhoglo channels Gogol's own views on the nobility of agrarian life:

> *A landowner never finds the days wearisome—he has not the time. In his life not a moment remains unoccupied; it is full to the brim. And with it goes an endless variety of occupations. And what occupations! Occupations which genuinely uplift the soul, seeing that the landowner walks with nature and the seasons of the year, and takes part in, and in intimate with, everything which is evolved by creation. . . . It will mean that the harvest is being sown, that the welfare of the world is being sown, that the food for millions is being put into the earth.*
>
> *And thereafter will come summer, the season of reaping, endless reaping; for suddenly the crops will have ripened, and rye-sheaf will be lying heaped upon rye-sheaf, with elsewhere, stocks of barley, and of oats, and wheat. And everything will be teaming with life, and not a moment will there need to be lost, seeing that, had you even twenty eyes, you would have need for them all. And after the harvest festivities there will be grain to be carted to byre or stacked in ricks, and stores to be prepared for the winter. . . .*
>
> *And if, in addition, one discerns the end to which everything is moving, and the manner in which the things of earth are everywhere multiplying and multiplying, and bringing in more and more fruit to one's profiting, I cannot adequately express what takes place in a man's soul. And that, not because of the growth of his wealth—money is money and no*

more—but because he will feel that everything is the work of his own hands. . . . In all the world, [Kostanzhoglo] repeated, "you will find no joys like these, for herein man imitates the God who projected creation as the supreme happiness, and now demands of man that he, too, should act as the creator of prosperity."[11]

Nineteenth-century Russian authors' and painters' search for meaning led them and others to the spiritual humus of an agrarian *sobornost,* a term not easily rendered in English that relates to the mutual "common-work" of close community. In a related sense, great art as reckoned by Leo Tolstoy (1828-1910) is comprehensible to the many rather than to the few who assume critical insight. Tolstoy's return to the family's Yásnaya Polyána countryside estate south of Moscow for the last thirty-five years of his life fostered such experience as he worked alongside field-hands to plow and sow, harvest and thresh. The famous "peasant school" he established there from 1859 to 1862 was as much an attempt to appreciate local children's imaginative stories and colorful art as to pioneer a new pedagogy of teaching reading and arithmetic to young minds deemed unworthy or incapable of learning in aristocratic circles. "The most important thing in life," Tolstoy wrote, ". . . is to feel the mysterious gladness of a communion which, reaching beyond the grave, unites us with all men of the past who have been moved by the same feelings and with all men of the future who will yet be touched by them."[12]

Tolstoy's novels contain vivid descriptions of memorable agrarian life borne of the author's personal experience. The family dramas so passionately depicted in his 1867 epic *War and Peace* mostly take place against a backdrop of battlefield chaos and palace intrigue derived from stories gleaned from Tolstoy's acquaintances and readings of period accounts. Two of the main characters, headstrong Nikolai Rostov and his wife, dignified Princess Mariya Volkonskaya, are based on Tolstoy's parents. Generous though sometimes stern Count Nikolai Tolstoy was devoted to crown, family, and the wellbeing of Mariya's ancestral farming estate. Yásnaya Polyana was located a hundred and thirty miles south of Moscow near Tula and is where Leo Tolstoy was born. There he wrote some of his most famous works including *War and Peace* (1867) and *Anna Karenina* (1877), both included in the Heron volumes, that contain evocative descriptions of the Russian rural heartland.

Count Tolstoy had been raised the son of a provincial governor so had known little of country life. After service in the Russian army and marriage in 1822 that brought security from his Mariya's wealth, Tolstoy sought to be a diligent landowner

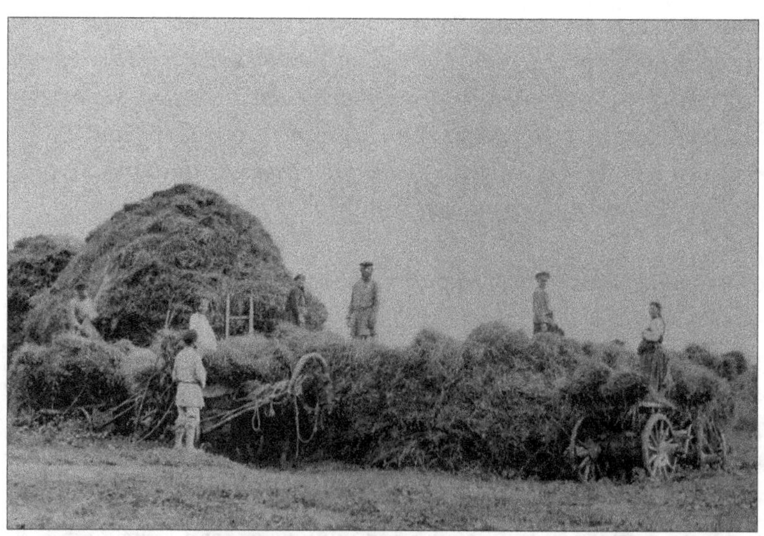

Tolstoy Estate Threshing Scene at Yasnaya Polyana, Russia (1899)
Leo Tolstoy State Museum, Moscow

and improve management of Yásnaya Polyána. Near the end of *War and Peace* Count Nikolai Rostov finally finds personal fulfillment in life on the land at Bald Hills, Leo Tolstoy's cipher for the family estate. He studied but did not embrace agricultural mechanization or the emerging field of soil chemistry. Instead, Rostov studied the culture of the serfs, endeavored to improve their lot, and consulted the best workers among them for advice on farming operations. He came to revel in Yásnaya Polyána's annual harvest and other labors:

> *Nikolai having thus been obliged to become a gentleman-farmer, had conceived a perfect passion for agriculture, and gave up almost all of his time to it. . . . He tried to understand [the peasants'] needs, to find out what they considered good or bad; and thus the orders he gave became a source of valuable training. . . . During fieldwork at sowing-time, haymaking and harvest, he gave the same watchful care to his own fields as to those of his peasants. Few landowners could boast of any in as good condition or equally productive. These occupations absorbed him so much that his wife felt a kind of jealousy. She was grieved that she could not enter into the pleasures and cares of a world apart from her own. Why was he content and happy when he came into tea, after rising at daybreak and spending the whole morning in the fields or the threshing-floor? Why this enthusiasm when he spoke of some wealthy peasant who had passed the whole night, with his family, in carrying his sheaves and making his ricks? Why that blissful smile when saw the fine, soft rain falling on the thirsty green oats, or watched the wind clear off a threatening cloud in haymaking time or harvest. . . ?*[13]

In similar spirit Tolstoy's novel *Anna Karenina* (1877) features virtuous aristocratic hero Konstantin Levin who decries social pretension and roguish behavior. He eschews the recommendations of modern agricultural economists to adapt to modern technologies and material worker incentives. His laborers may not be motivated by profit-sharing and innovation but are happy in their collective humanity. Levin nevertheless endeavors to improve his Pokrovskoye estate, another Tolstoy family property near Tula, and rises one morning to take part in the slow and rhythmic task of scything. Like Tolstoy himself, Levin was an independent thinker who came to find meaning in physical exertion, religious faith, and regard for the peasant agricultural commune, the *mir*. The scything episode is described in sentences of Tolstoyan length:

> *It was the time of year, the very top of the summer, when the prospects of harvest may be estimated, when the labors of the next year's planting begin to be thought of, and the mowing-time has come when the rye is already eared and sea-green in color, but still not fully formed; when the ears of grain swing lightly on the breeze; when the green oats, with scattered clumps of yellow grass, peep irregularly from the late-sown fields; when the early buckwheat is already up hides the soil; when the fallow fields, beaten as hard as a stone by the cattle and with paths deserted, on which the sokha, or primitive plow, has no effect, are half broken up; when the odor of the dry manure, heaped in little hillocks over the fields, mingles at twilight with the perfume of the "honey-grass," and on the bottom lands, waiting for the scythe, stand the protected meadows like a boundless sea with the darkening clumps of sorrel that has done blooming.*
>
> *. . . He thought of nothing, wished for nothing, but not to be left behind the peasants, and to do his work as well as possible. He heard nothing but the swish of scythes, and saw before*

> *him Tit's upright figure mowing away, the crescent-shaped curve of the cut grass, the grass
> and flower heads slowly and rhythmically falling before the blade of his scythe, and ahead of
> him the end of the row, where would come the rest. Suddenly, in the midst of his toil, with-
> out understanding what it was or whence it came, he felt a pleasant sensation of chill on his
> hot, moist shoulders. He glanced at the sky in the interval for whetting the scythes. A heavy,
> lowering storm-cloud had blown up, and big raindrops were falling. Some of the peasants
> went to their coats and put them on; others—just like Levin himself—merely shrugged their
> shoulders, enjoying the pleasant coolness of it.*[14]

Near the end of that memorable summer in 1976, Lois and I traveled to Germany and then continued on with the Deynekas to the Soviet Union. We visited numerous history and art museums
including the Vogelsberger Heimatmuseum in Schotten, Germany, where our ancestors had lived for
centuries before migrating to Russia in the 1760s. Among the exhibits was a substantial collection
of Old World harvest implements and a stylized painting by Frankfurt painter-lithographer Fridolin
Lieben (1853-1912), *Jesus the Bread and the Wine* (c. 1900, PLATE 15), that shows Christ as a boy
standing before ripe grain and grape clusters. Lieben's allegorical *Stufenalter* (*Stages of Labor*) depicts
nine workers standing on a massive stair-stepped stone arch above a field where a reaper and binder take in the harvest. Among those on each side of the bridge are a teacher, architect, pastor, and
shopkeeper. Prominently standing above all atop the structure next to a prodigious sheaf of grain is a
farmer carrying a scythe upon his shoulder. While the other figures tend to their respective tasks, the
rustic looks straight ahead to the viewer in modest affirmation of society's ultimate source of order
and prosperity.

This first trip to the USSR introduced us to the magnificent collections at the Hermitage in St.
Petersburg and Moscow's Tretyakov Gallery that included paintings by Ilya Repin, Isaac Levitan,
and Grigoriy Myasoyedov. A dozen subsequent journeys to Russia not only allowed further study of
their works but, in the spirit of T. S. Eliot's *Little Gidding* "end of all exploring," fostered newfound
appreciation for rustic roots. The Russian masterpieces set included a book of tales by Ivan Turgenev
(1818-1883), and I enjoyed short stories like his "Bezhin Prairie" from *A Sportsman's Sketches* (1852)
that seemingly offered commentary on the Tretyakov's agrarian canvases. It was as if I had traveled
halfway around the world to apprehend the significance of fields back home. What I had experienced
for years with little reflection took on special meaning through the exquisite details of Turgenev's in-
candescent prose. The early harvest mornings I felt had been endured were seen in Turgenev's words
as something marvelous:

> *It was a glorious July day, one of those days which only come after many days of fine
> weather. . . . [T]he sunrise does not glow with fire; it is suffused with a soft roseate flush. The
> sun, not fiery, not red-hot as in time of stifling drought, not dull purple as before a storm, but
> with a bright and genial radiance, rises peacefully behind a long and narrow cloud, shines
> out freshly, and plunges again into its lilac mist. . . . About midday there is wont to be, high
> up in the sky, a multitude of rounded clouds, golden-grey, with soft white edges. Like islands
> scattered over an overflowing river, that bathes them in its unbroken reaches of deep trans-
> parent blue, they scarcely stir. . . .*

Although a thorough cosmopolitan who spent most of his adult life in France, Turgenev provided vivid descriptions of the rural commonplace with lively character portraits and descriptions of peasant deprivation. Sketches painted serfdom with a human face and significantly contributed to its abolition in 1861 just as Harriet Beecher Stowe's writing had created preconditions for emancipation in America. Among the most meaningful insights from the daylong foray of Turgenev's hunter is the everyday drama that casts the dependent fieldworker—and humanity—upon the greater forces of nature. "It is for such weather that the farmer longs, for harvesting his wheat," muses the traveler, who then gets lost while hiking through the tangled brush and crags of a nearby hollow. As the sun sets he encounters the camp of five young peasant herders who offer welcome. The group then engages in fireside conversation about haunted places, "heavenly portents" (an eclipse), and other topics of interest to rural folk. But man is not the measure of all things in Turgenev's universe. The hunter and youths and "fields that stretched endlessly upon fields" exist beneath "the dark unclouded sky . . . in all its mysterious majesty." Farmers, and everyone else, have limited autonomy from nature, and would be wise to live within its boundaries.[15]

Turgenev enthusiastically approved of the autobiographical family tales of Sergey Aksakov (1791-1859), whose vivid childhood reminiscences included *A Russian Gentleman* (1855) and *Years of Childhood* (1858). These magisterial works are remarkable for detailed, unpretentious narratives of life on the family's rural estates in Orenburg Province southeast of Moscow. Writing at the end of an era untouched by industrialization and progressive social reform, Aksakov recounts traditional harvest scenes depicted in the art of nineteenth-century Itinerant masters I would later see at Seattle's Frye Art Museum and in European galleries:

> . . . *For a long time we drove along the strips between the fields. Then we began to hear voices and an odd noise some distance off. The nearer we drove, the louder the sounds became, till at last we could catch the flash of sickles through the standing rye and could see ears of cut rye lifted up by some invisible hand. A minute more, and we saw the bending backs and shoulders of the reapers, about twenty men and women. When we came out upon the field, the talk ceased, but the rasping sound of the sickles on the straw grew louder till it filled the whole field with a sound new to me. We stopped, got out, and walked close to the reapers. "God speed the work!" said my father. He spoke pleasantly, and at once all bowed low; some of the older men exchanged greetings with my father and me. Their sunburnt faces showed that they were pleased; though some were panting and others had fingers or toes on their bare feet bandaged with dirty bits of rag. . . .*

Aksakov notes his "great admiration" for the harvesters, and observations even as a boy led to his conclusion that serfs were "much more ingenious and adroit" than members of the nobility, ". . . because they could do things which ladies and gentleman could not." Aksakov's honest accounts are filled with everyday conversations and happenings in fields, villages, and cities and upon publication in the 1850s proved to be persuasive expressions against absolutism. His evocations of seedtime, harvest, and other agrarian endeavors subtly contrast with conditions that rob the dignity of both rulers and the ruled and disaffect humanity's relationship to nature.[16]

The Frye Art Museum

Since my first teaching job was near Anita Deyneka's tiny rural hometown of Plain, Washington, we became acquainted through her with the remarkable Dorsey family of artists. Anita's lifelong friends Jack and Ann Dorsey lived on scenic Camano Island north of Seattle. A noted painter of landscapes and portraits, Ann came from a prominent family of American illustrators. Her grandmother, Fanny Young Cory (1877-1972), was among the first women to attend New York's Metropolitan School of Fine Art. She went on to become one of the country's foremost fantasy illustrators for classics and periodicals including *Scribner's*, *Century*, and *The Saturday Evening Post*. Jack's many professional accomplishments included service as president of the Northwest Watercolor Society and he was selected in 1972 and 1979 to hold exhibitions at Seattle's renowned Frye Art Museum.

From Jack I learned that the Frye had been founded in the early 1950s by Walser Greathouse, executor of the estate of Seattle philanthropists Charles and Emma Frye. The couple had begun collecting art after visiting the World's Columbian Exposition in 1893 which featured galleries laden with outstanding European fine art. The fair holds special significance in the history of agrarian art and is discussed in the third volume of this series. Charles Frye, son of German immigrant farmers who had settled in the Midwest, was especially impressed with the art of Munich Secessionists like Fritz von Uhde (1848-1911), the group's first president. The term was respectfully coined by Georg Hirth (1841-1916) for dozens of painters like him and von Uhde who had recently broken from the tradition-bound Munich Artists Association. Hirth would go on to establish the illustrated German art weekly *Jugend* in 1896 that promoted modernist perspectives in art and politics. Hirth's journal was instrumental in launching the *Jugendstil* (Youth Style) movement as Germany's Art Nouveau

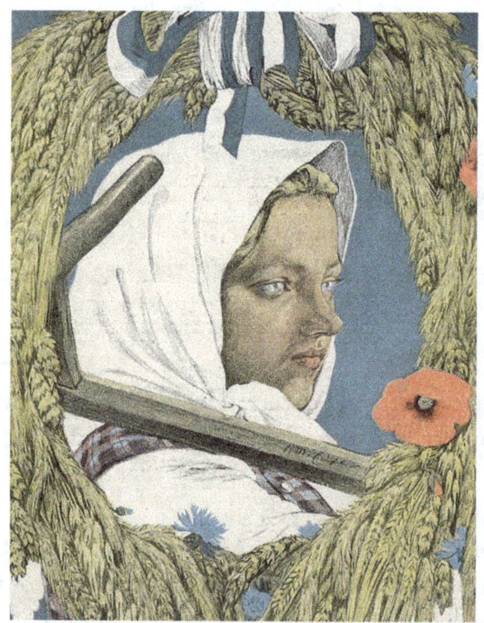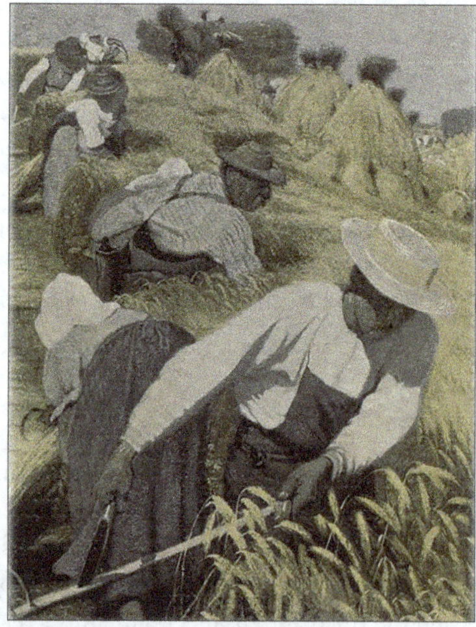

Left: Reinhold Eichner (1872-1947), *The Scyther*
Right: Walther Georgi (1871-1924), *Scythers*, *Jugend* 40 (Munich, 1901)
Columbia Heritage Collection

counterpart with fine art and decorative furnishings that combined twisting curves with geometric shapes. The style was significantly promoted by Grand Duke Ernst Ludwig of Hesse who founded the Darmstadt Art Colony in 1899 in the city's Mathildenhöhe district. Frye eventually acquired three of von Uhde's paintings including *Woman, Why Weepest Thou?* (1894), a depiction of the resurrected Christ's encounter with Mary Magdalene near the Garden Tomb. Whether of religious events or landscapes, von Uhde, a devout Lutheran, saw his art as witness to "the light of the Gospel." Many of his paintings are known for their masterful depiction of soft light that seems to emanate mystically from the canvas.

The Fryes also acquired placid *Landscape near Polling* (c. 1865) by Adolf Heinrich Lier (1826-1882) that shows two villagers on a makeshift dock as several ducks swim across the water nearby. Known for luminous canvases and loose brush strokes in his mature works, Lier was an acclaimed Munich School's landscapist and one Germany's most prolific painter of harvest scenes. Although influenced by the Classical-Romantic trends of the time, Lier traveled to France in the 1860s and became deeply influenced by Julien Dupré, Théodore Rousseau, and other Barbizon artists. He developed an oeuvre combining both styles characterized by muted tonal vistas that captured unusual if banal details of farm life. His earlier carefully constructed works like sunny *Harvest Near Munich* and *Grain Harvest in the Mountains* (1857, *Summer Day*) gave way to warmly atmospheric paintings with balanced coloring like *Harvest with Farmhouses* (1878). Lier traveled extensively throughout Germany to understand and depict everyday country scenes and from 1868 to 1873 operated a school in Munich that allowed for seasonal trips to the nearby Pang Art Colony near Rosenheim in the foothills of the Bavarian Alps. His bold study *Grainfield at Harvest Time* (1868) at Berlin's National Gallery shows bulging yellow-brown sheaves with a background of fieldworkers and rounded church steeples.[17]

Subsequent acquisitions by the Frye have also included significant examples of rural art including works by Andrew Wyeth (1917-2009), Winslow Homer's *The Wheat Gatherer* (1867), and his sketch from the same year on which it is based, *Woman in a Grainfield*. Homer's pair is remarkable not only for their stoic composition, but also for being completed during the artist's 1866-1867 visit to Europe. At that time he was introduced to the French Barbizon style of naturalist peasant and rural depiction magnificently expressed in works by Jean-François Millet (1814-1875) and Augustin-Léon Lhermitte (1844-1925). The latter's magnificent *Dans la vallée* (*In the Valley*, 1910, PLATE 18) is among the Frye Art Museum's largest and greatest treasures. Aldous Huxley's 1925 essay, "The Best Picture," bestows the honor on a fifteenth-century Italian masterpiece for its "natural, spontaneous and unpretentious grandeur." Lhermitte's substantial canvas is such a work for being majestic without theatrics. Its diaphanous presence depicts the heat and humidity of summer harvest as scythers and binders toil in a bountiful hillside of golden grain. Six years earlier, Lhermitte had painted the very similar *Harvest in the Valley*, presumably a scene from the Mont Saint-Pére district in northern France, which was subsequently acquired by the state to adorn the Palace of Versailles. A May 1904 issue of one of Paris's leading art journals, *The Builder*, described Lhermitte's paintings as "the most perfectly satisfactory" of all in the city's fashionable New Salon. The art of Barbizon masters like Lhermitte, Millet, and Breton elevated landscape art from the Academy's lower rank to a high form on par with historical depictions and traditional portraiture.

Following Walser Greathouse's passing in 1976, his wife, Ida Kay Greathouse—who lived to the remarkable age of 110, became the Frye's influential director and served for nearly two decades. She saw Jack Dorsey's work at an exhibition of the Northwest Watercolor Society and procured one of his paintings for the Frye's collection after his 1979 one-man show. Greathouse and Wenatchee artist William Reese (1938-2010) acquainted Jack more fully with the Frye's distinguished Permanent Collection of nineteenth and twentieth-century masterpieces. The museum's holdings had grown to include the Russian Itinerants Repin and Levitan, and Expressionists Nicholai Fechin (1881-1955) and his fellow exile and charismatic protégé, colorist Sergei Bongart (1918-1985). A native of Ukraine, Bongart had studied art in Kiev during his late teens with Mikhail Yarovoy, who had trained with Repin himself at the Imperial Academy of Arts, as did some of the early Munich artists. Fechin's art represented a powerful synthesis of continental Realism with a new spirit of Abstract Expressionism that profoundly influenced a new generation of artists on both sides of the Atlantic. (Bongart never personally met Fechin but considered him to be among the most significant figures in his life and bought Fechin's Santa Monica home and studio in the 1960s.)

Bongart began his association with the Frye in 1961 when Walser Greathouse was moved by his depiction of the Crucifixion (*Crown of Thorns*, 1962) and other works. He purchased the first of the Frye's eventual twenty-one Bongart paintings, *Brown Overcoat* (1962); Mrs. Greathouse acquired one of Bongart's many rural scenes, the splendidly sunny and brisk *Idaho Farm*, in 1973. Since the 1970s, the Frye's Permanent Collection has also acquired twenty-nine paintings by Fechin and such notable agrarian works as *Rest of the Haymakers* (1882) by Camille Pissarro and Levitan's *Two Haystacks* (1890). Being surrounded in the spacious Frye's Founding Collection Gallery by such densely clustered gold-framed masterpieces is to be transported back to the time of Europe's splendid salon exhibitions. Artists there like Lhermitte, Millet, and Pissarro used agrarian themes to question social privilege and explore worlds old and new. Their art and ideas raised considerations of aesthetics, meaning, and progress that remain as vital as ever in the wake of global cultural and environmental challenges.

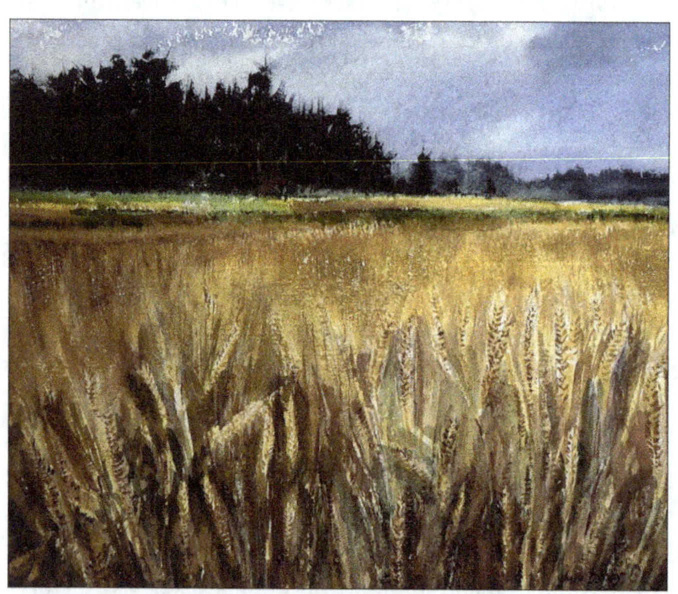

Jason Dorsey, *Whidbey Wheat* (2022)
Watercolor on paper, 15 ½ x 14 ¼ inches
Columbia Heritage Collection

Jack Dorsey went on to study with the Russian-Armenian artist Ovanes Berberian, another master of contrasting pigments. Like Reese, he had trained with Bongart at one of his Puget Sound summer workshops and at his beloved "Kievtshina" farmstead studio where he typically painted to the classical strains of Rachmaninov, Stravinsky, and other Russian composers. Bongart had established a reconstituted "homeland" residence on twenty rural acres near Rexburg, Idaho, soon after his first visit to the Northwest in 1970. Berberian subsequently made his home at nearby Rigby where he continues to conduct summertime plein air workshops. Through these diverse influences, Dorsey developed a distinctive style of "Impressionistic Realism" using watercolors as well as oil with looser strokes and brighter palettes for still lifes and landscapes. *Harvester at Rest* (2002, PLATE 16), from his 2002 Art of the Palouse Group series, was exhibited that year at Spokane's Chase Gallery and the Everett Center for the Arts. The painting shows an exceedingly weathered 1940s-era International Model 51 combine in the eastern Palouse that seems to float carefree in a sea of yellow grass beneath purple clouds.

"The masters paint from the heart, without ego or regard for commercial appeal," Jack explained regarding the Frye's remarkable collections. "They combine imagination with skilled technique to uplift and to depict beauty." Bongart's similar characterization of great art in his essay, "It's Time to Call Things by Their Proper Names" (c. 1980), excoriates the modern pop art of pink and black squares and plastic junk sculpture. At the same time he reaffirms high regard for the technique and creativity of Repin and Levitan, and American artists like the Wyeths. From this perspective, painters like Berberian, Jack Dorsey, and sons Jason and Jed Dorsey, can be seen as the present generation's East-West bearers of flame and brush bequeathed by masters like Fechin and Bongart. Their own styles in turn were influenced by such eighteenth-century canonical artists as Repin, Lhermitte, and Levitan.[18]

Socialist Realism and Modernist Slavic Art

Paintings by Berberian and Fechin are included in Springville, Utah's acclaimed Museum of Art Soviet and Russian Collection with notable works by Arkady Plastov (1893-1972), Yuri Kugach (1917-2013), Vyacheslav Fedorov (1918-1985), and other artists who featured genre scenes of Eastern European rural endeavor. Founded in 1937, the museum's holdings are among the nation's most extensive for rural subjects that include numerous harvest scenes painted in the 1890s by members of the Church of Latter-day Saints Paris Art Mission. Springville's Soviet and Russian Collection began unexpectedly in 1989 when museum director Vern G. Swanson first embarked on a series of trips to the USSR on behalf of the Grand Central Art Gallery Education Association. Swanson met Russian artist Vladimir I. Nekrasov (1924-1998) of Moscow's Surikov Art Institute who introduced him to important works of Russian Expressionism and Social Realism that led to a major exhibition at Springville in October 1990 and eventual artwork purchases for the museum.[19]

Successive assaults upon the Soviet Union's rural populace in the 1920s and '30s involved Stalin's brutal campaigns to collectivize agricultural lands and against religion that led to widespread violence and famine. Millions of peasants perished or were displaced from their native villages through the imposition of these policies to abolish private property and modernize the economy. Russia remained a major producer of grain until this period which witnessed the expropriation of commodities from landed peasants (kulaks) who had withheld harvests in order to boost prices. Stalin's push to industrialize the country at all costs required the provisioning Soviet cities, agricultural mechanization, the mass murder and exile of kulaks, and the exodus of vast numbers of younger rural residents to urban areas. The impact of these forces was devastating to traditional Russian village life and crop production. The nation was plunged further into cataclysm after war with Germany commenced in 1941.

As people and landscapes suffered, authors and artists sought memory for solace as well as lament. The glorification of communist principles through state-sanctioned Socialist Realism governed official Soviet art and literature from the 1930s to 1980s. Muscular representations of urban and rural life that lauded labor and socialist ideals generally characterized the approach, but later strains featured honest views of everyday life reminiscent of the French Impressionists and Taos Expressionists. Marx had viewed artists and writers as valued members of an intellectual vanguard promoting revolutionary change. Leon Trotsky later wrote in *Literature and Revolution* (1924) that their insights revealed the nature of society and if freely expressed would help guide the revolutionary struggle. Stalin, however, had no tolerance of art for art's sake. His authoritarian policies sought conformity and denigrated individuality—the basis of creativity.

Unless about earlier periods or other places, Soviet depictions of internal discontent and tragedy were forbidden in favor of sentimentalized worker characterizations of the proletarian dream. The character of Soviet monumental art was famously exemplified in Vera Mukhina's 80-foot steel sculpture *Industrial Worker and Collective Farmer* (1936) that was built to crown the Soviet pavilion at the 1937 World's Fair in Paris. Plated in radiant chrome-nickel, the massive female figure designed by Mukhina (1889-1953) grasps a sickle alongside her hammer-wielding companion in striding poses that symbolized the nation's aspirations. After the Paris fair, the sculpture was relocated to the entrance of Moscow's sprawling All-Union Agricultural Exposition on the city's north side where substantial halls showcased numerous aspects of crop, livestock, and food production. Architect Konstantin Topuridze (1905-1977) designed the enormous *Golden Sheaf (People's Friendship) Fountain* (1954) as one of the park's centerpieces that features a towering grain sheaf encircled by three colored glass cornucopias and sixteen bronze statues of young women who symbolized the Soviet republics. (In 1959 the complex's name was changed to the Exhibition of Economic Achievements and more recently to the All-Russian Exposition Center. Today it includes a grandiose amusement park, year-round trade shows, concert hall, and pavilions featuring space exploration and technological advancements.)

Idealized paintings of country life like Arkady Plastov's *Harvest Festival* (1937) and *Field after Harvest—Sheaves* (1954) by Yuri Kugach show bountiful fields and smiling brigades of *kolkhoz* (collective farm) laborers clad in red

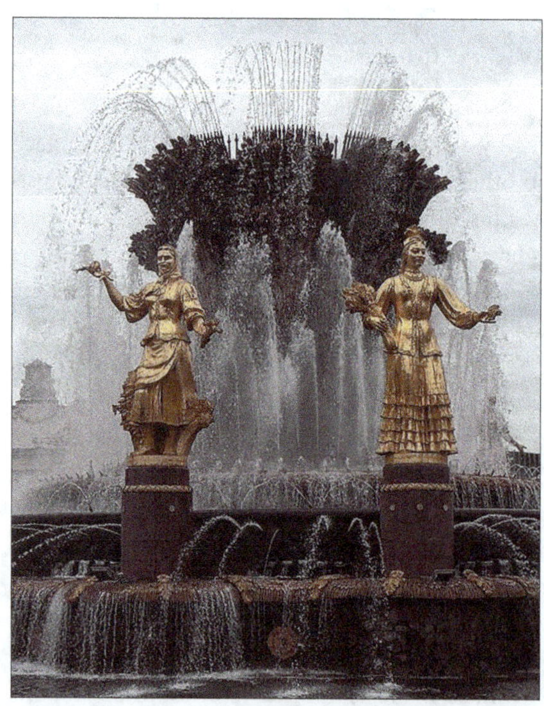

Konstantin Topuridze
People's Friendship "Golden Sheaf" Fountain (1954)
Exhibition of Economic Achievements, Moscow
Alix Saz Photograph; Wikimedia Commons

neckerchiefs and head scarves enthusiastically driving farm equipment or tending threshing operations. Plastov, born to a family of icon painters near Simbirsk on the middle Volga, also painted works like *Harvest* (1945) and *Spring* (1954) that risked official condemnation given his Impressionistic renderings of commonplace scenes devoid of political sentiment. *Harvest* is a discomforting view of an aged reaper sharing a meal in the field with three children scarcely old enough to shoulder such responsibilities. Completed in the last year of a war that had inflicted enormous suffering throughout Europe, the scene also inspires appreciation for the home front brigades of women, children, and the elderly who labored for years to sustain soldiers and civilians. Plastov's dynamic, colorful *Haymaking* (1945) shows a shirtless teen flanked by two elderly men and a woman who cut grass near a copse of birch trees.

Kugach, who settled in the Tver countryside after the war, went on to establish the Moscow River School in 1974 to revive the dramatic style of Repin, Levitan, and other Russian Realists. Ambidextrous painter Nikita Fedosov (1939-1992), Yuri Kugach's nephew, became a prominent member of the group and painted numerous country scenes including *Last Rays* and *Overcast Field* (1966). Muscovite Victor Ivanov studied with Kugach at Moscow's Surikov Institute of Art in the late 1940s and in the 1960s became a leading member of the Avant Garde Severe Style that depicted the grim austerity of post-war Soviet life in opposition the naïve depictions of Socialist Realism. Artists like Ivanov risked establishment censure but painted throughout the Khrushchev reform era in ways that recalled the 1910s Futurism of Kazimir Malevich and Kuzma Petrov-Vodkin. Ivanov painted numerous harvest scenes like *Harvesting near Ryazan, Men Resting at Harvest,* and *Women Harvesting* (1965). These spare, balanced compositions in irregular blocks of olive, mustard yellow, and chestnut contrast rural toil with the rustic beauty of the Russian countryside.

Contemporary Russian artists Ordjonikidze Izmaylov, Natalia Klimova, and Vasily Belikov have been drawn to the immense steppes of western Russia that have nourished Slavic body and soul since time immemorial. The mystical Russian affinity for the countryside transcends residence in city or town and is evident in the twenty-first century popularity of gardening, dachas and rural excursions, and influence of masterful works by the *Peredvizniki* Itinerants. Vladimir native Izmaylov's *Rural Motive* series show nineteenth-century Abramstsevo Colony influence with traditional Russian imagery but in a contemporary context. His *Summer* (c. 2000) casts the grandly swirling gold contours of a broad grain field beneath bulbous gray-white clouds and a deep blue sky. A solitary red combine plies a swath at far left as if in testimony to the patient, anonymous labors that provision mankind.

Natalia Klimova paints broad sinuous landscapes like *Fields* (2015) contrasting yellow and green stripes that are flanked by dark chernozem (fertile "black earth") fallow lands. Vitaly Gubarev, a native of Uzbekistan, has been drawn to rural scenes of the Russian North and Volga for masterful etchings and drawings that express human interactions with the environment. Through nuanced shading with filigree swirls, his velvet-tone etchings like *Ripening Grain* and *Haymaking* (1998) are remarkable for depth of field and impart a calm assurance of the heartland's enduring fertility. *Grainfield* (1994) shows slopes laden with wheat that seem to pulse in a warm summer breeze. Soft yellow highlights brighten the center of the black and white expanse that shows a small grove on the horizon that flanks a church. Vasily Belikov (1922-1994) was a direct heir to Itinerants as a student at the Penza Art Academy of Ivan Goryushkin-Sorokopudov (1873-1954), himself a protégé of the master

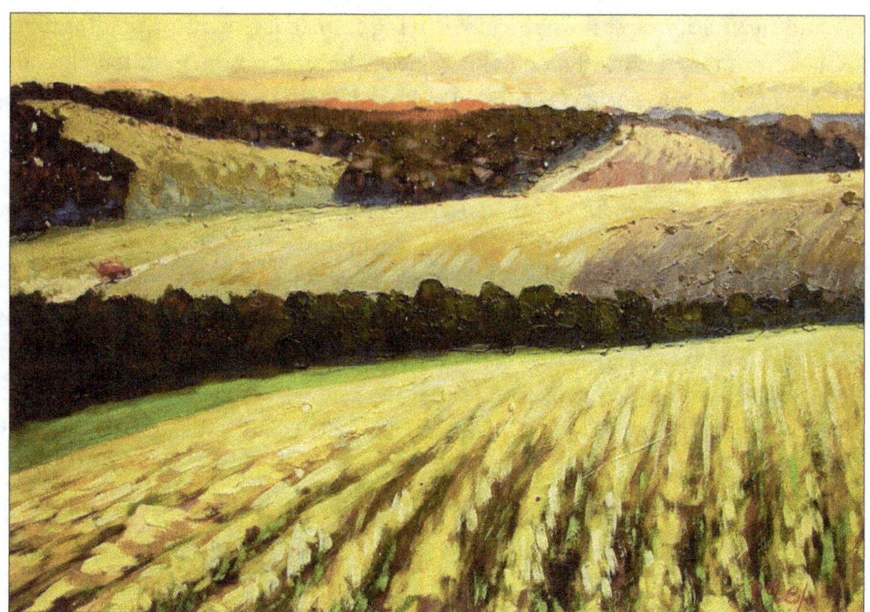

Dmitry I. Slobodin, Donbas Harvest Scene (1982)
Gouache on paper, 17 ¼ x 22 inches
Columbia Heritage Collection

Ilya Repin. Belikov endured horrific personal circumstances as a boy during the Russian Civil War and subsequent famine and served in the Russian navy before pursuing a career in art following World War II. Belikov characterized his luminous expressionistic Volga landscapes "autobiographical encounters with nature" to be experienced but that defy meaningful explanation. His paintings like *Grainfield* (1979), *Summer Evening* (1980), and *Path in the Field* (1980) cast protective if moody sunsets that glow with golds, pinks, and purples against dark blue skies.[20]

Russian landscapist Oksana Svinar of Murmansk travels throughout northern European Russia scenes she paints in colorful acrylic impasto. Raised west of the Ural Mountains in the Udmurtian city of Izhevsk, Svinar recreates familiar summer scenes from that area that are alive with agrarian bounty. One of the artist's personal favorites, *At the End of the World* (2020) is a kaleidoscope of seasonal hopefulness showing a prodigious field of wheat flanked by a country lane. A native of the Donbas region in eastern Ukraine, Dmitry I. Slobodin (1929-2005) graduated from the Art College of Lugansk and became a master of Impressionistic palette-knife paintings in tempura and oil earth tones that depict the quiet beauty of his native land while resisting the artificiality of the regime's officially sanctioned Socialist Realism. Slobodin's untitled Donbas harvest scene (1982) shows mottled field rows of tawny cream with shadowed forest greens beneath a gleaming orange ribbon of setting sun. At far left in the distance beyond a darkened tree-lined swale one can almost hear the hum of a late model Rostelmash self-propelled combine throwing a roiling cloud of yellow-white chaff. The red machine appears to be opening up a field of ripened grain at day's end near at base of a broad gentle slope in a scene of bounty and peace.

Serbian Expressionists Milan Konjović (1898-1993) and Fedja Soretić (1930-1023) also faced charges of individuality from the authorities for unpeopled agrarian landscapes with twisting clouds and fields in intense colors. The fields of Konjović's 1938 masterpiece *Harvest* swirl with vibrant summertime energy reminiscent of van Gogh's fevered Auvers wheatfields. A cluster of leaning sheaves in the right foreground painted with thick daubs of yellow-green, brown, and orange shows the vital, primitive forms seen in other Konjović harvest scenes. A native of Novi Sad in Serbia's Vojvodina's breadbasket, and graduate of Yugoslavia's Academy of Fine Art in Belgrade, Soretić lived during and after the Soviet grip on the Balkan nations. He returned time and again to paint rural Vojvodina scenes where farmers followed traditional agrarian ways throughout most of the twentieth century. In 1965 Soretić completed his large-format Summer series of oil paintings as part of his acclaimed Four Seasons collection. Most of the summer paintings are hopeful midday views of standing grain and shocks against broad horizons and billowy white clouds. The artist's works like *Harvest* (1990) are even more serene with copper-brown stubble, shocks, and sky divided by a lush band of indigo. The glowing sun seems to be at once rising and setting as that part of the world seemed to experience at the time of its creation. Soretić established a household in Florida in 1992 and a major Northwest exhibition of his work was held at the time through support of Oregon's Portland and Eugene Serbian communities.[21]

Harvests of Sorrow

Tales of Andrew Lund's immigration to America from Denmark and wartime memorial art remind me of Grandpa and Grandma Morasch's reminiscences on their 65th wedding anniversary in December 1973. They told about the day he decided to leave their Russian village seven decades earlier: "Harvest was with a scythe and cradle and cut by hand," Adam Morasch remembered. "My father was dead; my oldest brother, he says, 'Take your choice. If you want to go to America, then alright, or if you want to serve about three or four years in the army. . . .' So I went out, thrashin' by hand, and we made one round and here come a big rain. I took my outfit and threw it in a corner . . . and went to the depot."[22] The obdurate structures on wind-blown rural landscapes and decayed barn relics represented by Bruch's creations embody such past hope and achievements, as well as failed enterprise and exploitation. But they also serve as reminders that generations of strangers from all corners of the globe sought a better life in America, and to be responsible citizen-stewards. Untold numbers still do, and Kogan's bronze stand of grain attest to the ultimate sacrifice many like the Lund brothers have been willing to make.

In connection with work establishing exchanges between U.S. and Soviet universities as the U.S.S.R. teetered on the brink of collapse in the fall of 1990, I received permission to visit our grandparents' Russian ancestral village of Yágodnaya Polyána (Berry Meadow) on the lower Volga. The region had been off-limits to Western tourists since the 1920s but I had few travel restrictions apart from a three-day limit once departing Moscow by rail in early October for what seemed a journey back in time. In the regional Volga capital of Saratov I visited the State A. N. Radishchev Art Museum, founded in 1885 as Russia's first public art museum by artist and patron Alexander Bogolyubov (1824-1896) and named for his maternal grandfather, social critic Alexander Radishchev (1749-1802). I had more than incidental interest in Radishchev since he had been raised on a farming estate near Saratov and was a controversial contemporary of my ancestors who immigrated from Germany to Russia in the 1760s. Radishchev authored *A Journey from St. Petersburg to Moscow* (1790) in which

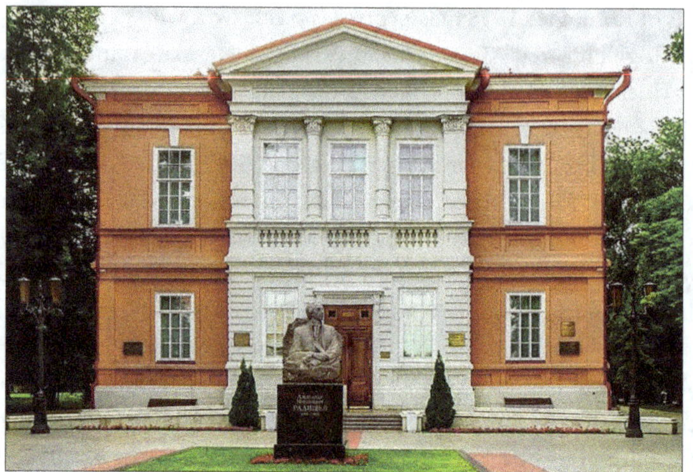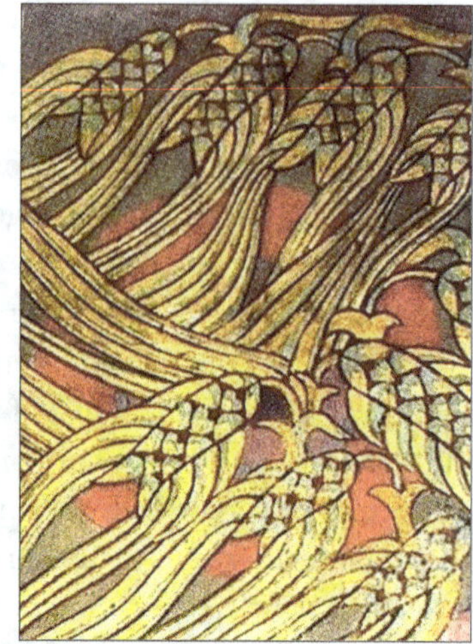

**Left: Alexander Radishchev Monument and Art Museum; Saratov, Russia
Right: Mikhail Vrubel, Grain Stalks Fresco Ornamentation (1889); St. Volodymyr's Cathedral, Kyiv
Columbia Heritage Photographs**

he dared suggest the Enlightenment idea that autocracy and serfdom were violations of Natural Law. For this heresy Catherine the Great condemned him to death, though the sentence was later commuted to lifetime exile. Publication of Radishchev's passionate book placed him in the first rank of Russia's earliest radical writers, and he invoked numerous references to the harsh realities of agrarian experience to decry the abuses of the established order. In Radishchev's travel account of life in mythic Vyshny Volochuk, he rails against the greed of landowners whose harvest prosperity leads to their further strangulation of the public good.[23]

Alexey Bogolyubov had studied and painted throughout Europe in the 1850s and took classes in Düsseldorf where he met the progressive *Malkasten* landscapist brothers Andreas (1815-1910) and Oswald Achenbach (1827-1905)—known in their day as "the A[lpha] and O[mega] of landscapists." Their enduring friendship and Bogolyubov's regard for the brothers' work led him to begin collecting *Malkasten* as well as Barbizon art. Bogolyubov's own painting changed from Romantic formalism to natural Realism and after returning to Russia in 1860 he became associated with the aesthetically kindred Russian *Peredvizhniki* (Wanderers). His extensive collection representing these artists and others became the nucleus of the Saratov museum which includes Oswald Achenbach's sultry family scene *Harvest* (c. 1890) and lush *Harvest in the Surroundings of Düsseldorf* (1867) by Academy landscape professor Eugene Dücker (1841-1916). The gallery also holds important paintings by Russian Symbolist Pavel Kuznetsov (1878-1968) and Mikhail Vrubel (1856-1901). Kuznetsov's placid Rest (1925) captures a slumbering, barefoot woman in dreamy reverie beneath a bower of golden sheaves. Museum docents shared pridefully that Vrubel, who also worked with maiolica, textiles, and sculpture, took his first art lessons in Saratov in the late 1860s before studying at the Imperial Academy of Arts in St. Petersburg and Odessa's Richelieu Lyceum. During his years in Ukraine, the versatile artist immersed himself in Byzantine Christian art and completed exquisite frescoes for cathedrals in Odessa and Kyiv which I later visited. Vrubel's frescoes in Kyiv's St. Volodymyr's Cathedral include sinuous grain stalks representing Christ as Bread of Life in an idiosyncratic style that combined medieval Orthodox ornamentation with Art Nouveau.

After my Saratov city tour and spirited discussion of Radishchev's relevance for our day, a Russian acquaintance arranged a car rental and we headed northwest in the direction an old map indicated that Yágodnaya Polyána still existed. I knew that any distant relatives who might have inhabited the village before World War II had been exiled to Siberia and Central Asia when Germany declared war on Russia in 1941. But I still hoped to see where Adam Morasch had once scythed grain and Mollie Bafus had scrutched flax, and to find the houses where our grandparents had lived. We passed stubbled grain fields and an immense congregation of ochre sunflowers bowed with heavy heads for a late harvest offering. The murky gray skies we had known for several days, however, soon dissipated into a ceiling of lucent turquoise. Climbing the kind of gentle rise for which the colonists named the west side of the Volga the *Bergseite*, we spotted a sign for Yágodnaya Polyána that directed us eastward off the main highway onto a severely cratered road.

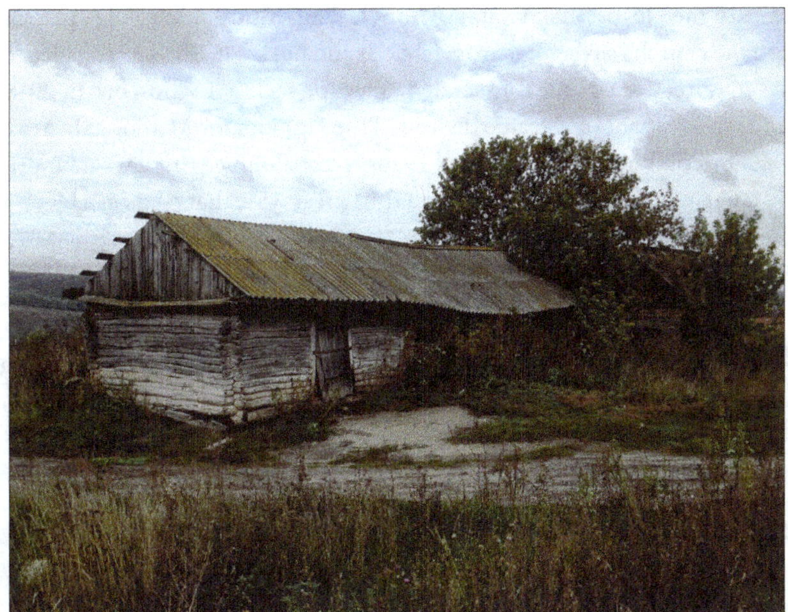

Colonial Era Log Granary (*Ambar*) in Yágodnaya Polyána, Russia (2013)
John Clement Gallery Photograph

The route soon further deteriorated into a deeply rutted lane as we began descending steeply through an autumn gilded cowl of quaking birches protecting the tranquil valley. Visible among the peeling white *beryozky* was an endless aisled carpet of humble moss and frosted woodland flora where family elders had long ago gathered wild mushrooms and strawberries. In the distance appeared the outskirts of the village which still had many log homes from the era of colonization under Empress Catherine. We learned that an elderly couple had recently returned there from Kazakhstan after wartime exile and an absence of half a century—Georg and Maria Scheuerman. I ventured an introduction in my best Russian and was warmly welcomed into the tidy home where they shared stories not only of trial and tribulation, but also memories about the beautiful land and life that had beckoned their return. Other family members later gathered to show us the timbered remnants of the village's original granary, location of the old outdoor threshing floors, and where the flour mills once stood.

Having experienced recurrent personal and cultural trauma from the time of the 1917 Bolshevik Revolution to modern times, the aged couple related their experiences in the aftermath of civil war, collectivization, religious oppression, famine, and exile. During the 1930s when they sent letters appealing for help to relatives in North America, Stalin dispatched brutal NKVD security service units to villages along the Volga and in Ukraine where they surrounded granaries and forced locals to load the contents onto trucks and wagons for transport to northern cities. Expropriation of grain stocks by militia unfamiliar with farming left many villages without seed for future crops. Family storehouses were not spared as soldiers broke them open to requisition grain and flour, and also helped themselves to equipment that might be sold for cash like sewing machines and brass samovars. Rev. Arthur Pfeiffer was among the last openly practicing pastors in the region until his arrest in 1934 and exile to the Novosibirsk region. Before leaving his destitute Volga parishioners, Pastor Pfeiffer wrapped the church's silver altar cross in cloth and buried it. The sacred object remained hidden for decades until recovered and consecrated anew for service in Moscow. In 2015 the National Park Service and Government of Ukraine dedicated the Holodomor Memorial in Washington, D.C. to honor the memory of the victims of Stalin's 1930s terror-famine. The stark memorial near Union Station features a bas-relief "Field of Wheat" wall thirty feet wide that rests on a black granite plinth.

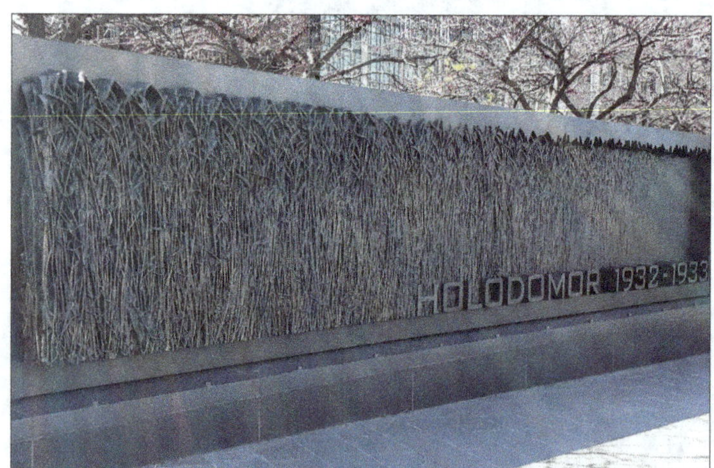
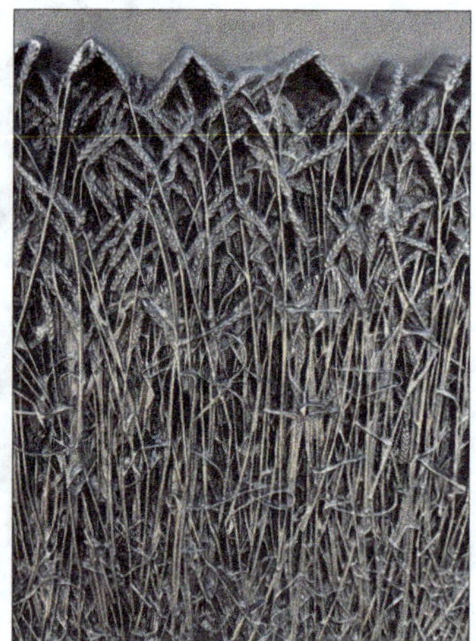

**The Kurylas Studio, Holodomor Memorial "Field of Wheat" (2015); Washington, D.C.
Columbia Heritage Photograph**

After Stalin's 1930s terror-famine, Germany's invasion of Russia in June 1941 ushered in a new series of calamities for the peoples of Eastern Europe. Stalin charged the pacified Volga German villagers with being "spies and diversionists" sympathetic to the invaders. In August military trucks arrived to carry them to area train stations and transport east to sparsely settled districts in northern Kazakhstan and Siberia. Given a single day's notice of their impending fate, Georg and Maria along with others from Yagodnaya Polyana soon crowded into cattle cars packed for the two-week journey that eventually led to the Trans-Siberian Railway station of Mosklenki near Omsk. Here they were

fable-like World War I-era story is set in the South, filming took place in 1976 on the Montana-Alberta prairie borderlands near Lethbridge. Malick's award-winning film chronicles the experiences of lovers Bill (Richard Gere) and Abby (Brooke Adams) who are on the lam with Bill's young sister following his inadvertent killing of a Chicago steel mill foreman. The threesome embarks on a quest for sanctuary and eventually find seasonal work in the extensive grainlands of prosperous but dying Farmer (Sam Shepherd) as harvest commences. Solitary, taciturn Farmer inhabits the lonely landscape in an incongruous Victorian mansion that seems to represent a passing if entrenched old order. Circumstances set up a dramatic love triangle against contrasts of rural agrarian and urban industrial cultures. Director Malick, who had worked as a youth in Texas harvest fields, also noted the influence of different worldviews: "[The characters only] see before them another season, another harvest, and feel unable to build a life. Though this is familiar to a European, it may seem puzzling for Americans. Americans feel entitled to happiness, and once they manage to find it, they feel as if they can own it."[28]

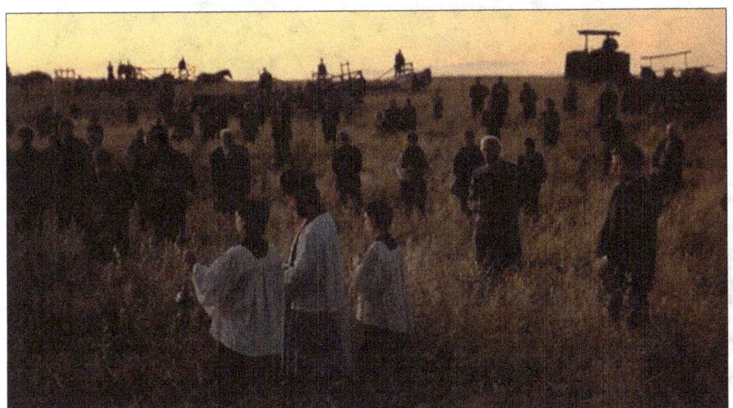
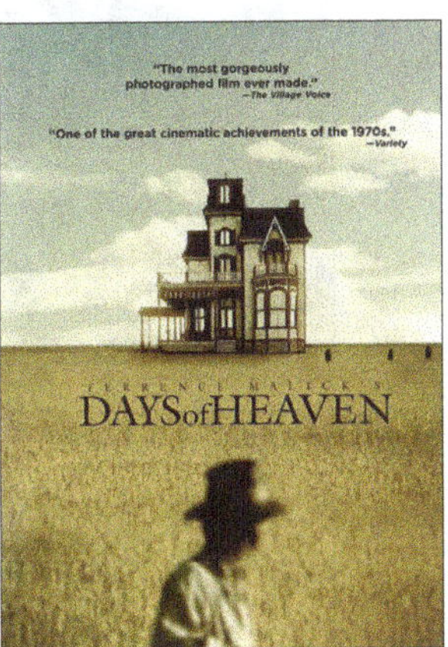

***Days of Heaven* Blessing of the Field Scene and Poster (1978)**
Paramount Pictures Corporation

Frequent narration by the young girl veers from the banal to profound and unites the plot's elliptical path. The story has been likened to biblical allegory suggesting David and Bathsheba, Ruth and Boaz, and the bondage and plagues of Egypt. (The title suggests a place of security and is from the King James Version of Deuteronomy 11:21.) Considerable footage in *Days of Heaven* is devoted to reaping, binding, and threshing operations with period equipment and substantially filmed in natural light during the ethereal "golden hour" of dusk. Ennio Morricone's (1928-2020) soundtrack features recurring themes from the haunting orchestral melody "Harvest" and rhythmic "Threshing." The vast panoramas and wistful music are punctuated with machinery clatter and spare dialogue for emotional depth without epiphanies; genders and cultures collide as elemental forces of nature shape unintended outcomes. Malick and Morricone craft a primal epic of disrupted souls experiencing love and loss with artful use of ephemeral elements—frolicking workers, birds on the wing, pulsing crickets, and silhouettes of swaying grain against the setting sun. The sumptuous amalgam of struggle, sight, and sound garnered the 1979 Academy Award for the painterly cinematography of Néstor Almendros (1930-1992) that he credited to the influence of art by Vermeer and Andrew Wyeth. Morricone's soundtrack was also nominated for best original score and won the British Academy Award.[29]

The Basket (1999) by Spokane's North by Northwest writing team of Don Caron, Rich Cowan, and Frank Swoboda, won the International Family Film Festival Gold Award. The film is an elegant homage to early twentieth-century life on the golden-hued harvest fields of the Columbia Plateau. Starring Peter Coyote, Karen Allen, and Robert Karl Burke, the movie opens with period scenes from the Palouse Empire Threshing Bee near Colfax. Considerable action follows in the one-room Fishhook Standard school near "Waterville," but actually located northeast of Pasco, Washington. The plot centers upon a newly hired teacher from the East, Mr. Conlon (Coyote), and orphaned German siblings Helmut (Burke) and Birgitta (Amber Willenborg) who must adapt to the suspicions of peers and parents against the backdrop of World War I-era prejudice. Conlon schemes to obtain a threshing machine for area farmers by hoping to coach the school's older boys to unlikely victory in a basketball tournament.

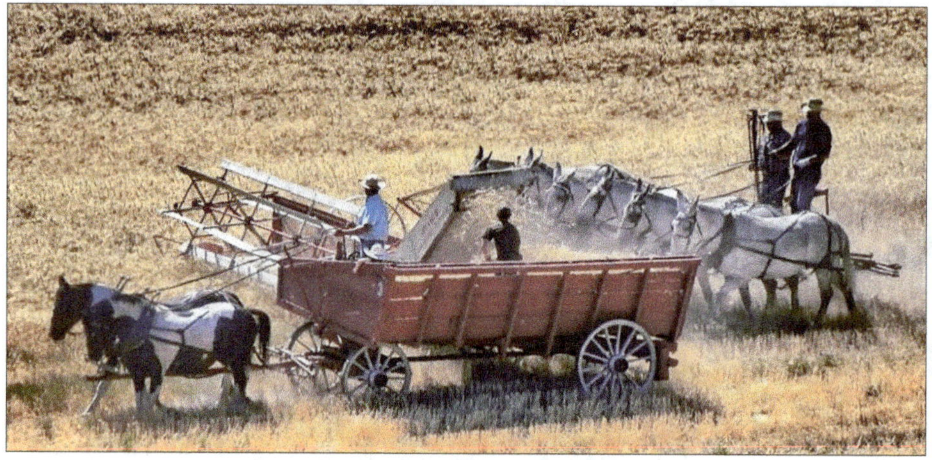

Palouse Empire Threshing Bee Header Crew (2011)
Digital print on paper, 8 x 10 inches
John Clement Gallery Photograph

The lush, full orchestral score by Spokane composer Don Caron was performed by the Hungarian National Symphony and based upon a mystical, recurring four-note theme inspired by the music of the great Austro-Bohemian Romantic composer Gustav Mahler (1860-1911). Eight separate pieces comprise *Der Korb* (The Basket in German), an imaginary opera supposedly written at the turn of the century by "Gottlieb Müller." The name is Caron's nod to the influence of Mahler, a subject of his musical studies while majoring in music at Whitworth College in Spokane. The opera's lyrics, interpreted in stages for Conlon's students by Helmut, tell of a mysterious stranger in medieval times who appears in an isolated village and challenges the residents to stand firm against invading barbarians. He offers them nothing more than the contents of his *Korb*—two stones and a sling, and during his visit romances the daughter of the local duke. Conlon uses the tale as a metaphor for teamwork among the boys.

One of the nineteenth century's most notable composers, Gustav Mahler fused popular European literature with complex musical scores for magnificent creations imbued with pathos. A great lover of pastoral life and folk traditions, Mahler based his first four symphonies and other works on texts from

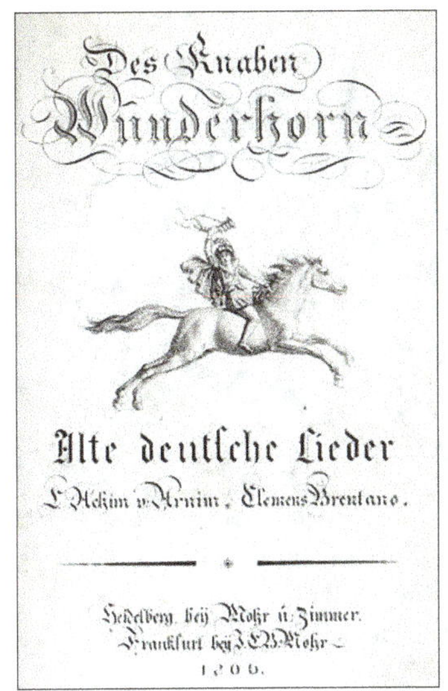
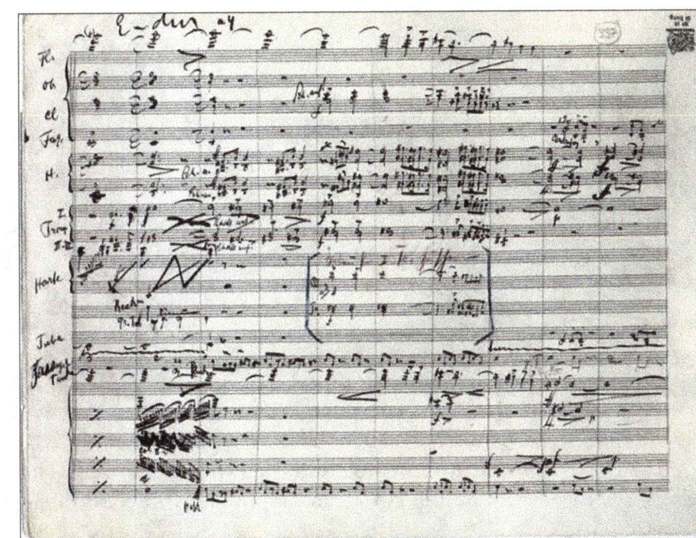

Left: *Des Knaben Wunderhorn* First Edition Title Page (1806)
Right: Gustav Mahler, *Fourth Symphony* Score Fragment
Hans Moldenhauer Collection (Whitworth College), Library of Congress

Des Knaben Wunderhorn (*The Boy's Magic Horn*), a popular collection of German folktales published in three volumes from 1805 to 1808. Mahler's *Lied von Des Knaben Wunderhorn* (1899) contained his mournful melody "*Das Irdische Leben*" ("The Earthly Life") to accompany the stark Magic Horn poem. It tells of a child begging his mother for bread, only to perish just before the grain was harvested. The music's whirring rhythm imitates both the flailing of the grain and burden of life itself.

Contemporary Northwest Agrarian Aesthetics

Northwest state legislatures established publicly funded arts and humanities commissions in the 1960s that have significantly contributed to regional aesthetic experience though promotion of public art, writing, and education projects. Formed in 1961, the Washington State Arts Commission was among the first in the nation to enact legislation requiring and funding art for public spaces and similar support was forthcoming with the organization of commissions in Oregon (1967) and Idaho (1967). A wide range of artist backgrounds, media, and subjects including agrarian art have enriched cultural experiences in Northwest city and rural schools, libraries, and other public spaces with commissions advancing their missions through relationships with the National Endowment for the Arts, state trusts and private foundations, and colleges and universities.

The three Northwest commissions eventually began acquiring art in a variety of forms for the state's permanent collections and temporary exhibits with many reflecting harvest themes (see Appendix). Among the first to be so honored by the Washington Arts Commission were photographs by Genesee, Idaho, photographer Bill Woolston—*Harvest* (1976) and *Combine* (1976). Woolston had come to the region from Chicago in 1973 to teach art at the University of Idaho in Moscow after challenging times dealing with family illness and military service in Vietnam. In the rolling Palouse he found "green, gold and brown—the colors of life," and through area photography and

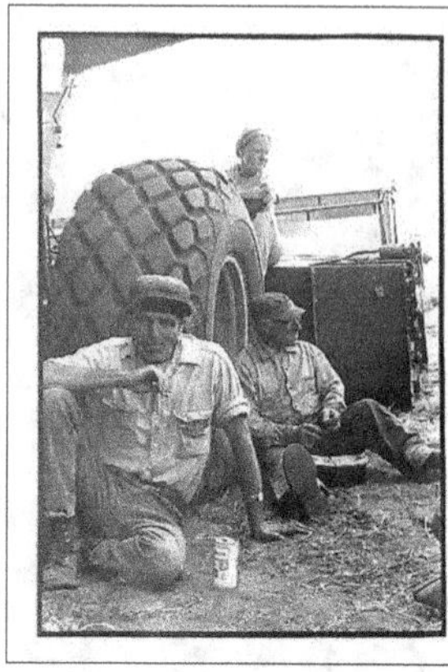

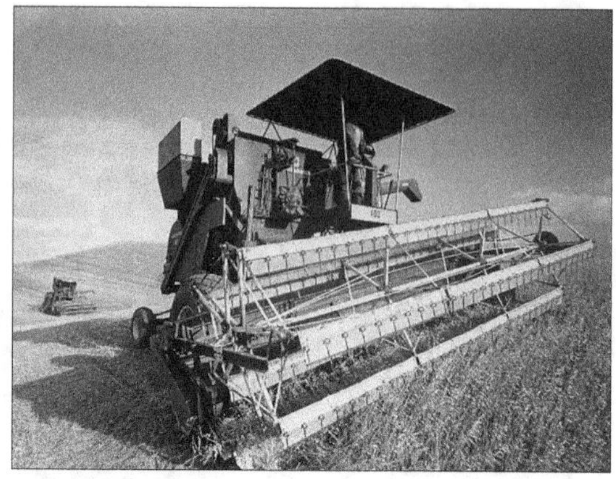

Left: Bill Woolston, *HARVEST: Wheat Ranching in the Palouse* cover (1982)
Cover Photograph: "Lunch in the shade of a combine" (1976)
Right: Bill Woolston, *Harvest* (1976); Washington Arts Commission

newfound friends discovered "a means of expressing my renewed faith in life." He devoted three years to the very personal experience of visually documenting the experiences on one farm by one farmer—Cliff Wolf of Uniontown, Washington. Woolston characterized their relationship through this peculiar project that yielded thousands of stark black-and-white images as moving from complacency to annoyance and eventual acceptance. Woolston went on to a thirty-eight-year career at the Moscow school and published *Harvest: Wheat Ranching in the Palouse* (1982) that chronicled his time on the Wolf farm.[30]

In 1982 the Washington Commission acquired *Summer Wheat Harvest* (PLATE 17) by George Stillman (1921-1997), which was one in a series of large-format paintings depicting the rich colors of Columbia Plateau farmlands. A prolific member of the influential San Francisco Bay Area "Sausalito Six" who influenced first-generation American Abstract Expressionism in the 1940s, Stillman taught art at Central Washington University in Ellensburg from 1972 to 1988 where he was surrounded by Kittitas Valley agrarian views. Like other notable works by Stillman, *Summer Wheat Harvest* (1982) evocatively straddles the boundary between expressionistic and representational art. Two modern Massey-Ferguson combines ply through whitened grain in the flatland distance with one accompanied by tractor and dump-wagon. The bright red of the machines enlivens the harvesters' activity in contrast to the muted yellow-brown of the stubbled foreground and immense gray-blue overcast sky.

Among Oregon Art Commission acquisitions are the serene Robert Weller paintings *The Combine* (1979) and *Fields of August* (1981) that show harvest scenes in the fertile Tualatin Valley southwest of Portland. *The Combine* is a wide-format acrylic painting in warm ochres, siennas, and umbers that shows an older, dark red self-propelled combine with raised header. Parked in front of a weathered barn, the machine seems at rest after completion of another season. Tree-lined hills and mountains appear in the distance beneath low, fading sunlight. In *Fields of August* Weller masterfully captures a radiant "white-out" of swirling summer dust near day's end that momentarily floats between distant drifting clouds a prodigious foreground stand of bending prairie grass.

Based in the famed Willamette Valley heartland at Corvallis, Oregon State University's Art About Agriculture program maintains one of the nation's most enduring and extensive collections devoted to contemporary agrarian art. The program's principal founders, Gwil Evans and Ken Kingsley, directed communications for the OSU Extension Service in 1982 when College of Agriculture Dean Ernest J. Briskey encouraged them to address growing concern by farmer groups that consumers and policymakers had but superficial awareness of their dependence on family farms. Evans and Kingsley responded by seeking a "'common denominator' that would have inherent value and meaning" for anyone, regardless of location or profession. To this end, they formulated a contemporary agrarian art program launched the following year to encourage creative stimulation on a wide range of media, involve the general public across the region, and purchase exceptional juried works from participating artists. The pair enlisted Tom Allen, a professor of botany and watercolorist himself, and university graphic artist Tom Weeks to organize promotional efforts for Art About Agriculture's first exhibition at the OSU LaSells Stewart Center Giustina Gallery.

The remarkable collaborative of artists, gallery partners, jurors, and educators has led to participation by more than 2,000 artists and a permanent collection of over 300 paintings, prints, photographs, sculpture, and mixed media pieces. Traveling exhibits have shared the mission with dozens of venues in Oregon, Washington, and British Columbia. Notable examples

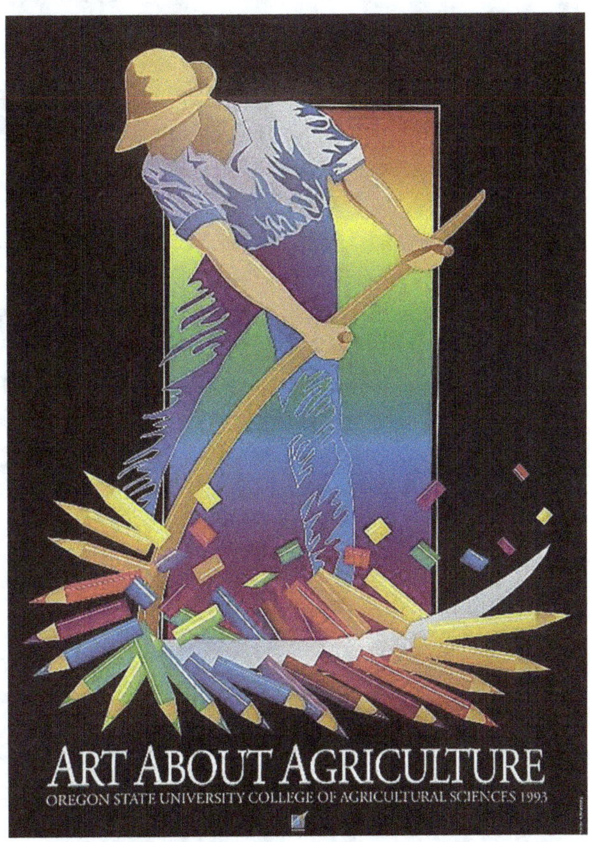

Tom Weeks, Art About Agriculture Scyther poster (1993)
Lithography on paper, 17 x 24 inches
Oregon State University Publications, Corvallis
Columbia Heritage Collection

of harvest-themed art in the collection include Tom Weeks's annual poster series that includes the whimsical 1993 scyther of richly colored pencils, Doris Frischknecht Opager's graceful oak sculpture of a sickle, *Tradition in Flight I* (1982); the abstract oil paintings *Wheat Fields* (Berkeley Chappell, 1985) and choppy, ebullient *Peoria Wheat Fields* (1985) by John Rock (1919-1983), and Donna Trent's adventuresome *Winter Wheat Growing* (1999). Other notable works include Christie Froese's elegant monochromatic drawing *Persevere* (1996) that depicts several generations of farmers, the evocative wax painting *Threshold* (2002) by Jeana Edelman, and contemporary grainfield scenes by Robert Weller and JoAnn Gillis. Gary Fields contributed the *Harvest Synergy* (1996) photographs and Sally Finch's colorful abstract pattern *Dryland Farming* series (2012) consists of four paintings based on National Climate Data Center Records for locations in Oregon, Washington, and Idaho. In an essay on OSU's landmark program, critic Lois Allan writes of an underlying theme throughout the

Pacific Northwest Art Commissions and Oregon "Art About Agriculture" Grain Harvest-Related Works (1976-2019)

Bill Woolston, *Combine and Harvest* (1976), Job Service Center, Walla Walla, Washington[1]

Robert Weller, *The Combine* (1979), Oregon State Office Building, Pendleton, Oregon[2]

Robert Weller, *Fields of August* (1981), Department of Revenue Building, Salem, Oregon[2]

George Stillman, *Summer Wheat Harvest* (1982), Washington State Investment Board Building, Olympia, Washington[2]

Jo Fyfe, *The Train* (through a wheatfield, 1984), Cashmere High School, Cashmere, Washington[5]

Doris Frischknecht Opager, *Tradition in Flight I* (AAA, 1984), Strand Hall, Oregon State University, Corvallis[4]

John Rock, *Peoria Wheat Fields* (AAA, 1985), Strand Hall, Oregon State University, Corvallis[2]

Berkley Chappell, *Impending Summer Storm, Wheat Fields* (AAA, 1985), Office of the Registrar, Oregon State University, Corvallis[2]

Janette Hopper, *Palouse Panorama* (1986), Ridgeview Elementary School, Yakima, Washington[2]

Suzanne Duryea, *Oregon Vignette* (1988), Heritage Hall, Western Oregon University, Monmouth, Oregon[5]

Susan Bennerstrom, *Grain Elevator Near a Ravine* (1989), Washington State Patrol Office, Ritzville, Washington[2]

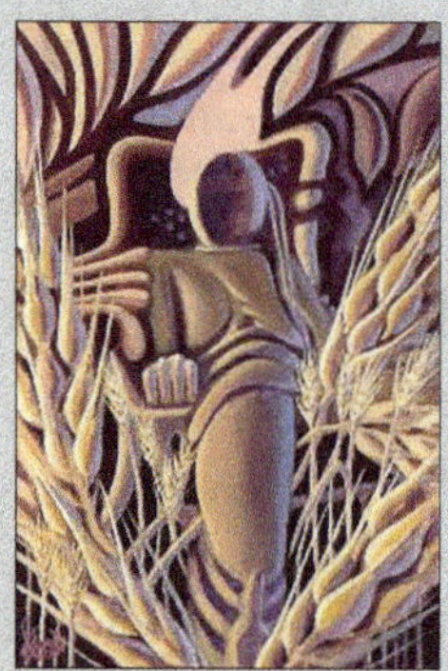

Jo Fyfe
The Train

Bob Pool, *Helix Wheatland* (AAA, 1992), Agricultural Research Foundation, Oregon State University, Corvallis, Oregon[1]

Mark Allison, *Virtual Wheat* (AAA, 1993), Dryden Hall, Oregon State University, Corvallis[2]

Judith Findley, *Bird's Eye View of Agriculture* (AAA, 1993), Horticulture Building, Oregon State University, Corvallis[1]

Justen Ladda, *Wheat* (1993), Department of Labor and Industries Building, Olympia, Washington[3]

Claudia Cave, *Palouse Wheat* (AAA, 1994), Strand Hall, Oregon State University, Corvallis[5]

Gary Fields, *Harvest Synergy* (AAA, 1996), Strand Hall, Oregon State University, Corvallis[1]

Loren Nelson, *Silos, Meeker Seed and Grain, Amity, OR* (AAA, 1999), Strand Hall, Oregon State University, Corvallis[1]

Mark Allison
Virtual Wheat

Donna B. Trent, *Winter Wheat Growing* (AAA, 1999), Gig Harbor, Washington[2]

Jeana Edelman, *Threshold* (AAA, 2001), Strand Hall, Oregon State University, Corvallis[1]

Don Kirby, *Wheatfield, Preston Ranch, WA* (AAA, 2002), Academic Programs Office, Oregon State University, Corvallis[1]

Phillip. V. Augustin, *Silo Shadow* (AAA, 2002), Strand Hall, Oregon State University, Corvallis[1]

Rich Bergeman, *Boston Mill, Shed, OR* (AAA, 2002), Strand Hall, Oregon State University, Corvallis[1]

Nancy Peterfreund, *Swirled Tracks* (AAA, 2003), Food Innovation Center, Portland[1]

Jan Boles, *Granaries at Soda Springs* (AAA, 2008), Strand Hall, Oregon State University, Corvallis[1]

Laura Bender, *To Nurture* (2009), Winters Building, Western Oregon University, Monmouth, Oregon[2]

Erik Hall, What Grows Here (2012), Connell Elementary School, Connell, Washington[2]

Sally Finch, *Dryland Farming Series 3: Moro (4: Pullman, 5: Spokane, 6: Moscow)* (AAA, 2012), Wheat Marketing Center, Portland, Oregon[5]

Kathleen Frugé Brown, *Wheat Country* (2014), Davenport Middle School, Davenport, Washington[3]

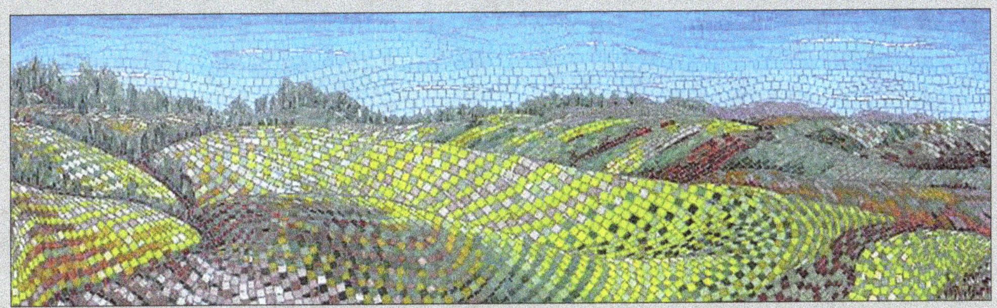

Kathleen Frugé Brown
***Wheat Country* mosaic**

Mark Abrahamson, *Winter Wheat* (AAA, n.d., acquired 2016), Strand Hall, Oregon State University, Corvallis[1]

Alexis Gregg and Tanner Coleman, *Wheat to Wheel* (2019), Colton High School, Colton, Washington[4]

Linda McMillan, *Agriculture* (2019), Idaho State Capital Building, Boise, Idaho[3]

Carol Chapel, *Grass of Gold* (AAA, 2019), Wheat Marketing Center, Portland, Oregon[6]

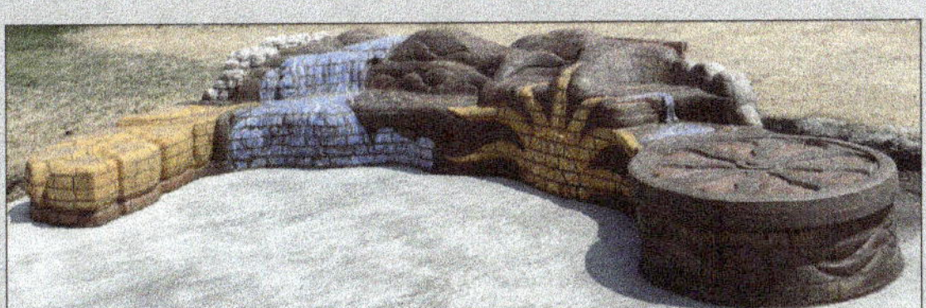

Alexis Gregg and Tanner Coleman, *Wheat to Wheel*

[1] Photograph
[2] Painting
[3] Mural
[4] Sculpture
[5] Mixed Media
[6] Drawing

eclectic collection the founders would surely appreciate. Devoid of the sullen rancor that characterizes much modern art, these works stir hearts and minds to "the necessity of good stewardship of the
area's natural resources so that they can be used judiciously and enjoyed in perpetuity."[31]

Although his undergraduate studies emphasized modern abstract approaches, Steve Henderson
of Dayton, Washington, felt drawn intuitively to the figurative styles of Rembrandt and Spanish landscapists before he discovered Isaac Levitan and his Russian Itinerant contemporaries. He found their
art to beautifully capture the emotional essence of outdoor scenes as well as of portraiture. In the
fall of 2012 Henderson was conducting a painting workshop in White Bird, Idaho, and mentioned
his regard for the Itinerants. One of the participants told Henderson that his paintings reminded her
of Nicholai Fechin, and she later presented him with an illustrated biography of the artist that had
been signed by his daughter, Eya. The book not only told of Fechin's life in Russia and the American
Southwest, where Henderson had also spent his youth, but led him to the work of Ilya Repin and
Abram Arkhikov. "The Russian masters showed profound appreciation for matters of the heart and
the peasantry," Henderson observes, "by mixing emotion with paint" for a personal style he also
characterizes as Impressionistic Realism. Henderson's lush harvest scenes include *After the Harvest
Rain, Golden Field*, and *Harvest Time* (2018, PLATE 19). The colorful swirls of field and path seen in
Highland Road, named for a lonely stretch of elevated southeastern Washington ridgeland, inspired
verse by Illinois poet Michael Escoubas:

I'm on the Highland Road,	*a burning maple sighs*
the air is sweet with autumn aromas.	*As I go by, harvest-ready*
My boots collect a cake of mud	*wheat lets go its fragrance.*[32]

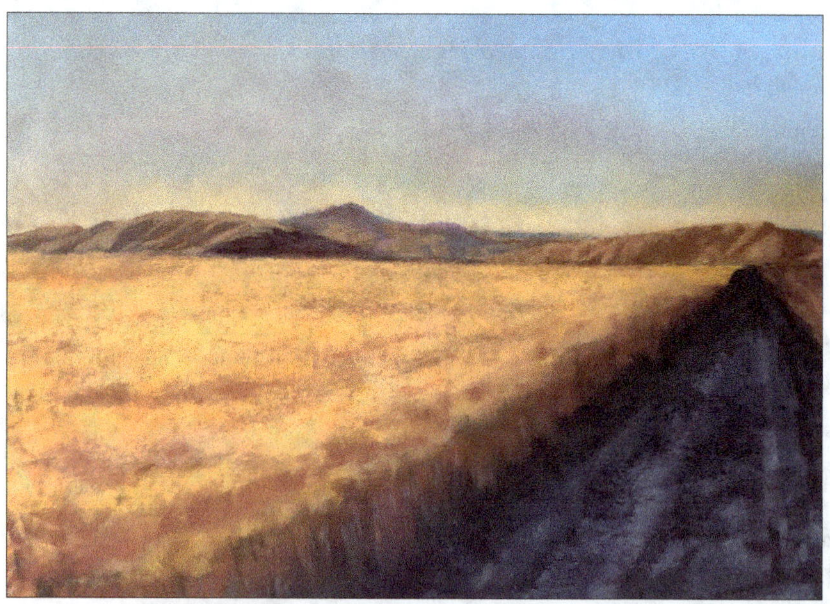

Sue Nash, *Horse Heaven Wheatfield III* (2019)
Pastel on paper, 10 x 14 inches
Columbia Heritage Collection

The Frye's paintings by Fechin, Winslow Homer, and the Wyeths also made a deep impression on Northwest artist Kathleen Hooks who first visited the museum as an art student in the 1970s. The Pasco, Washington, native went on to develop a Tonalist "art of rural life" inspired by her own agrarian experience and the broad landscapes of the Columbia Plateau. She began painting professionally in the 1990s and created highly detailed watercolors of draft animals and other livestock. She shifted to oils at the end of the decade for contemplative plein air depictions of shadow and chromatic harmony. In the spirit of Hudson River School Tonalists a century ago, Hooks's subdued canvases emphasize the formal elements of style using color, line, and shape for depictions that exist mysteriously between realism and abstraction. Soft atmospheric effects of twilight are seen in *Season of Harvest* (2002), and *Dryland Wheat* (2002, PLATE 20) with a field of golden grain pulsing before purple-tinged fallow slopes of the Horse Heaven Hills. Hooks is drawn to the "mature landscapes of summer" as fulfillment of nature's metamorphosis. "We are surrounded by harvest where we live—cutting wheat, digging potatoes, picking fruit," she observes, "and the isolation of rural life appeals to me."[33]

Rural scenes in southeastern Washington's Palouse and Horse Heaven Hills have similarly inspired Richland, Washington, artists Sue Nash and Maja Shaw. Nash moved the Northwest in 2010 after a cross-country career in art conservation. She characterizes her favored pastel medium as "the best of painting and the best of drawing" that makes possible array of "seductive colors." Nash's 2019 Harvest Series includes serene *Horse Heaven Wheatfield III* (2019) and explores the shapes and changing light across the sparsely inhabited plains of southcentral Washington. She finds the creation of agrarian pastels a "thoroughly liberating" experience that makes possible appreciation of a range of field patterns and tonal contrasts. Paintings by watercolorist Maja Shaw depict regional agrarian landscapes throughout the seasons. *Palouse Harvest* (2017, PLATE 22) is Shaw's stunning impressionistic work that captures Touchet Valley days when late spring sunshine washes swirls of grain from greenish amber to lush russet and golden brown beneath cerulean skies. Shaw, who values natural color and location above social commentary, also fashions "re-purposed" cut paper assemblies from earlier works for *Blue Mountain Harvest* (2019) and other collages of Northwest landscapes.[34]

In 2016, the Frye hosted "Cris Bruch: Others Who Were Here," an exhibition of contemporary agrarian sculpture and installations inspired by the Bruch family's twentieth-century experiences as dryland grain farmers in eastern Colorado. Featuring a diverse assembly of wood, glass, aluminum, and other utilitarian materials, the Seattle-based artist sought to address "the durability of hope in the face of hardship, obsolescence, and decay." His creations are informed by personal and historical narratives recalled through stories and letters of his relatives and others. These narratives further testify to resilient rural values born of necessity in "a wheel of adversity, scarcity, perseverance, humility, gratitude and love." Bruch expresses this aesthetic of attachment to the land and others through transformed and rescaled objects displayed as a "memory palace" of shrouded grain elevators, recycled church windows, and elegant wooden forms of farm hand tools.[35]

Country Life literature remains popular throughout rural America through more recent contributions by popular regional author-artists like Nona Hengen of the Pacific Northwest's Palouse Country. Following retirement in 1981 from a career in education and media, Hengen returned to the Northwest. She established a studio and spacious gallery in a restored barn on the property of her grandparents' farm home south of Spokane, Washington, near the hamlet of Spangle. Surrounded by wheat fields in all directions, Hengen draws inspiration from the landscape's seasonal color changes and local memories shared by community elders. Acclaimed as one of the country's finest painters of horses, many of her illustrated short stories and books feature Clydesdales, Belgians, and other massive draft animals as leading characters in lesser-known sweaty affairs of the 1920's and 30's as seen in her three-canvas Palouse Harvest series (2011). Her book *Plodding Princes of the Palouse* (1984) tells of local farm family experiences during the early twentieth century when horses pulled combines and heavily laden grain wagons to cavernous wooden storage flathouses along the rail sidings of area communities.[36]

Prominent twenty-first-century writers Stephen Turner (1937-2012), Ivan Doig (1939-2015), William Least Heat Moon, and Timothy Egan have explored agrarian themes in books of various genre and settings. These include Doig's and beloved Big Sky Montana stories, Least Heat Moon's *Blue Highways: A Journey into America* (1982) and *PrairyErth: A Deep Map* (1991) travelogues, and Egan's *The Worst Hard Time* (2006). Seattle native Egan's chronicle, winner of the National Book Award for Nonfiction, explores the darkest years of the Great Depression when choking dust storms roiled across the Great Plains in one of the gravest environmental disasters of recent times. In "High Plains Deutsch," a chapter about the Volga German farmers who had settled in the Cherokee Strip of northwestern Oklahoma, Egans tells of their thick, yeasty beer, *Felstiefel* felt boots, and harvests of famed hard red "Turkey" wheat. The Old Country grain revolutionized American milling and bread-making. But overproduction in American harvests led to forty cents per bushel wheat in late 1929, or one-eighth of its value a decade earlier. Farmers throughout the country plowed up still more native grassland in a futile attempt to make ends meet.

During West Coast travels as a young man in the 1950s, Vermont native Stephen Turner signed on as a harvest hand in Washington's dryland grain Big Bend country and returned to same area a half-century later to renew old acquaintances and explore changes in country life for *Amber Waves and Undertow* (2009). Turner and many of his journalistic contemporaries have written of an enduring rural America exceptionalism evident in close family and community ties, but one that also confronts decline and complex challenges through dependency on unpredictable national farm policies and global markets. In "Bringing in the Sheaves," his chapter about harvest today on a 2,200-acre farm near Lind, Washington, Turner opens with a line from van Gogh's letter to his brother Theo in the summer of 1889 about the Arles grain fields he enjoys "like a cicada." Turner tells of values that endure amidst economic uncertainties:

> *Van Gogh, of course, bespoke the intensity of color and texture that his increasingly fevered brain was converting to paintings. But the voluptuousness of a crop ready for cutting conveys itself also to the Adams County farmers who reap this bounty they have grown from the earth. You can't avoid the esthetics of subtly undulating ripe wheat when you drive a combine through it for literally miles with nothing else to obstruct your view.*
>
> *. . . Later we sit on the veranda with drinks, talking leisurely as the day begins to cool. Sophie, a bundle of energy, cavorts on the green spread of lawn that fans out past flowerbeds and some of Annie's outdoor sculptures toward shade trees, the child herself a blossom in a floral play dress as she romps. At the table for dinner then, we hold hands in a circle while Annie asks a blessing for friends and food. And I find myself truly grateful for this day, my welcome into experiencing once again, however briefly, the elemental satisfaction of bringing in a harvest.[37]*

Writer Ralph Beer grew up on the broad arid grasslands south of Helena, Montana, where his ancestors had eked out a living raising livestock and crops since the late 1800s. He writes in muscular prose to express in elegiac tones the inherent paradox of viewing land as both contemplative source of inner life and contentious, sometimes brutal host disinterested in allowing it inhabitants to make ends meet. In "The Big Dusty," the opening short story of his collection *In These Hills* (2000), Beer recalls part-time harvest work driving a grain truck for a seventy-five-year-old neighbor who works late into the night pressing a newer John Deere 7700 combine to get his wheat crop in before autumn rains: ". . . [T]he machine ahead seems otherworldly indeed, the twenty-four-foot header gnashing wheat into its maw, the straw chopper in back flailing stems into a mist that haloes the yellowed superstructure lights, then spills away into darkness. . . . But as I ease the truck in below the spout and lean to close the passenger window against a storm of noise and dust, I can see an actual man at the controls above me, guiding this ungodly complexity of augers and gears, drive chain and belts over ground that must have taken decades to learn." Elsewhere Beer tells of growing up in a place where one generation bequeaths to the next the fragile treasure of "remembered experience" involving good and bad times through the seasons and with neighbors, many times inconsequential but occasionally momentous. To Beer, farm and ranch stories passed down through generations impart a sense of wonder, meaning, and love.[38]

Architect-artist John Rankin has drawn from experiences as youthful farm laborer in the grain fields of the Northwest's Columbia Plateau to inform color and design of stunning silk screen serigraphs. For *Home Place Harvest* (1995, PLATE 23), Rankin superimposed a 1920 black-and-white harvest photograph of the Hugh Phillips threshing outfit near Cunningham, Washington, across a broad wash of deep yellow. The enormous combine and substantial team of horses seem to float across a ripened sea of wheat beneath of placid turquoise sky. The scene's mesmerizing starkness is reminiscent of art by pop pictorialist Andy Warhol, whose provocative compositions Rankin credits along with Hatch poster art for influencing his own printmaking.[39]

"I love distance, openness and endlessness," explains Washington artist Kim Mathews Wheaton, "and want my paintings to have a feeling of looking into infinity." She credits the abstract aerial landscapes of Richard Diebenkorn and Russell Chatham for influencing her austere style of saturated

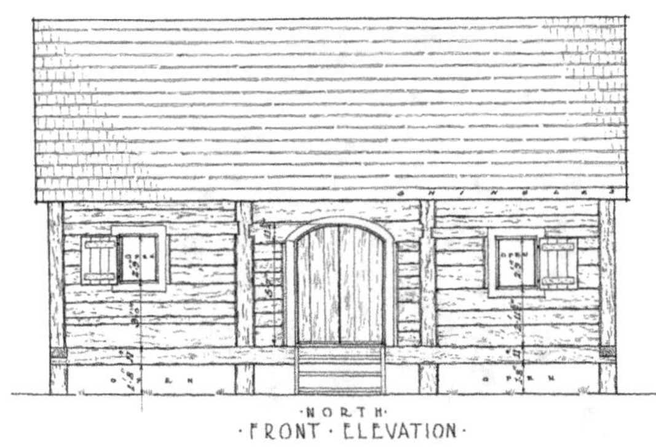

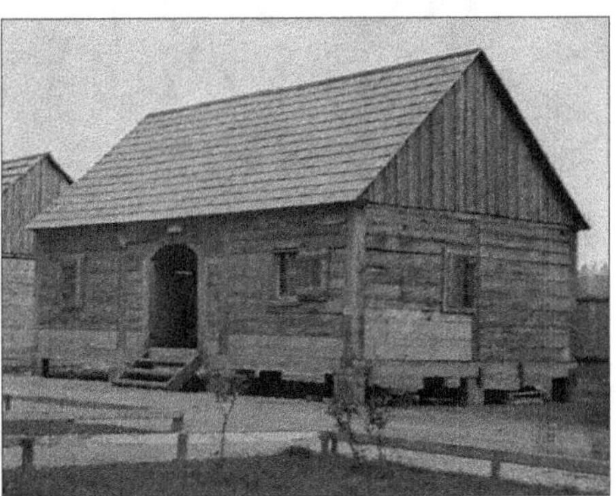

Fort Nisqually Granary (1851) Architectural Plan and Restoration
Historic American Buildings and Landscape Surveys **(1937)**
Prints and Photographs Division, Library of Congress

tonal bands depicting grand vistas of grain fields, summer-fallow, and distant natural landforms. In works like *Heading Over the Hill* (2010), *Palouse Gentle Hills* (2011), and *Sweeping Shadows* (2012), Wheaton applies muted shades of orange, brown, green, and white to horizontal canvases several feet wide. Her paintings impart an introspective, spiritual quality without sacrificing the reality of such majestic views.[40] Spokane native Ricco DiStefano studied at Rocky Mountain College of Art and Design in Denver and worked as a graphic designer before co-founding the Issaquah, Washington based artEAST gallery consortium of regional artists and craftspeople. Residing on a small farm in scenic Maple Valley southeast of Seattle, DiStefano seeks to express ambient impressions of the countryside through visual imagery. In his large format acrylic painting *The Winds of Harvest* (2012, PLATE 24), a shadowy Demeter-like figure gazes toward the sinuous expanse of luminous gold, russet, and green.

Living History and Remembrance

In 2014, Mike McGuire of Tacoma, Washington's Fort Nisqually Living History Museum sought help to help reestablish demonstration plots of the area's early eighteenth-century Hudson's Bay Company cereal grains. This project and generous access to the fort's archives better acquainted me with the 1930s WPA efforts to document and restore such historic sites associated with the nation's agrarian past, and with the substantial collections of the WPA Farm Security Administration. (The Fort Nisqually Granary, built in 1851, is the region's oldest wooden structure.) Roy Stryker, WPA/FSA Historical Unit Director, was especially interested in pictures of farmworkers, and designated with taxonomic obsession a vast range of such storytelling themes as "harvest," "barns," and "church." He hired experienced photographers including Dorothea Lange, Arthur Rothstein, Russell Lee, and Marion Post Wolcott who traveled vast distances across the country to document "Grapes of Wrath" era of American farm life. The visual record amassed by the unit from 1938 to 1942 numbers some quarter-million negatives of which approximately 170,000 FSA images have been digitized and publicly available.[41]

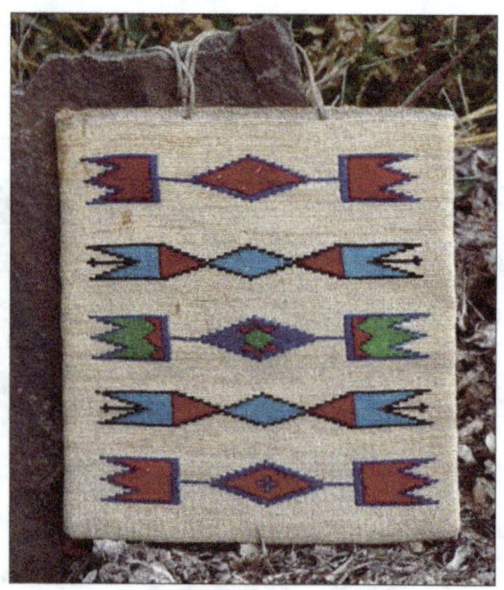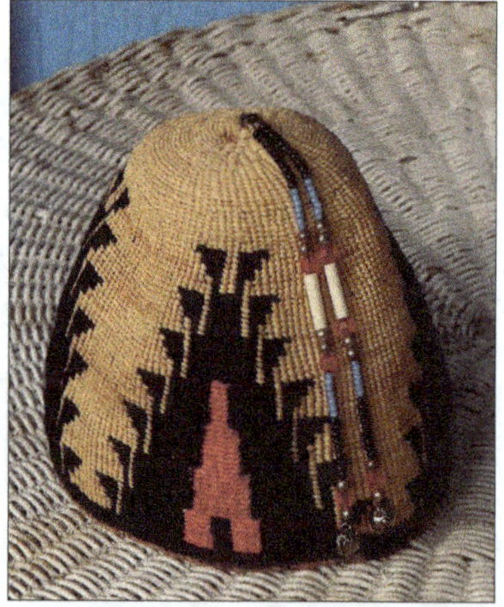

Snake River-Palouse Corn Husk Root Bag and Woman's Hemp Twine Hat
Carrie Jim Schuster Collection
John Clement Gallery Photographs

Introduction to the Fort Nisqually Living History Museum story led to publication of *Harvest Heritage: Agricultural Origins and Heirloom Crops of the Pacific Northwest* (Washington State University Press, 2013), coauthored with Alexander C. McGregor. The book describes Hudson's Bay Company Governor Sir George Simpson's epic 1825 journey from Great Britain to the Northwest that led to the establishment of Ft. Vancouver and Ft. Colville. Simpson sought trader self-sufficiency on the far frontier which led to extensive grain, fruit, and livestock production. The HBC's influential Puget Sound Agricultural Company began operations at farming stations around Ft. Nisqually near present DuPont in 1841 under the capable management of the fort's indefatigable master of many interests and skills, Dr. William Tolmie. A later view of one of the farms is the likely subject of a magnificent landscape (c. 1881) by Luminist Raymond Yelland (1848-1900) that shows log structures and adjacent field with snow-laden Mt. Rainier in the background.

The fur trade farmers and missionaries shared their seed with area Native Americans who were tending substantial acreages on both sides of the Cascades by the 1840s. Anthropologist Alan Marshall has written that Columbia Plateau First Peoples practiced natural horticulture prior to Euro-American contact that facilitated their adoption of agriculture. Their extensive tending and gathering of plant resources (e.g., camas, bitterroot, kouse) did not involve rectilinear plots or fences so was not recognized as agriculture by the newcomers, but the distinction was largely cultural and symbolic. Grains came to flourish on the Columbia Plateau where Yakama-Palouse Chief Kamiakin of the Yakama, Moses of the Columbia-Sinkiuse, and Chief Seltice of the Coeur d'Alene, and many others raised crops long before American settlement. In 1859 frontier roadbuilder John Mullan reported Palouse and Nez Perce Indians were raising several hundred acres of "corn and wheat" on seven tracts of fertile bottomlands along the lower Snake River. Columbia Plateau women became

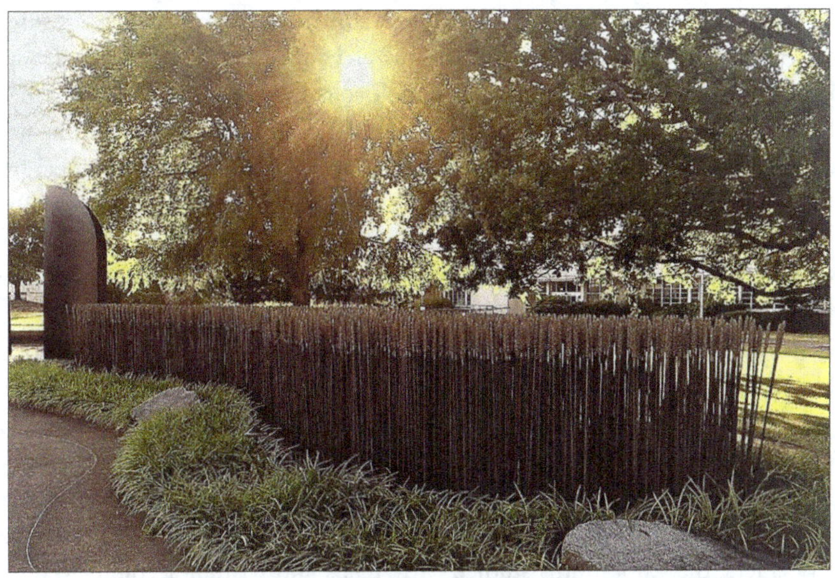

Simon Kogan, *Washington World War II Memorial* (1999)
Bronze, steel, granite, and concrete installation
Washington State Capitol Campus, Olympia
Columbia Heritage Collection Photograph

adept at incorporating fibers from the newly introduced plants and native species into exquisitely crafted designs for colorful root bags, baskets, and other items.[42]

Chief Factor Tolmie reestablished company operations in British Columbia after the Treaty of 1846 and a reinvigorated PSAC was soon producing grain and livestock on four substantial farms in the Victoria area as well as on Tolmie's own 1100-acre Cloverdale Farm near Saanich. Immigrant settlers recruited in Great Britain to operate area PSAC farms recorded their adjustments to Northwest frontier life in verse, diaries, and works of art. Poetry by James Deans, a young man assigned to Victoria's Craigflower Farm in 1853 on a five-year labor contract, expresses the homesickness experienced by such newcomers with hints of hope seen in nearby fields:

<table>
<tr><td>Cheer up, sad heart and while I stray,</td><td>Then lay aside, your hopes and fears,</td></tr>
<tr><td>Sadly o'er Craigflower woody brae</td><td>A change shall come with passing years</td></tr>
<tr><td>Come fancy paint each glade and glen,</td><td>A time will come when you will say,</td></tr>
<tr><td>With waving grain and homes of men.</td><td>This is my home, here I will stay.[43]</td></tr>
</table>

To visit quiet places like Ft. Nisqually and Craigflower, and to view Cris Bruch's silent, draped models of grain elevators brings to mind small towns of the Washington-Idaho Palouse and elsewhere that struggle to survive. That my grandfather's family came to the Northwest from Russia after three years of farm life in Kansas, and my wife's grandfather, Adam Morasch, immigrated from the same Volga village to our hometown via the Colorado beet fields makes Bruch's Frye exhibit "Others Who Were Here" deeply personal. So does sculptor Simon Kogan's grain-based 1999 World War II memorial installation fashioned from metal, stone, and greenery for the state capitol campus in Olympia. A native of Russia who trained under renowned sculptor Isaac Brodsky in Moscow, Kogan's solemn creation features 4,000 bronze wheat stalks fashioned from melted torpedo railings of decommissioned American battleships. The meandering field represents state residents like our own relatives who perished during the war. A half-dozen young men from farms near my hometown lost their lives including Andrew and Francis Lund's sons from Benge who were killed in Pacific battles—Otto in 1943 and Victor in 1945. (Compounding such loss was the wartime passing of the Lunds' daughter, Irene, from natural causes.) John B. Carter was one of five brothers from the tiny hamlet of Hooper who volunteered for service. Lt. Carter was killed on June 6, 1944 while attempting to capture an enemy pillbox overlooking Normandy's Omaha Beach.

The first stanza of *America the Beautiful* inspired Kogan's war memorial, and a bronze plaque carries a wistful inscription: ". . . Reach out now, across the years and through the tears. . . . Remember me." In his classic study of the English landscape, *The History of the Countryside* (1986), Oliver Rackham identifies memory and meaning as critical cultural attributes that are jeopardized when rural landscapes are abandoned or despoiled. Places like the Lund farm near tiny Benge have something to tell us, and longtime residents like Roy and Karin Clinesmith, who later acquired the property, exemplify a generation of responsible land stewards who have sought to understand and share the meaning of place. In addition to recording the Lund family saga through the recollections of community elders, the Clinesmiths studied early maps and written accounts to document area agricultural history and the wartime sacrifices of local farm youth like Otto and Victor. They found that the original buildings were located at the confluence of two notable frontier trading routes—the Colville Trail and Mullan Road—still prominently etched across area rangeland and grainfields.[44]

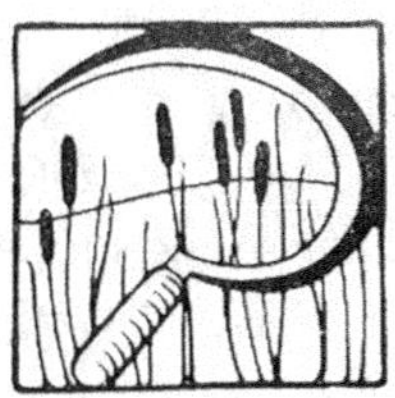

IV

Threshold
NEW AGRARIANS FOR RENEWED COMMUNITY

French filmmaker Agnès Varda's documentary, *Les glaneurs et la glaneuse*, winner of the Mélès Prize for Best French Film of 2000, offers controversial interpretation of Jean-François Millet's iconic painting *The Gleaners* (1857). Distributed in the United States as *The Gleaners and I*, the movie shows how poverty need not deprive individuals in any age of dignity and humor. They may be compelled, however, to overcome significant social and economic obstacles to eke out an existence. The film has contributed to a broader, contemporary definition of gleaning to include the gathering of unwanted foods of all kinds—bread, fruit, vegetables, and fish, as well as other castaway resources. Varda's sobering images of oppressed, vulnerable, and often young souls, illustrate the disturbing trend of income inequality in modern societies like France where "gleaning" remains a salient reality for many, and its potentially harsh consequences. Her work also suggests possible solutions in the food service sector through the stewardship of surplus distribution via urban pantries and community food banks.[1]

This more broadly defined concept of gleaning was described in *The Other America* (1962), Michael Harrington's influential study of hunger and homelessness that shaped Lyndon Johnson's 1960s War on Poverty. In the wake of growing public awareness, social service and religious groups have formed new partnerships in recent decades to develop food security programs to distribute perishable produce and processed foods. At least one-third of food produced annually today in America—as much as 40 million tons valued at approximately $75 billion, is wasted due to spoilage and inefficient storage and distribution. Applying the idea of gleaning to such lost resources, a group of Phoenix activists organized the country's first urban food bank, Second Harvest, in 1975 (known as Feeding America since 2008). Similar humanitarian efforts followed in Portland (Interagency Food Bank, 1975), Chicago (Food Depository, 1978), Seattle (Food Lifeline, 1979), New York City (City Harvest, 1982), and spread to many other large cities. Some of these endeavors have been affiliated with denominational benevolent ministries including the Society of St. Andrew Gleaning Network (United Methodist Church), Evangelical Lutheran Church of America World Hunger, and Catholic Relief Services Hunger Campaign.

Brad Bailie of Lenwood Farms near Connell, Washington, produces organic grain and vegetables, and regularly works with local churches and crews of Feeding America gleaners to supply Second

129

Harvest and other regional food banks. He explains his and other farmer-contributors' motivations in both practical and moral terms: "Sometimes growers have surpluses because commercial buyers have certain commodity specifications by size or weight. This can leave a considerable amount of quality produce in the field, and we don't like seeing such waste. We also believe that the blessing of a bountiful harvest brings responsibility to share with others."[2] The opportunities and responsibilities that come with abundant harvests are also evident in revivals of the ancient Passover Festival among religious fellowships throughout the world. Israel's celebrated and prolific composer, Matityahu Shalem (1904-1975), wrote numerous folk songs for contemporary Jewish worship including Passover celebrations when the first sheaves of barley are cut for presentation at the temple. His popular *Shibbolet Basadeh* (Ears of Grain in the Field) is sung and danced to traditional choreography shaped by Shalem's experiences on a kibbutz in western Galilee where he tended flocks and fields after relocating to Palestine before World War II.

<table>
<tr><td>

Shibolet basadeh kor'ah baru'ach
me'omes gar'inim ki rav.
Uvemerchav harim
yom kvar yafu'ach.
Hashemesh ketem v'zahav.

Sdei se'orim tama
 zer chag oteret
shefa y'vul uvracha.
Likrat bo hakotzrim
b'zohar mazheret
cheresh la'omer m'chaka.

</td><td>

Ear of grain in the field, bowed in the wind
From the weight of its seed, which is great.
And in the expanse of the mountains
The day already rises.
The sun is fine gold.

A pure field of barley is crowned
with a holiday wreath,
an abundance of produce and blessing.
Just before the coming of the harvesters,
with shining brilliance,
silently, it waits for the sheaf.[3]

</td></tr>
</table>

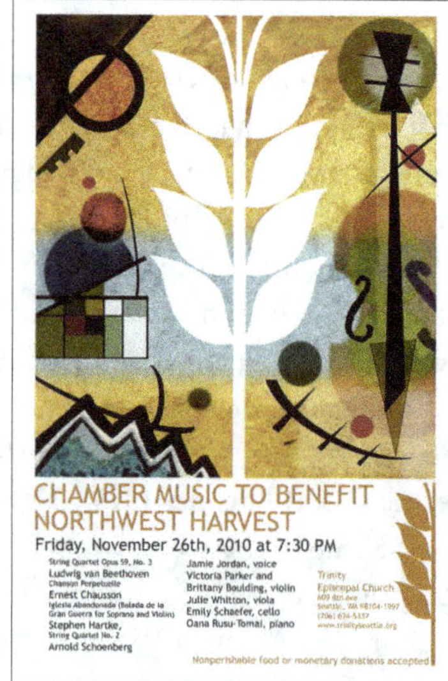

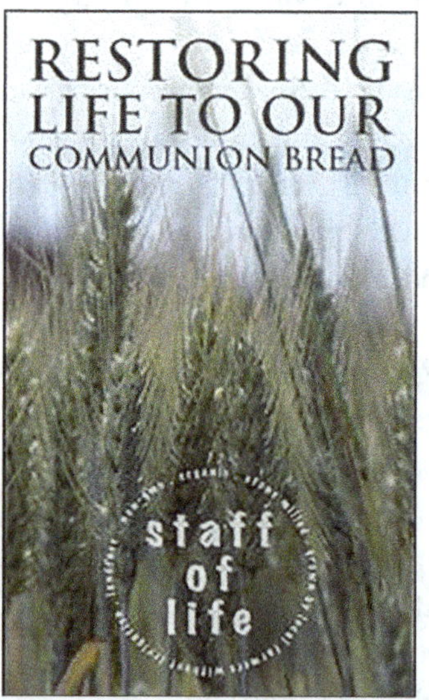

Left: Jeff Whitton, Northwest Harvest Poster Art (2010)
Right: General Convention of The Episcopal Church Banner, Salt Lake City (2015)
Columbia Heritage Collection

For religious thinkers like Shalem, meaning still retains a supernatural sanction derived from humanity's simultaneous temporal and spiritual nature. Contemplation of the harvest labor and its bounty can be perceived in the particularities of agrarian experience whether along a Galilean shore or Dakota slope.

Heartland, KareLift, and Harvest Hope

Pope John Paul II (1920-2005) made explicit reference to farmers' charitable obligations to the poor during an unprecedented papal visit to the American heartland in October 1979 hosted by the Diocese of Des Moines, Iowa, and the National Catholic Rural Life Conference. The pope celebrated an open-air Mass where a vast crowd of some 300,000—the largest in Iowa history, had assembled on a broad hillside at Urbandale, Iowa's Living History Farms. Local St. Mary's parishioner Joseph Hays had sent a hand-written letter to the Pope inviting him to witness the church's "Community in the Heartland" ministry of rural study and outreach. The pontiff's decision to visit the Iowa countryside led to weeks of preparation by members who broke from customary harvest routines to host the special ecumenical event. Surrounded by area church and civic leaders, the pope led the service from a massive platform fashioned of white oak from a century-old corn crib. The temporary sanctuary was draped with an enormous quilted banner designed by Fr. John Buscemi of Madison, Wisconsin showing a cross with four colorful contoured field patterns symbolizing the seasons. From this peculiar setting, Pope John Paul II delivered a homily urging his hearers "in the middle of the bountiful fields at harvest time" to embrace "three attitudes . . . for rural life"—humble gratitude, land stewardship, and generosity toward the poor.

In an address ten years later commemorating the church's "Declaration *Nostra Aetate*" regard-

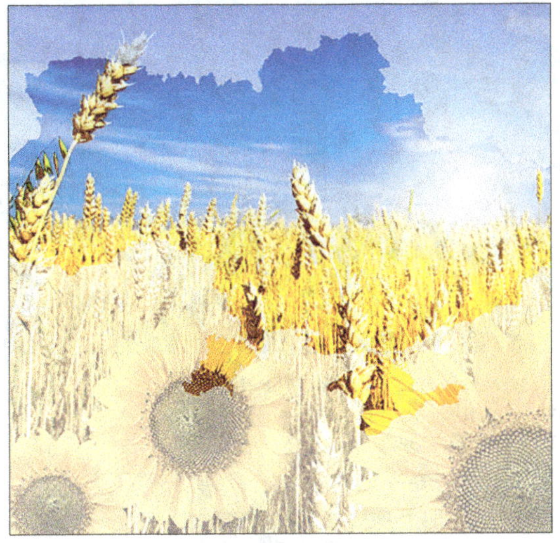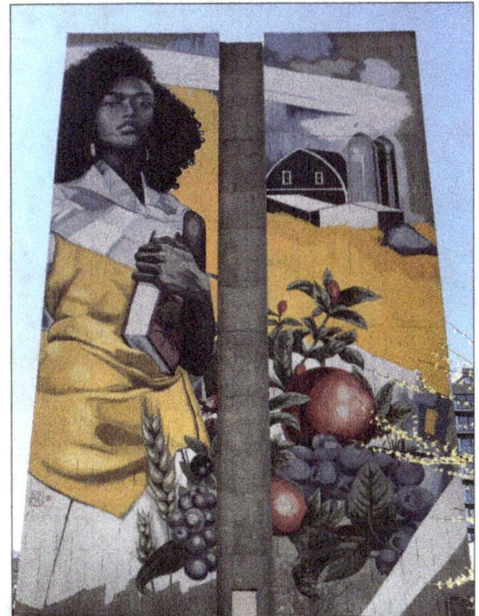

Left: Sara Quinn, *We'll Still Be Here When This Is Over* Cover Montage
Tumbleweird 7:4 (April 2022), Courtesy of the Artist
Right: Lula Goce, *Zero Hunger* (2021)
I-195 "Mystery Silo" Vent Mural; Washington, D.C. / Columbia Heritage Photograph

ing mutual respect and cooperation among world religions, John Paul II mentioned the notable contributions of American Trappist monk Thomas Merton (1915-1968) to interfaith dialogue. A longtime resident of Gethsemani Abbey near Bardstown, Kentucky, Merton also fostered fellowship with prominent Asian and Native American spiritual leaders and formulated a corpus of ecological writings permeated with contemplative appreciation of nature and agrarian endeavor. In his poem "Trappists, Working" (1942), farming is likened to a liturgy of worship amidst outdoor sanctuaries of divinely bestowed sun, wind, and "walls of wheat." "Landscape: Wheatfields" (c. 1950) likens faithful soldiers of the faith to shocks of grain sheaves awaiting transport in holy service of others.[4]

American farmers participated more directly in domestic gleaning programs in the 1980s as well as in similar global aid projects. In the wake of the collapse of the Soviet Union, a group of Pacific Northwest growers formed WestWind Ministries in 1991 in response to appeals from newly independent Russian leaders to provide food and medical assistance to schools and orphanages in the Russian Far East. A coordinated "Operation Karelift" effort involving the National Association of Wheat Growers, Washington-Idaho Pea & Lentil Association, and The McGregor Company of Washington, Idaho, and Oregon led to delivery of over one thousand tons of aid to areas in greatest need. Farmers hauled truckloads of wheat for processing into flour while Northwest barley, lentils, and beans were combined into nutritious soup mixes.

When Russian President Boris Yeltsin made an unprecedented visit to Seattle in September 1994 to report on newly normalized relations between the two countries, he cited "this help in our hour of need" in the context of the food campaign as a key factor in his historic decision. Yeltsin's gala reception was hosted by Washington Governor Mike Lowry, himself a native of the Palouse Country hamlet of Endicott, Washington, where his father, Robert, had managed the local grain grower

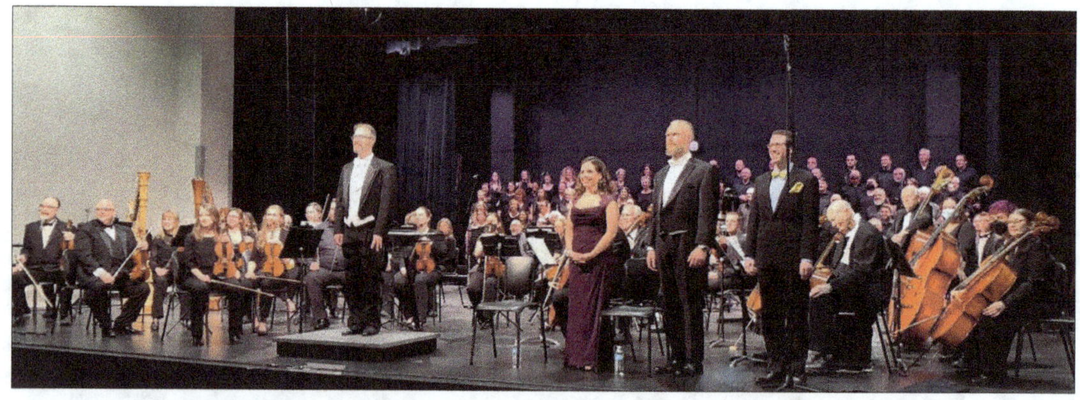

**Mid-Columbia Symphony and Mastersingers Ukraine Benefit Concert,
Kennewick, Washington (March 2022)**
Johannes Brahms, *A German Requiem to Words of Holy Scripture, Op. 45*

Sie gehen hin und weinen   *(They go forth and weep,*
und tragen edlen Samen,   *bearing precious seed,*
und kommen mit Freuden   *and come with joy*
und bringen ihre Garben.   *bearing their sheaves.)*

—Psalm 126:5-6
Columbia Heritage Collection Photograph

cooperative in the 1950s. Lowry's dedication to humanitarian and migrant farm worker causes was the subject of many tributes following his passing in 2017. Officiant Kacey Hahn of St. Matthew's Lutheran Church in Renton opened the late governor's memorial with explicit reference to moral responsibility from Leviticus 23:22: "And when you reap the harvest of your land, you shall not reap your field right up to its edge, nor shall you gather the gleanings after your harvest. You shall leave them for the poor and for the sojourner."

With the outbreak of war of Ukraine in February 2022, many of the original KareLift partners joined with other groups through "Operation Harvest Hope" raise funds and send Northwest commodities to help feed the several million refugees who fled the conflict to safe havens throughout Europe. The war between Russia and Ukraine—nations that provide some 30% of world wheat and barley exports, destabilized global grain markets and put at significant risk the wellbeing of millions living in the Middle East and North Africa who depend on imports and subsidized bread.[5]

Midwest farmer Oren Long contributed for decades to agrarian periodicals and his local Kansas paper, the *Valley Falls Vindicator*, to offer insight on topics ranging from food security and social unrest to seed rates and meaning in art. In a 1983 *New Farm* article, Long underscored the vital understanding that rural experience is at once terrestrial and transcendent. "My farm is my refuge from the deception and hopelessness that haunts this intrusive commercial world. . . . I am an inseparable part of a great biological scheme of things and the greater contribution toward the complexity and harmony of that scheme, the greater will be the beauty of my world and the greater my significance to it." In this way rural experience is understood to impart beauty to life in ways long expressed by agrarian painters and writers who have shown the abiding value of sowing, reaping, and other "cooperative arts" practiced with attention to land care and the less fortunate.[6]

Good Scythes and Thresholds

Jim Crace's 2013 dystopian novel, *Harvest*, transports readers to a sixteenth-century English village to experience a week of celebration, intrigue, and disturbance that marks the end of harvest. Area residents gossip and gather in the barley field but are more concerned with the recent arrival of several vagrants than the momentous events about to engulf them. The story is told from the standpoint of Walter Thirsk, who after residing there for a dozen years is himself a relative newcomer to a place. "We should face the rest day with easy hearts," he muses, "and then enjoy the gleaning that would follow it, with our own Gleaning Queen the first to bend and pick a grain. We should expect our seasons to unfold in all their usual sequences, and so on through the harvests and the years."

The strangers who camp nearby are refugees from enclosure of open lands, and their coming coincides with that of a man of uneasy silence the villagers call Mr. Quill for the peculiar instrument he carries for his work: "We mowed with scythes: he worked with brushes and quills. He was recording us, he said, or more exactly marking down our land." Quill is making a map and compiling numbers, measuring locations of streets, houses, and fields. He informs his rustic hearers that such work is about "improvements" being done on behalf of the manor estate's absentee heir who is zealous for improvements to enlarge the estate by enclosure and replace fieldworkers with sheep which will also render gleaning obsolete. "We know enough to understand that in the greater world," Quill explains, "flour, meat, and cheese are not divided into shares and portions for the larder, as they are here, but only weighed and sized for selling." The old order of Enough is being displaced by More. To be sure, pre-enclosure landscapes were not idyllic spaces since commoners depended on hard labor and the vagaries of the seasons for their welfare. But conditioned by faith and custom, daily anxieties poignantly expressed by Crace were moderated by community fellowship and shared resources from the commons.[7]

One of Kentucky farmer-philosopher Wendell Berry's first public addresses on trends in consolidation of family farms and land care took place in July 1974 at the "Agriculture for a Small Planet" symposium held at Gonzaga University in conjunction with Spokane's "Expo '74." The world's fair was promoted as the first international ecological exposition and Berry's passionate talk, delivered from scribbled notes on a large yellow pad, included a call for "a constituency for a better kind of agriculture." The presentation inspired organization of the Northwest Tilth movement for sustainable farming and became seed for of Berry's best-selling book *The Unsettling of America: Culture & Agriculture* (1979).

In his essay "The Good Scythe" (1979), Berry grapples with the meaning of progress in modern times. He recalls buying a "power scythe" for cutting grass on a steep hillside near his home, but soon found that the anticipated advantages of reduced labor were offset by the machine's temperamental motor and considerable racket. The turning point came when a neighbor showed him an old-fashioned scythe that was comfortable to handle and efficient. "There was an intelligence and refinement in its design that make it a pleasure to handle and look at and think about," Berry observed, and he promptly replaced the powered machine and gas can with a wooden-handled Marugg scythe and whetstone. Berry does not dismiss mechanical innovation; the scythe, after all, is an improvement on the sickle. But he found the episode to have "the force of a parable" about life, labor, and definitions of progress. He advocates a time-honored approach for judging claims of saved labor and short cuts, and warns against the embrace of solutions that tend to bring longer working hours with greater equipment expense, and further move the balance between nature and needs.[8]

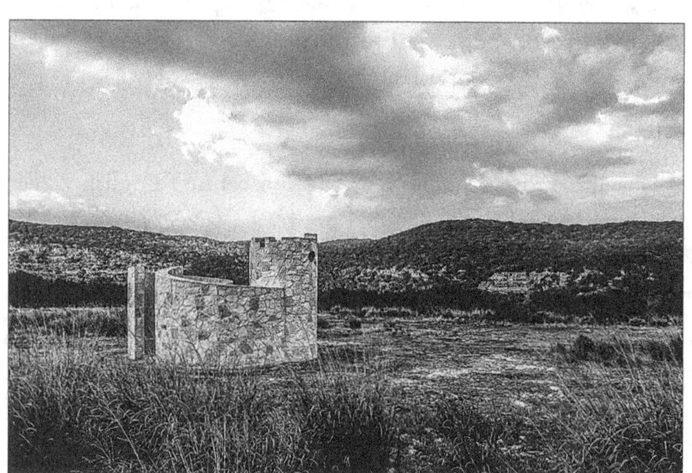

Roger Feldman, *Threshold* (2013)
Laity Lodge near Leakey, Texas
Courtesy of the Artist

Lewiston, Idaho artist W. Craig Whitcomb has painted rural scenes for a half-century in watercolor and acrylic with subject matter ranging from isolated Northwest grain elevators to English thatched cottages and Japanese landscapes. His *Amber Waves* (2008), finalist for the first annual "H'Art of the Palouse" Banner Competition, shows an immense abandoned grain elevator in vivid rusty reds and blues rising from a field of ripe grain. Vibrant watercolors of Northwest grain and legume fields scenes by Andy Sewell of Viola, Idaho, have appeared on posters for the Pullman-based National Lentil Festival. His dramatic *Doubletime Before the Storm*

(2021, PLATE 24) shows the skillful choreography of two John Deere combines moving in tandem with tractor-pulled grain carts in the face of threatening clouds and lightning. Sewell, a graduate in fine arts from the University of Idaho in Moscow, spied the late afternoon scene near his eastern Palouse home. The golden browns and dark shadows of land and sky express Sewell's appreciation for the primal forces of nature that make harvests possible. Other richly colored agrarian landscapes by Sewell include *Palouse Summer Glory* and *Palouse Country Summer.*

Works by Seattle's Roger Feldman, winner of the 2005 Prescott Award in Sculpture, reflects his study of theology and art education. Raised in the Palouse Country community of Rosalia, Feldman has created large site-specific sculptures in the United States, Canada, and Europe. He meticulously plans each installation by visiting the location to "dream about the possibilities" before rendering a small 3-D scale maquette from mat board before fashioning a larger, more refined model from wood. For *Threshold* (2013) at Laity Lodge, an ecumenical retreat along the Frio River in Texas's Hill Country, Feldman conceived of three interconnected chiseled limestone monoliths including a 15-foot-tall tower to represent the three-in-one concept of the Trinity. The work's title is derived from Hebrew words used in the Old Testament (*saph, miptān*), a raised beam at the edge of a threshing floor, to signify the boundary between the outside world and sacred space for contemplation and worship.[9]

Tradition and innovation have presented cultural tensions since the dawn of civilization, and responsible influence from each has contributed to humanity's wellbeing. Like van Gogh paintings of gleaners and reapers with factory smokestacks on the horizon, agrarian fine art and literature foster better understandings of tensions that involve emotion and reason, and local and universal values. Among other recent developments in grain production, the advent of minimal tillage operations using specialized power equipment has greatly reduced soil erosion on American farms while increasing yields. The emerging New Agrarianism of the twenty-first century moves beyond nostalgic romanticism to moderate use of industrial energy within the context of natural systems for soil fertility. Wise approaches to innovation respect stewardship of land and the longterm wellbeing of others. Duke Divinity School environmental theologian Norman Wirzba writes of a New Agrarian ethic that honors modern science as well as ancient religious appreciation for the transformative mystery of soil, water, and grain for human sustenance. Implicit acknowledgment is also made of fair compensation for farmers and other workers. "How we make bread, how we share and distribute it, are of profound moral and spiritual significance," he writes in *Food and Faith: A Theology of Eating* (2011). "[E]very loaf presupposes decisions that have been made about how to configure the social and ecological relationships that make bread possible."[10]

Tim Dearborn of Fuller Theological Seminary and author of *Taste & See: Awakening our Spiritual Senses* (1996) tells of Jesus' reference to bread in the context of material well-being and spiritual strength. During his temptation in the Wilderness (Luke 4:4), Jesus quotes the familiar Old Testament passage, "[M]an does not live by bread alone" (Deuteronomy 8:3), which recognizes legitimate needs for "daily bread" physical sustenance (Matthew 6:11) provided through divine provision and sacrifice. Sharing food and faith goes hand in hand with prayer ("grace") and communion with family and friends for the vital, sensuous experience of daily feasting. In this way, meals can transform mundane consumption into enriching spiritual experience that honors grains, greens, and other foods, but recognizes their material essence, cultivation, harvest, and preparation as rooted in meaningful service. The tragedy of religious piety is not materialism Dearborn writes, "but that in a particular way we are not materialistic enough." By dividing aspects of human existence into sacred and secular realms, one can also render possessions, physical needs, and the land into domains separate from their divine source and protection.[11]

Frustrations with equipment repair and long hours of solitary fieldwork may appear scarcely related to religious faith. But farmers and other members of St. Macrina's Episcopal Church near San Francisco regularly meet to share the challenges of twenty-first-century farming with area millers, bakers, brewers, and consumers. All contribute perspectives on grain as a "community crop" and how each group can participate in consequential efforts to strengthen cultural ties and serve as stewards of the land. In 2015, St. Macrina co-founder and Agricultural Chaplain Elizabeth DeRuff established The Bishop's Ranch Field on Russian River Valley church property near Healdsburg, California. Young and old gather there throughout the year to plant, till, and harvest heritage grain that is milled for communion bread and distributed throughout the diocese. "We want to see local farmers succeed and be part of local communities," explains Rev. DeRuff, "and to learn with them about 'belonging' as well as 'having.'"[12]

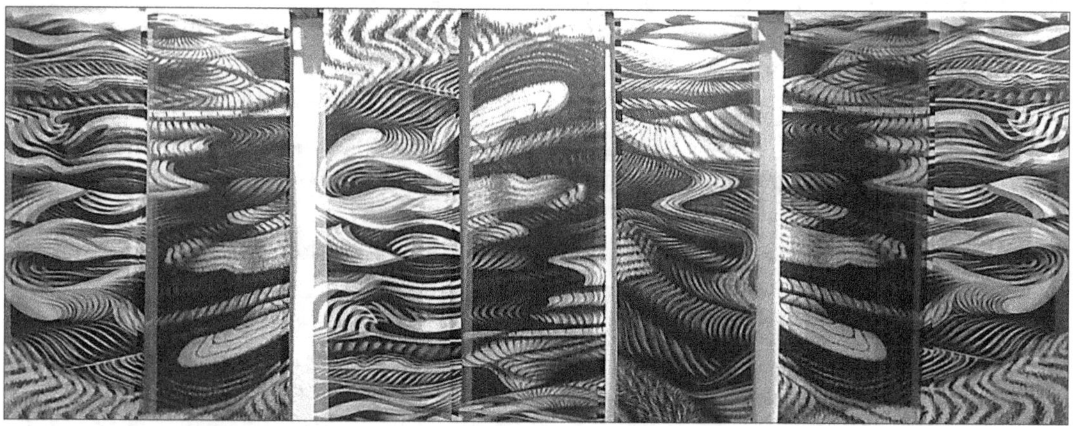

Katherine Nelson, *Ideas About Infinity* (detail, Grainfields from Steptoe Butte, 2018)
Charcoal and dye sublimate on opaque and sheer fabric, 3 x 9 feet
Courtesy of the Artist

Although based in Baltimore, landscapist Katherine Nelson has regularly traveled cross-country since 2001 to the Palouse's undulating grainlands. Her fluid charcoals and dye sublimates capture the summertime chiaroscuro of swirling slopes, saddles, and swales laden with grains and legumes. Nelson has also contributed to Oregon State University's Art About Agriculture program and to Glen Echo, Maryland's Yellow Barn Gallery exhibitions. She traces threads of her fascination with the region to her diplomat father's interest in Turkish rugs: "I remember their luxuriant textures and shapes which influenced my affection for rolling landscapes. The Palouse is a tapestry of woven connections among seasons, fields, and people. The effect is thoroughly spiritual and provides a place of reflection, solace, and beauty that overcomes the noise of the outside world." To emphasize the rhythmic effects of light for line and shadow, Nelson works entirely in black-and-white which evokes heightened awareness of layering, texture, and movement. "My 'Portraits of the Palouse,'" she explains, "are metaphors for the human prospect. 'Harvests' to me are exhibitions that depict the land as hallowed space through views of heritage farm architecture and landscape vistas. Implicit rural values relate to the natural environment, hard work, and community, and are relevant anywhere."[13]

Perilous Bounty vs. *Golden Wheatfields*

Twentieth-century British folklorist George Ewart Evans remained sanguine about contemporary smallholder and rural community prospects. He recognized the possibilities of new cooperative relationships by which growers could pool resources to buy machinery and share storage and marketing facilities, and characterized these arrangements as "a return on a higher level to the structure of the Middle Ages." The situation was not unprecedented in Evans's view, as he cited the introduction of the heavy Saxon *carruca* plow to Britain in early medieval times and the enclosure movement as changes that necessitated innovative cooperative practices. The "break" in appreciation of the old ways of labor, thrift, and economy, Evans wrote in the 1960s, "has chiefly been in the oral tradition: a farm-worker of the old school, a horseman for instance, had latterly no apprentice to take up his lore; and the young—the true bearers of the tradition—have in this respect been receiving a speedily diminishing heritage. It is not so much that they are not interested . . . ; they have now so few points of reference against which to measure it."[14]

Mutual dependence among neighbors and community members was more than virtue. It was necessity when harvest-time was essential endeavor and ritual for all able-bodied persons including field laborers, cooks, and craftsmen. The rise of mechanization that has reduced exhausting manual labor and technologies to facilitate communication and transportation will not abide nostalgic appeals to preserve the old ways. Evans characterizes such doomed efforts as "misguided romanticism" that is impossible in practical application and ignorant of the abiding dynamics of rural life through the ages. Aspects of social cohesiveness evident in harvest operations of former days have also diminished an isolated parochialism that limits wider multicultural understandings as well as individual opportunity in life. Moreover, a host of political and environmental conditions that threaten the wellbeing of farmers and rural communities cannot be understood apart from participation in global solutions.[15]

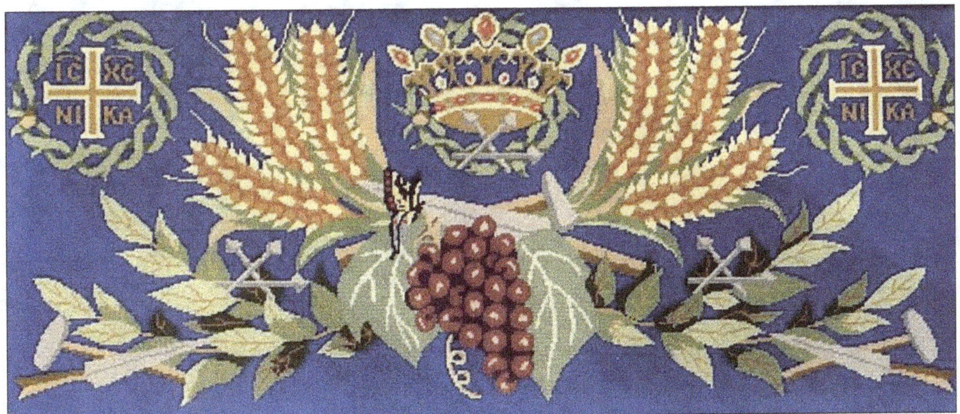

Needlepoint Grain and Grapes Altar Kneeler, National Cathedral (2019)
Columbia Heritage Collection Photograph

Public awareness of land stewardship takes on special significance in a day of unprecedented industrialization and centralization of the food supply given world population growth and pressure for land use. The number of farm residents declined during the twentieth century from 42% of the nation's population in 1900 to just 1% in 2000. (Global rural population decreased over the same period from 84% to 53%.) After peaking in 1935 at 6.8 million, the number of U.S. farms and ranches fell sharply until the early 1970s with about two million operating today. But just 5% of farms currently produce approximately two-thirds of the nation's food supply. Science writers now contribute to a new literary genre of environmental despair in the wake of global warming and food insecurity with such troubling titles as *The End of Plenty, Perilous Bounty, The New Age of Catastrophe,* and cultural critic Brian Watson's big-picture *Headed into the Abyss.* (The phenomenon started with publication of *The End of Nature* in 1989 by mild mannered Methodist Bill McKibben, who warned in Falter [2019] of significant disruption to world crop production and decrease in grain protein levels due to global warming.) Contemporary science fiction has likewise shifted in tone from the fantasy upheaval of alien invasions or asteroid impacts to speculative dystopian thrillers and television series with names like *The Last of Us.*[16]

American-Canadian writer William Gibson's books *The Peripheral* (2014) and *Agency* (2020) depict a menacing state of corporate control and online existence substantially disconnected from the natural world. Instead of a single make-believe threat, Gibson's characters face a convergence of intractable problems exacerbated by climate change, pandemics, and challenges to global solidarity on the issue. More disturbing if absurdly entertaining are novels by Joy Williams like *The Quick and the Dead* (2000) and *Harrow* (2020) in which characters vainly navigate through modern social upheav-

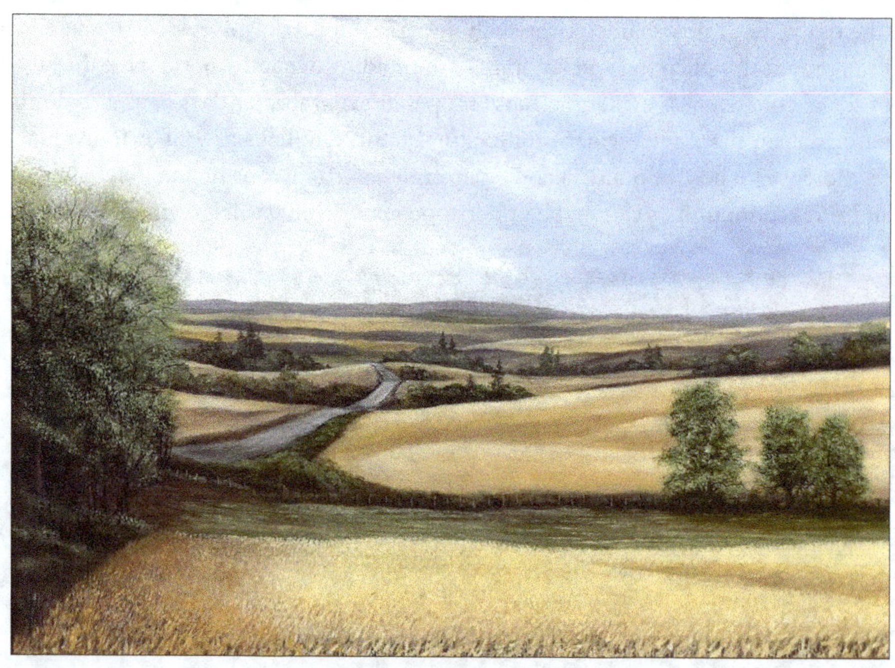

Vicki Broeckel, *Fields of Gold* (2022)
Oil on canvas, 18 x 24 inches
Collection of the Artist

al. Williams's latest title alludes to the ancient farm implement as cipher for humanity's relationship to nature, and recalls a passage from Job (39:9-10) about the foolishness of tethering a wild ox to a harrow. This varied literature disdains publicly invoked cultural pieties about responsible living. Such stories often invoke ancient myths bearing the common assumption that the wellbeing of humanity is inextricably linked to respect for the natural world's titanic potential.

Societal expectations for tomorrow are strikingly varied. As a boy I experienced our family's 1962 cross-state trip from the Palouse Hills to Seattle's optimistically titled "Century 21" World's Fair. Visitors were dazzled by exhibits on space travel and consumer abundance. A half-century later Milan, Italy, hosted the 2015 "Feeding the World" Fair with themes related to the problems of food security, sufficiency, and safety. A UN-sponsored session discussed the disturbing flatline of world grain yields since 2000, and how one billion developing world inhabitants were at risk of chronic malnourishment after decades of decline. Medieval era population peaked at approximately 300 million inhabitants but rose to a billion by about 1800, doubled to two billion in 1927, and reached three billion in 1960. Demographers at Milan predicted this exponential growth rate would result in ten billion by 2050 and bring attendant challenges for food resources, species diversity, and stewardship of soil.

Recent Palouse Country fine art exhibitions that have explored these themes have been sponsored by the Pullman Arts Council, Moscow Arts Commission, and Colfax Arts Council. They have featured works by George Bedirian, Jacqueline Daisley, Vicki Broeckel, and Anna Blomfield. Colfax-area native Broeckel is known for the luminous dimensionality of rural landscapes throughout the seasons. *Fields of Gold* (2022) shows a harvesttime view in the LaCrosse-Winona district that beautifully captures the delicate white sheen characteristic of fully ripe wheat. Daisley lives on a farm near Pullman where the surrounding countryside inspired her curvaceous masterpiece, *Palouse Hills— End of Harvest* (2018, PLATE 25). Anna Blomfield embarked on an artistic farrago in the fall of 2006 to explore and depict Palouse grain production throughout the seasons. The British newcomer had studied painting and printmaking at St. Martin's School of Art in London and in Liverpool before moving to California in 1987 to work as an animator and muralist. Her popular online "frogblog cartoondiary" commenced on November 7, 2006 when she observed, "If I wanted to buy a combine harvester there are 3 dealers in town to choose from." Blomfield, whose family had operated a farm implement business, first applied her special interest in the venerable forms of manual labor to paintings of foundry and tannery workers.

Relocation to the Northwest brought unexpected agrarian vistas and serendipitous friendships with area farmers, combine dealers, shop mechanics, and others engaged in agricultural pursuits. "I was entranced with the landscape, the people, and the various crops" she remembered; "each one had a story to tell and show." Over the course of her five-year residence in the region, these encounters yielded dozens of such colorfully illustrated posts as "Out Amongst the Wheatfields (8.28.07)," "Threshing Bee (9.3.08)," and "A Combine Chorus Line (4.25.2010)" with commentary provided by the artist's alter ego Froggie who traveled year-round in Studio Subaru. Characterizing her captivating genre as "visual storytelling," Blomfield combined illustration and information to transport and amuse readers through expeditions to harvest fields, grain elevators, inspection stations, rural railroad sidings. Public "gallery" showings of her popular work were appropriately held in re-purposed silos, barns, and the farm equipment dealerships she had first noted in her earliest online posts.[17]

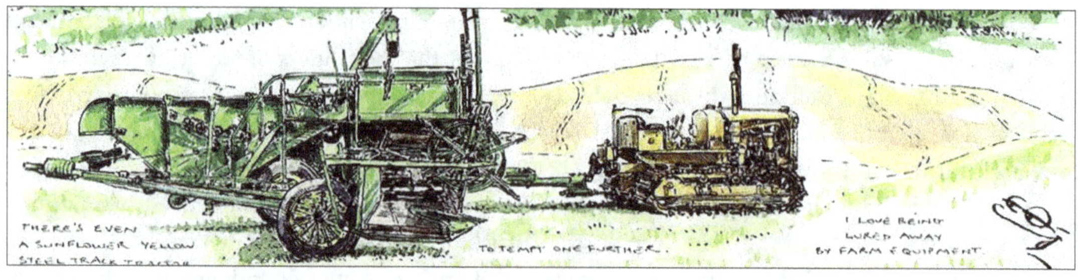

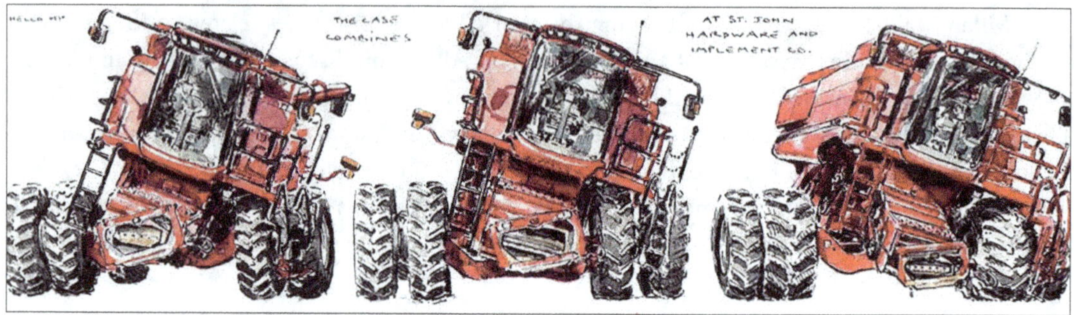

Anna Blomfield *Farming Temptations* and *Combine Chorus Line* (detail, c. 2010)
Digitized watercolor reproduction on paper
Courtesy of the Artist

Pullman, Washington, artist Henry Stinson has painted beautiful representational canvases of harvesting combines and other modern farm equipment in action, but is especially known for whimsical views that reflect his lifelong fascination with gadgets and electricity that attest to modern American society's ubiquitous connections to technology. An enormous untitled Stinson 2019 exterior wall mural looks as if *American Gothic* met a 1960s episode of *Lost in Space*. The painting reflects the artist's interests and concerns in a world where people and livestock increasingly share rural landscapes with drones and computer-monitored field equipment.

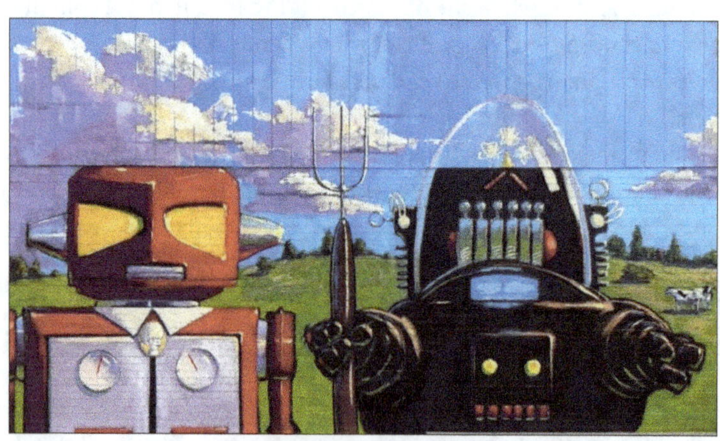

Henry Stinson, Untitled Wall Mural (2019)
Fonk's Store Mural; Colfax, Washington
Columbia Heritage Collection Photograph

Provocations have long been essential work for artists and authors who question popular assumptions about the benefits of innovation, deregulation, and globalization while trying to preserve what should not change. Shapes and names of the surreal artwork of contemporary Canadian painter Jo-Anne Elniski reflect rural environmental concerns. *The Last Harvest* depicts a fulminating sky in vivid swirls of yellow, purple, and white that rain down upon rows of grain that wave in the same garish colors. Other works by Elniski like *Field of Gold* and *Prairie Harvest* appear as flaming fields of abundance that rise to confront brightly lit horizons of pink, orange, and yellow. The depictions are awesome if sometimes unsettling.[18]

Changes on the Land

The Columbia Plateau had scarcely been explored by European-Americans when Constable completed his harvesttime masterpiece, *The Haywain* (1821). That year also marked the emergence of the Hudson's Bay Company's virtual monopolistic control of regional trading posts where Northwest agriculture first gained a foothold. As mentioned previously, HBC Governor Sir George Simpson journeyed to the Pacific Northwest in 1825 and brought the first seed wheat to promote self-reliance among the far-flung company engagés. Three decades later Washington achieved territorial status and its first governor, Isaac Stevens, undertook extensive surveys to promote development. In June 1855, Stevens led a party that included German-born interpreter-artist Gustavus Sohon (1825-1903) northward from the Walla Walla Valley to Coeur d'Alene country. Sohon sketched the first-known view of the rolling Palouse Country as seen from the eastern slopes of Moscow Mountain. "...[W]e have been astonished to-day at the luxuriance of the grass and the richness of the soil," Stevens noted in his report at the time. "The whole view presents to the eye a vast bed of flowers in all their varied beauty. The country is a rolling tableland, and soil like that of Illinois."[19]

Comparison of Sohon's scene (later colored by expedition painter John Mix Stanley [1814-1872]), to contemporary views shows a landscape without the industrial impact more proximate to Constable's Sussex. Five large tribes of the grass family (Poaceae) representing at least eighty-five native species blanketed the virgin Palouse in a rippling expanse of prodigious fecundity as seen in the Sohon-Stanley depiction. In addition, some forty varieties of grass-like sedges, rushes, and reeds grew primarily in wetland areas. The vicinity of Sohon's perspective hosts the region's only growth of leafy, white-flowered Piperia elegans, spiky-podded entanglings of the tall grass *Calamagrosis macouniana*, and multicolored beard-tongue and pink-petaled Claytonia. The pine-covered ridges of the Thatuna Hills near present Moscow, Idaho, shelter the delicate, blue-blossomed bellflower amid one of the region's few stands of the bromegrass *B. eximius*. In addition to the aesthetic worth of paintings and photographs, they can also provide important documentation of changes to landscapes. Among the most disturbing trends in the eastern Palouse has been the loss of native habitat and estimated fourteen tons of preventable annual soil erosion where conventional summerfallow operations were used during the century from settlement to substantial introduction of conservation practices in the 1980s. Use of minimum tillage and divided slope cropping along with sidehill and streamside grassland enhancement have been shown to significantly reduce the trend.[20]

A scenario of consequences resulting from environmental change is the subject of *Uncertain Harvest* (2021) by Vermont sociologist Charles Simpson (1941-2021). The novel relates American journalist Ed Dekker's hazardous global quest after uncovering evidence at a European biotech conference of environmental and food security risks wrought by transnational business interests. Dekker finds that the company "Naturetek" and others exert unprecedented control of the university and government research sectors and develop genetically engineered terminator seeds that produce ster-

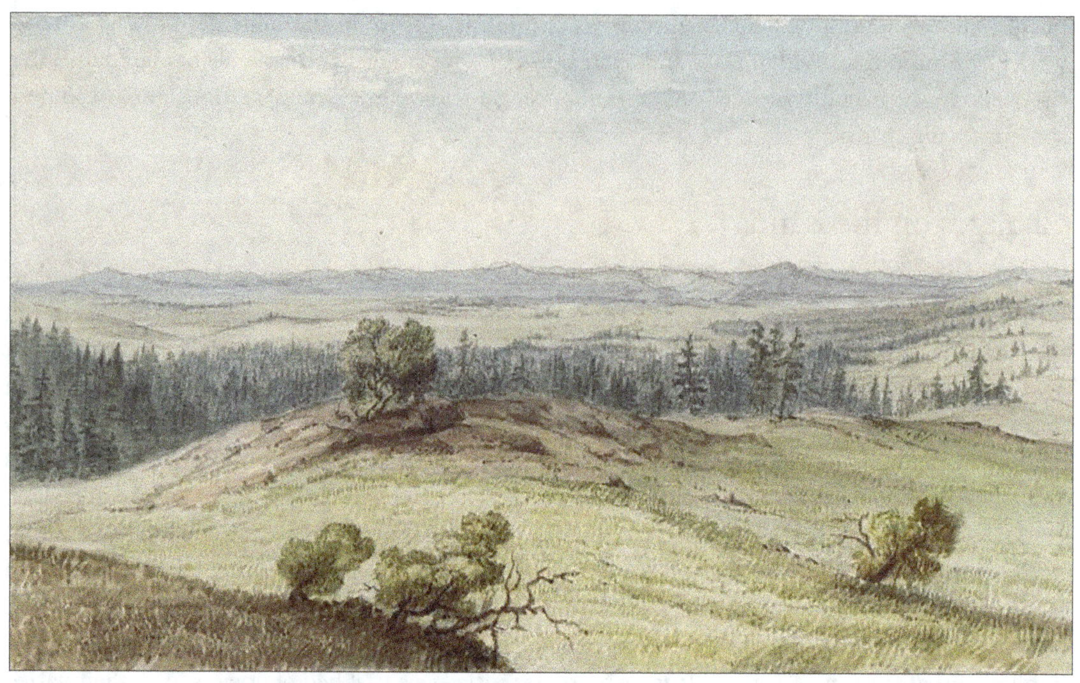

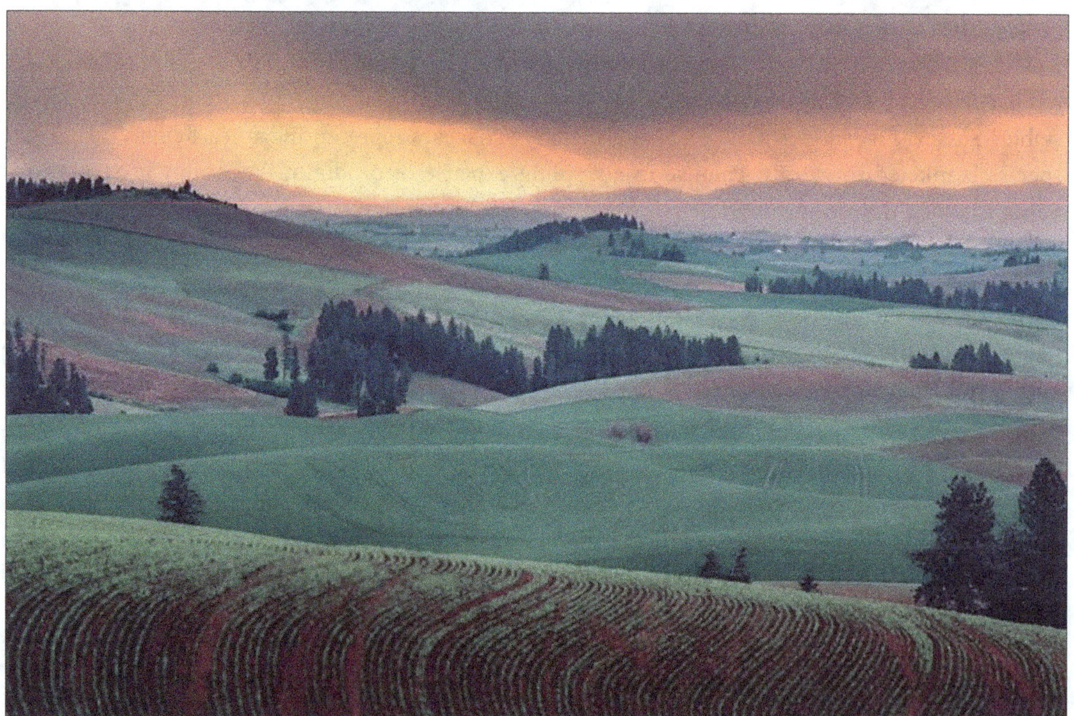

Above: John Mix Stanley (after Gustavus Sohon, June 1855), *Source of the Peluse* (c. 1860)
Watercolor over pencil, 7 x 10 ¼ inches; Paul Mellon Collection, Yale University Art Gallery
Below: John Clement, *Moscow Mountain Morning Squall* (2010)
Digital image on paper; John Clement Gallery, Richland, Washington

ile offspring and compel grower reliance on their patented germplasm. The engaging story draws on Simpson's own experiences working with farmers and indigenous peoples in Mexico and Guatemala and investigation of U.S. government subsidies, lax regulation, and global trade policies that are inimical to small-scale farming and in developing world settings.

I first learned about the scope of GMO commercialization in 2012 while traveling from Seattle to a global orphan care conference in Indonesia. While on the Tokyo to Bali connection, I noticed the logos of prominent U.S. seed suppliers on the shirts and backpacks of several other passengers. As we waited for our bags at the airport they told me about the Asia & Pacific Seed Association Conference taking place that same week at the city's posh Legian Seminyak Resort. During a break in the meetings I was attending in more humble surroundings, I ventured across town to the seed conference and met security befitting the visit of a head of state. After considerable effort and the intervention of delegates I had recently met, my admission was granted. A combination trade show with large group presentations, the forum provided an opportunity for transnational firms like Sygenta, Bayer, and Monsanto to showcase proprietary seeds and technologies that promised "transformation of Asian agriculture."

One of the influential contemporary voices on small-scale farming and rural culture is farmer-author-painter Lynn R. Miller of Sisters, Oregon, who has written over twenty books of nonfiction, poetry, and fiction. In publication since 1976, Miller's quarterly *Small Farmer's Journal* reaches 20,000 readers and features articles on a wide range of topics for farmers and ranchers devoted to the "craft" of farming and husbandry and for those interested in the life and literature of the countryside. The *Journal* has featured numerous harvest-related articles including Esther M. Jensen's c. 1940 Willamette Valley memoir, "The Day the Threshers Came," equipment reprints like H. R. Trolley's "The Efficient Operation of Thrashing Machines" from a 1918 USDA *Farmer's Bulletin,* and detailed accounts by twenty-first century operators of horse-drawn reaper-binders and stationary threshers like Khoke and Ida Livingston of Davis City, Iowa ("The Harvest of Grain"). Miller's photo essay "Anatomy of a Threshing" (2021) is a prelude to forthcoming books, *Threshing Machines* and *Grain Binders and Reapers,* which will be the culmination of his years-long mission to glean practical information from out-of-print books, manuals, parts lists, and catalogs.

Miller describes the *Journal* as more of a "community odyssey" than periodical. In a recent editorial titled "Seedbeds & The Rooted Mend," he recalls visiting a local high school where he was asked to speak about farming. "I told them, no I tried to show them, that the small *s* big *F*—small Farming—the one where the farmer is the measure—contains every single element of life, most every attainable pedestal of adventure, and the best chance at everlasting health." Perhaps spurred by the immediacy youthful hearers, Miller then mused about the significance of their high desert homeland: "To take root here is to find and see the elements, wind, sun, trees, wildlife and our own meddlesome footprints as belonging. You do that in part by keeping sight of what is near at hand, by being 'short-sighted.' If you are looking off to imagined holiday vacations, to deep seedy city environs, to video fantasies, to cosmetics in pursuit, to worrisome accounting, you will NOT see yourself rooted, you will NOT see the other elements near you as rooted." In closing Miller expressed hope that his enduring effort at "community journalism" had provided "a soft drumming of information, shared adventure, and kinship" to support, celebrate, and gather the vital assembly of "human-scale" farmers.[21]

Rural Landscapes Then and Now

Stories and paintings that relate a range of interpretations regarding contemporary and future existence add voice and visibility to diverse perspectives on land use. Consolidation of family farms in recent decades into larger corporate enterprises and the commodification of grain—William Cronon's "transmutation of one of humanity's oldest foods," warrant high regard for stewardship of the land. Reinvigoration of Americans' deep-seeded social memory and cultural capacity can guide landowners and public officials who contend with environmental challenges and finite production acreage. When Conrad Blumenschein left Russia for America just before the outbreak of World War I (as recounted here in "Grain Scythes and Altar Cloths"), ten families lived on a dozen farms of about 320 acres each scattered along the road between my hometown of Endicott and the Palouse River some seven miles to the north. (The other two landowners lived in town.) Numbering some fifty people, most attended one of two Lutheran churches in the area—the Missouri Synod in the country, and the Ohio in town, and two country schools enrolled the area's children through the eighth grade. Many of these families were related to each other, and regularly gathered for summer harvest labors, fall butchering bees, and various ceremonies and celebrations.

A half-century later in the 1960s when I began interviewing first generation immigrant elders like Conrad Blumenschein and Mary Morasch, the number of farms had fallen to nine with some consolidation of property holdings among the seven families of thirty-two individuals who remained. The size of area farms had increased to an average of 550 acres, and both country schools had consolidated with the larger town district that offered instruction through grade twelve. Our father was able to complete the month-long harvest on about that much acreage by keeping in good repair the old D6 caterpillar tractor and International pull-combine that had teamed up for at least a quarter-century to make the annual run. I remember well not only his pleasure at seeing a bumper crop, but his momentary agony prior to harvest when the county USDA officials armed with surveying equipment measured out bountiful hillsides for destruction. To comply with federal farm "allotment" limits for stabilizing grain prices, we periodically had to hitch the D6 to a heavy disc for the sad task of tearing out some of the ripening grain.

The price of a bushel of wheat rarely rose to $2 from 1960 to 1973, when a controversial U. S. trade deal to supply the Soviet Union with grain boosted prices to as much as $6.25. The long-sought optimism felt by growers ushered in a season of equipment upgrades and land purchases encouraged by Agriculture Secretary Earl Butz's 1973 "get big or get out" slogan. But favorable Russian harvests the following year coupled with reduced federal subsidies contributed to America's 1980s "farm crisis" of rural economic stagnation. The 1973 Farm Bill championed by Butz represented a sea change in federal agricultural policy as New Deal legacy price supports were replaced with production incentives that boosted commodity supplies. The decade witnessed a 40% increase in wheat acreage as the 49 million sown in 1970 swelled to 81 million acres in 1980. Grain prices plummeted and food processors scrambled to find new purposes for surplus wheat, corn, and other crops including high-fructose syrup, biofuels, and massive grain-fed cattle feedlots and hog and poultry factories.

As rural life historian Steven Conn points out in *The Lies of the Land: Seeing Rural America for What It Is—And Isn't* (2023), crop production has always represented an efficiency dilemma for farmers since increased supply depresses prices. Agriculture as viable business enterprise for rural renewal is a function of a balanced "middle way" involving government protections coupled with sound agronomic practices and reasonable market fluctuations. But unlike most businesses that operate within a range of managed variables, farmland is an economic constant with continual labor, tax, and other obligations. Replacement of crop production and soil conservation policies that had been heir to 1930s New Deal legislation with the Nixon-Butz Farm Bill of 1973 created new challenges to life in the countryside.[22]

Many smaller farms failed or were sold under the pressure of rising input and equipment costs, which led to an unprecedented period of land consolidations. A spate of mergers also took place among farm input suppliers and food processors, and by the end of the century a half-dozen retailers controlled a majority of the American grocery market. In 1985 Congress enacted the Farm Security Act to curb commodity surpluses and reduce soil erosion. The legislation authorized the USDA to establish the Conservation Reserve Program (CRP) that qualified up to 25% of eligible county farmlands for direct payments to farmers based on historic land use and highly erodible criteria. With a goal of idling up to 45 million acres, CRP land by 2023 totaled about half that amount.

Earl Butz's "get big" remarks in 1973 were coincident with publication of British-German environmental economist E. F. Schumacher's *Small is Beautiful: A Study of Economics as If People Mattered* (1973). The contrast in perspectives regarding the wellbeing of farmers could not have been starker. Schumacher (1911-1977) considered federal promotion of larger production acreages and related needs for more expensive equipment to be problematic. Inspired by thinkers like Tolstoy and Gandhi, Schumacher understood the forces of modernity to be complex but fueled by preoccupation with short-term solutions that ignore ancient ways of sustainable living and soil care. He invoked a sovereignty of reason to advocate for enterprising farmers who value the longstanding natural and social commons that provided economic benefits well worth protecting. Without such public policies more younger farm families would be at risk of being driven from the countryside.[23]

In the fifty years that have passed since my 1960s high school FFA speech on world hunger, global population has increased by four billion and the world has consumed some 1.3 trillion barrels of oil. (Approximately one and a half trillion metric tons of carbon dioxide have been released into the atmosphere since the dawn of the Industrial Revolution.) The result has been a warming of the earth's surface by 0.6° C, disappearance of a million square miles of spring snowpack in the northern hemisphere, and 30% rise in acidification of the oceans. The pace of change has led to devastating impacts on biodiversity and prospect of neo-Malthusian food security crises. Regional trends can only be discerned over time but already some disturbing patterns are evident. Since 2020 annual precipitation has declined across the Columbia Plateau and in the summer of that year the only firestorm in living memory made national headlines. It devastated the Palouse hamlets of Malden of Pine City and destroyed vast areas of standing grain.[24]

My 1960s boyhood rural neighborhood of several thousand acres that had been home to about fifty souls in 1900 along the seven-mile stretch from Endicott to the Palouse River is comprised today of just eight farms. All but one are portions of larger family-owned operations whose members also own or lease other cropland in the area. (The average Palouse Country farm size in 2020 was approximately 1200 acres.) Only four households are located on the same route that supported those ten families a century ago which is populated today by just five adults whose grown children live elsewhere. In between these habitations one can now see lonely clusters of skeletal locust trees, broken fences, and

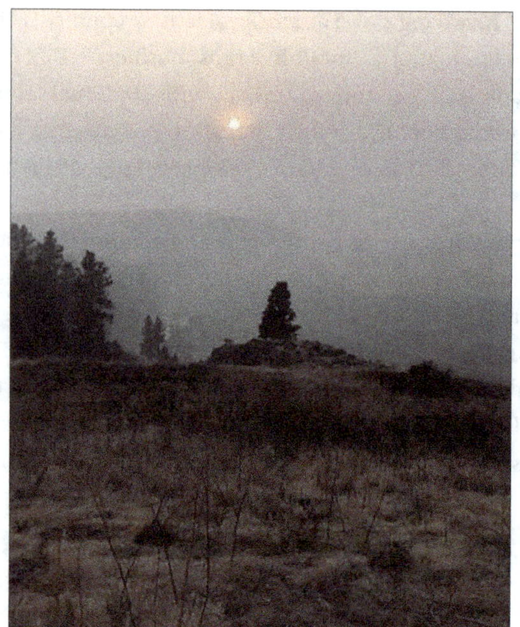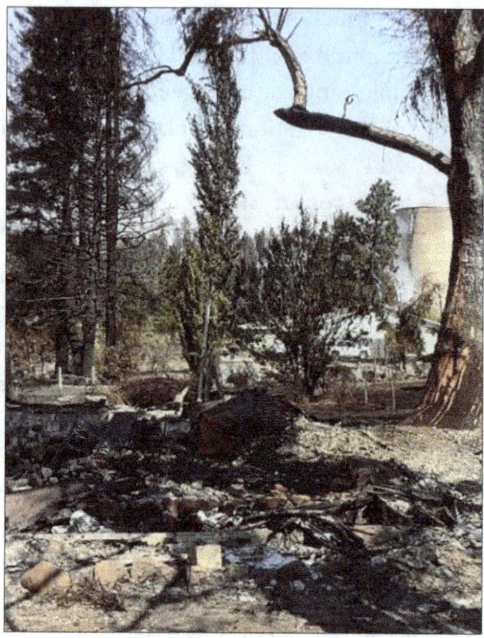

Palouse Harvesttime Smoky Sunset and Pine City Grain Elevator Firestorm Aftermath (September 2020)
Columbia Heritage Collection Photographs

rusted equipment guarding abandoned buildings. The percentages of young people—mostly males, who remained in the Endicott and St. John area to farm or returned to do so after post-secondary study have remained fairly constant—11% in my father's high school generation (1939-1942) compared to 10% in that of our son's (2000-2004). But the total number of graduates during those years fell by 57%. The numbers are consistent with figures from the wider region where areas farmed by about the same number of individuals over three generations have more than doubled in size. The trend has brought challenges to rural communities that contend with closures of local stores, banks, and public services.[25]

The broader demographic impacts on rural life and labor reflect trends over the past two centuries that have changed the nature and necessity of worker communities. In 1840s pre-industrial America, for example, a farmer could produce an acre of hand-broadcasted wheat yielding about twenty bushels from approximately fifty hours of annual work using simple implements like a single-shear plow and scythe. (Soil exhaustion and other factors in early nineteenth-century France and Germany contributed to average yields of less than half that amount, or about ten bushels per acre. Yields on unmanured fields in England were in the range of fifteen. Continental Europeans commonly faced substantial crop failure and famine at least once every ten years.) A single day's harvest by an able-bodied scythe-wielding reaper could cover up to one acre. By 1900 an American farmer equipped with horse-pulled gang plow, harrow, and mechanical drill still produced about twenty bushels but in about ten hours of annual per-acre labor. An experienced crew operating a reaper-binder and steam-powered thresher at that time could cut about forty-five acres a day for some 1,200 bushels (31 tons) of grain. A farmer in 1940 using a gas-powered tractor, three-bottom plow, and combine with 12-foot header further reduced annual per acre labor to 3.5 hours.

Dryland grain yields increased three-fold nationally during the twentieth century and Palouse

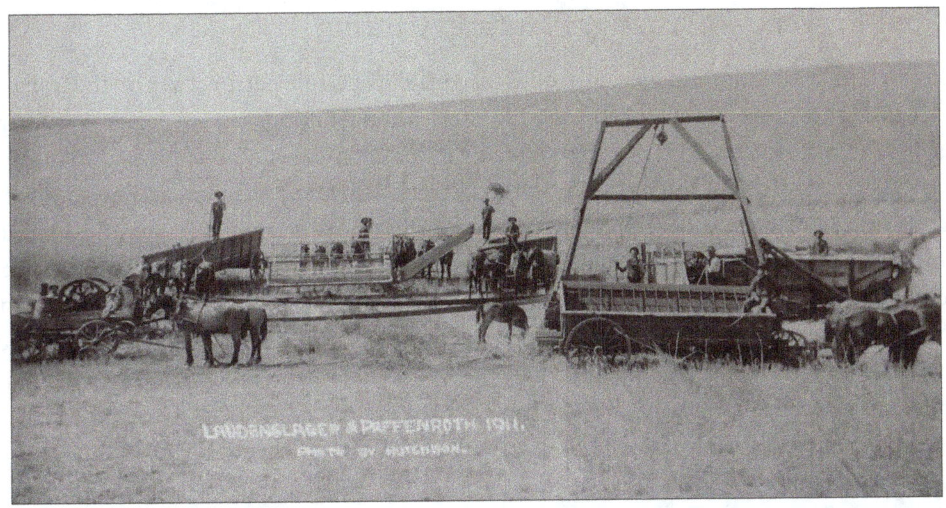

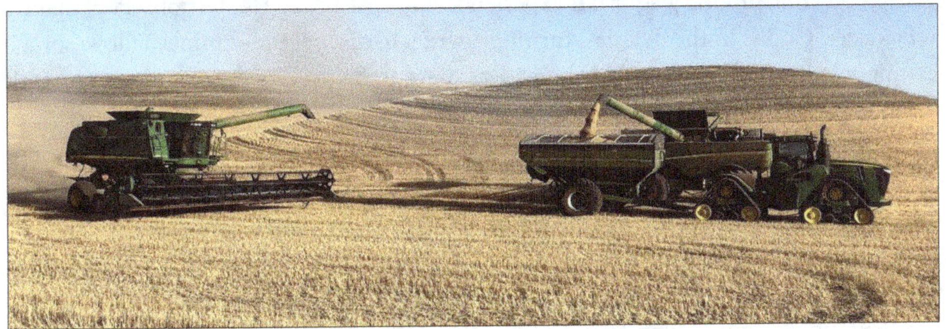

Harvests Yesterday and Today—Different Times, Same Location
Lautenschlager & Poffenroth (1911) and Klaveano Brothers (2019) Threshing Outfits
Four miles north of Endicott, Washington
Columbia Heritage Collection Photographs

Country yields of eighty bushels per acre are common today along with diesel-powered, satellite-guided equipment that makes crop rows of linear perfection. High-capacity combines now cost as much as a million dollars and feature sidehill leveling, cruise control, and electronic monitoring of threshing functions that automatically adjust to crop load. Modern farmers invest about fifty minutes in total annual per-acre labor and using a combine header forty feet wide can harvest 250 acres in a ten-hour day to yield up to 30,000 bushels (900 tons) of wheat, or 72,000 bushels of corn. Sickle-section cutter-bars endure. Modern versions feature four-inch-wide chromed, serrated triangular sections arranged toothlike in a row that run the length of the header and move back and forth at lightning speed. Such mechanical marvels represent the output of a thousand reapers and twice as many binders who labored in harvest fields before the Industrial Revolution. (Substantial numbers of others were tasked with carting unthreshed stalks to barns, flailing grain, tending livestock, and other related tasks.) A phalanx of these modern behemoths cruising through a field of golden grain evokes appreciation for techno-mechanical ingenuity, and still stirs ancient feelings of gratitude for agrarian bounty. Dayton watercolorist Paul Strohbehn's dramatic *Sorghum Hollow Gold* (2018), for example, dramatically captures three immense John Deere Titan combines rising from a Palouse hillside chaff cloud.[26]

"Ancient Prayers"—Discing, Planting, Cutting

Contemporary agrarian art and literature offer insightful if variable perspectives on the transformation of rural landscapes and ways of country life. Stories and paintings can be elegiac and abstract as well as hopeful, with expressions of agrarian renewal evident in newfound appreciation for regional heritage and stewardship of the land. These considerations juxtapose values related to the natural world with those of private development. There is little to regret about archaic rural prejudices, grinding aspects of exhaustive dawn to dusk farm labor, and highly erosive tillage practices that once characterized areas like the Palouse. Small town redevelopment efforts are examining in new ways how local stories, specialty crops, and other resources might be shared to better contend with shifting labor patterns and demographic change. Harvest bees and church benefits still aid neighbors in times of special need. The annual harvest experience still requires seasonal urgency, and the instinct for necessary provision unites humanity worldwide through rituals of planting and harvesting and thanksgiving.

Reflecting on contemporary agrarian experience, novelist Bruce Holbert observes that in places like the Palouse, "God is in the details—turning a wrench, discing the summer fallow, spraying and rod-weeding, planting and cutting." He considers these to be prayers "of the ancient sort, the ones you offer not for an answer as much as to be heard. Their reward is the opportunity to perform similar acts tomorrow and the next day. Their faith is not invested in an end; it is the opposite, a prayer to continue and in it is a kind of patience with the fates that few outside this place share." The inhabitants of Holbert's stories are not portrayed to explore the classic American theme of personal freedom amidst the conformist mainstream. Instead, they seem to take for granted a life of mystery and misery amidst economic hardship and the vagaries of nature and speak perceptively from deep within as they move about in clouds of uncertainty. Holbert explores the abiding toil and periodic terrors of country life in *Hour of Lead* (2014), winner of the Washington State Book Award for Fiction, and in other novels and writings. His short story "Ordinary Days," in which the Mason Hills are cast for the Palouse, features an exploration of rural change and meaning-making as a farmer takes his fourteen-year-old daughter on a drive through the countryside:

> *The wind through the wheat stalks alternated like the back and forth of the sea, green or brown or golden or tawny swells depending on which direction you looked and which time of year. My father had always claimed its authority, like God's, was omnipresent. Omnipresent not omnipotent. The difference mattered to him. We resided in Nighthawk. Grain siloes paralleled the railroad lines like castle keeps. An enormous American flag flapped upon the tallest. The town is only a few cross hatched roads lined with well-kept but unassuming houses and lawns, a school tucked against a hill.*
>
> *The Mason Hills were once peppered with such communities but farming requires fewer people and more equipment and land so kids who twenty years ago inherited ranches now attend college and opt for city jobs. Shrinking towns bled children to those a little less tenuous through yellow school buses until they were only wide spots along the highway and even freight trains no longer blasted their whistles when they passed.*
>
> *In Nighthawk, the grocery closed at 5:00 but the owner distributed keys to the locals. After hours customers hung post-its on the cash register to update their bill. Each year, the Red*

Cross circulated a large donation envelope from house to house that held hundreds of dollars in cash. Neighbors and acquaintances claim to know you but so do those with whom you just exchange a nod in town. When a person gets sick it plucks a string in a spider's web and all the other strings vibrate.[27]

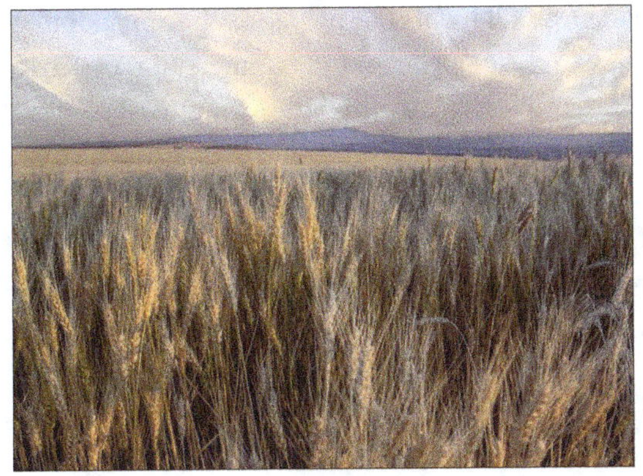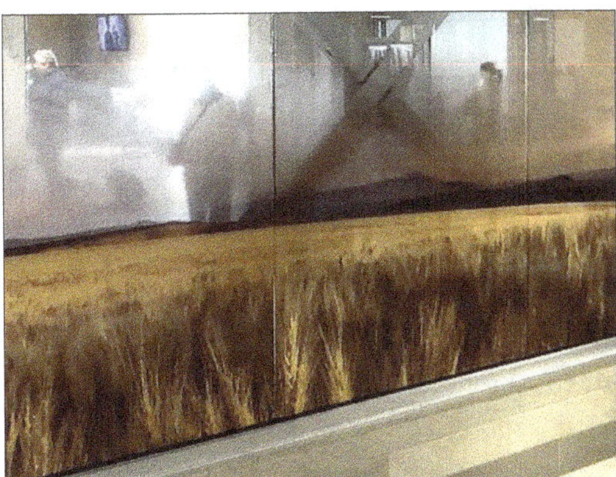

John Clement, *July Sunrise—Horse Heaven Hills* (2011)
Digital image and dye sublimate on plate glass, 4 x 8 feet panels
John Clement Gallery and Pasco Airport; Pasco, Washington

Nostalgia for some halcyon past contributes to the popularity of rural art but tempered with consideration of what has been lost and what has been gained. These contrasting themes are considerably explored in contemporary photography and are the special interest of Richland, Washington, native John Clement (see Preface) and Southwesterner Don Kirby. Ambivalent considerations about such trends are expressed in "Palouse" by Lewiston, Idaho, poet William Johnson:

There is always an empty house
by the road at the edge of town,
its windows whiskered with lilac
and letting in rain. Nearby,

a barn drags itself home,
and in May, daffodils trim the yard
against an ocean of wheat
that rolls in on a slow inexorable tide.[28]

The stark, mysterious black-and-white photography of Kirby's *Wheatcountry* (2001) shows unpeopled agrarian vistas from Texas to Washington. Essayist Richard Manning writes of the contrast between the imaginative West of the national consciousness—reshaped since settlement and largely uninhabited, and landscapes tended by farmers who contend with the vagaries of weather and maneuver through an array of government programs to provision the masses. Kirby's monochromatic views bear the titles of nearby locations scarcely known or seen by outsiders, but that conjure memory and meaning to locals. His Palouse series includes Diamond and Lancaster (places where farmers long came from miles around to procure seed wheat), Harrington and Pomeroy (once home to area flour mills), and the Snake River port of Central Ferry, which remains one of the Northwest's largest grain exporting terminals.

Having grown up in the vicinity, I immediately recognized Kirby's "Wheatfield III, Repp Road, Endicott, Washington," which shows a prodigious stand of ripening wheat cloaking an enormous swirl of sloping summer-fallow beneath a stack of cumulus clouds. The Repp family had the only pool for miles around before the town built one in the 1960s, so kindly taught a generation of us how to swim. The matriarch of the clan recently passed away at age 104 and her nephew Mike Lowry—our state's twentieth governor who contributed to normalized US-Russia relations in the 1990s, may well have helped harvest that very hill. Trends in the depopulation of the countryside are found throughout the nation, even as affordability of houses in small towns has helped keep some populated with newcomers to sustain local schools, churches, and clubs. Shrinking numbers of farmers remain as vital carriers of intimate knowledge about the land and growing conditions, and of practical skills that keep bringing forth the crops.

Themes of change upon the landscape mixed with agrarian wonder characterized many poems by Pulitzer Prize winning author Howard Nemerov (1920-1991). Although the New York native spent much of his life in academia, he traveled widely through the New England countryside and with publication of his 1955 collection *The Salt Garden*, Nemerov's refined, contemplative verse took on more practical tones in defense of the land. Poems like "Midsummer's Day" and "The Winter Lightning" reflect upon the timelessness of the seasons and consider a consilience with humanity's ephemeral presence. In "A Harvest Home" an abandoned vehicle stands in a recently harvested field:

> *So hot and mute the human will*
> *As though the angry wheel stood still*
> *That hub and spoke and iron rim. . . .*

Marvelous creatures of the wing appear throughout the day—jays "proclaim" dawn, afternoon crows "arise and shake their heavy wings," and an owl "complains in darkness."[29]

Sacred Harvests

Cultural tensions rise with proliferating perils of climate change, concern about impacts on soil biomes and wildlife, and global food security needs. Establishing balance involves mediation of the ancient urges for veneration and exploitation, and consideration of technocratic limits and trade-offs in agricultural improvement. Agronomists estimate that no-till farming has reduced erosion on American farms by 40% since widespread use in the 1980s but this has also created conditions for decreased soil microbiome diversity. Yet less tillage in such systems reduces American diesel consumption by two-thirds for some 280 million gallons of annual savings. For Promethian bio-entrepreneurs CRISPR gene modification and related technologies offer prospect of crop improvement although prominent biologists acknowledge that cellular arrangements can be altered in ways that are poorly understood.[30]

Re-engineering life forms through transgenesis raises questions about the complicated relationships between the natural and unnatural, and theological distinctions between obeying God and playing God, or between earthly tenancy and proprietorship. Physicist Richard Feynman urged caution by likening the complexities of modern science to attending a cosmic chess game where we are observers with limited capacity to predict outcomes regardless of intentions. Wendell Berry explores these tensions and offers recommendations for how anyone can apply the useful habits of memory, attention, and reverence. He recognizes the primacy of earth care to support all other human endeavors, and likens this to holy service: "The most exemplary nature is that of the topsoil. It is very Christ-like in its passivity and beneficence, and in the penetrating energy that issues out of its peaceableness. It increases by experience, by the passage of seasons over it, growth rising out of it and returning to it. . . . It keeps the past, not as history or memory, but as richness, new possibility." In this sense the urgent partisanship in political and economic affairs becomes secondary to stewardship of land and life, and the function of healthy community for enduring fulfillment.[31]

Whether on a Columbia Plateau farm or in a Seattle high-rise, one may live with fidelity to place by learning and practicing domestic arts and community building. Family care and homemaking need not require moving "back to the land," though new paradigms for remote working are facilitating rural residency. In any setting folks can summon moral courage to eat together, shop locally to support practitioners of local crafts, connect young people to worthwhile endeavors, and affirm the values of environmental care. Policies and practice of self-reliance and promotion of the common good that characterized republicanism in the ancient world are relevant more than ever in an era of threatened landscapes, endangered species, and marginalized labor. Ethicist Julie Crouse characterizes Berry's ongoing literary conversation about these matters as a "sacred harvest" for renewing the common good by promotion of connections among people, landscapes, and spirituality. In the context of farm work for participation in the market economy, or mental work for healthy imagination and discernment, Berry calls the external standard of such endeavors the "Great Economy," or the "Kingdom of God." The goal is to promote abiding, abundant harvests and the longterm wellbeing

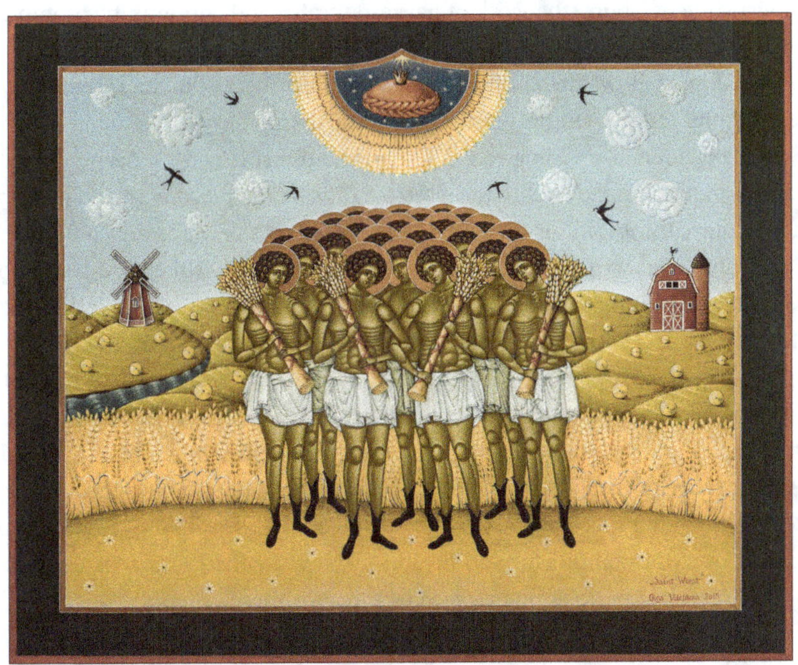

Olga Volchkova, *Saint Wheat* (2015)
Acrylic and gesso on wood, 20 x 24 inches
Courtesy of the Artist

of individuals in community. In spite of dreams of space colonization, exploration of extraterrestrial neighbors near and far has shown that, in the words of science-fiction writer Kim Stanley Robinson, "There is no Planet B."[32]

Berry has explored ideas of earth care most fully in his novels and short stories about the Coulter, Catlett, Branch, and other families. They dwell together in imaginary Port William, for which the author's native Port Royal, Kentucky, serves as a touchstone. The place is not a utopian metaphor as its residents navigate with failure and success through cross-currents of old ways and modernity's encroachments and benefits. In "That Distant Land" (1965), a rural neighborhood crew gathers for tobacco harvest in the sweltering August heat, and Berry casts a scene that resembles field laborers of ancient times: "They dove into the work, maintaining the same pressing rhythm from one end of the row to the other, and yet they worked well, as smoothly and precisely as dancers. To see them moving side by side against the standing crop, leaving it fallen, the field changed, behind them, was maybe like watching Homeric soldiers going into battle. It was momentous and beautiful and touchingly mortal." The classical allusion is not incidental. In "The Agrarian Standard" (2002), Berry invokes lines from Virgil's Fourth Georgic about "an old Cilician" who cares for a small plot of land that produces abundantly because of an ethic "rooted in mystery and sanctity" that values giving back to the health of the soil—an affectionate agrarian stewardship for the sake of present and unborn generations.[33]

Meaning-making in classical thought came through an honorable paideia of civic engagement and reflection. Intellect detached from action risks loss, and empathy apart from action is purposeless. Apparent in literature and art from Virgil and Horace to British Georgics, French Rustics, and Russian Itinerants, a holistic life of labor—the work of harvests, crafts, and community, promotes personal as well as cooperative wellbeing. A related education grounded in distinctly local connectedness through stories and art, mealtime fellowship and field study, offers prospect of cultural renewal.

Mentalities of limits and moderation foster individual identity as well as social cohesion to sustain both culture and soil. French sociologist Émile Durkheim criticized modern society's "malady of infinite aspiration" that trades cultural renewal for a fixation on replacement and perpetual commercial agitation to acquire things. In his foreword to *Of the Land and the Spirit* (2008), an anthology of Walter, Lord Northbourne's writings on ecology and religion, Berry (like Lord Northbourne) points to perennialist matters of ultimate purpose as expressed by Jesus to, "[S]eek first the kingdom of God, and his righteousness . . . ," in order to be provided food, drink, and clothing. Berry suggests that mastery of agriculture, husbandry, weaving, and other skills are assumed in this context, and that Jesus demonstrated faith by prayerful service to others in feeding and healing them. In these ways, literary, artistic, social, and vocational endeavors in any age represent harvests in which anyone can participate. A wide range of sustainable farming initiatives have been advocated and implemented in recent decades with varying success. Those with prospect for enduring progressive change incorporate core elements of crop rotations for soil health; improved conservation through reduced tillage and terracing of slopes; and policies that reward farmers for these practices.[34]

Amidst the mistral winds of June 1888, Vincent van Gogh painted his Harvest (Wheat Fields) series of ten paintings which provided opportunity to experiment with color and technique. He worked quickly, "just like the harvester, . . . intent only on the reaping," and blended striking hues of gold, copper, and bronze with yellow, red, and brown. The lush expressionistic masterpiece *Wheat*

Field with a Reaper he painted in July 1889 gleams with a swirling sea of grain in impastoed layers of yellow-orange with white highlights seen in many of his Saint-Rémy paintings. A benevolent sun stands against a sky of aquamarine and seems to shine from the canvas uninterrupted by shadow or shade. The view is from the upper story of the building where van Gogh had sought recovery and shows the field's gray-white boundary wall without any sense of confinement. The artist wrote of its "vague figure toiling away . . . in broad daylight with a sun flooding everything with a light of pure gold," as a modernist expression of "sacred realism" with calm, religious hope in the face of death and his own demise.[35]

Association of grain with life suggested humanity's vulnerability and resiliency in passionate reds and golds composed in balanced synthesis with the greenery and browns of verdant earth. These colors van Gogh complemented by mysterious orange and blue tones of the cosmos—a palette strikingly similar to his *The Resurrection of Lazarus* (1890). Despite grain's susceptibility to ruin from fire and wind, or harvest with scythe and sickle, the fields serenely endured as tangible evidence of incomparable beauty, fecundity, and energy of the divine order. The death reaper fulfilled the sower's purpose in the grand mysterious cycle of life. Tolstoy's exposition on aesthetics in *What is Art?* (1897) embraces a wide range of creative expression from painting and sculpture to literature, folklore, and liturgy. He characterizes their highest forms as conveyance of the makers' regard for human dignity and the natural world in soul-inspired ways that astonish, mystify, and benefit the common good. That Tolstoy points to artists like Millet and Lhermitte, and authors Gogol and Chekhov as aesthetic exemplars is significant for their use of agrarian themes to present universal values. Great painting and writing express struggle and beauty through memorable associations with places, people, and ideas.[36]

Such understanding is at risk in contemporary society at once smart and ignorant. Information technologies can be exploited to distraction for endless browsing and displace contemplation of relationships to the earth and to one another. Unprecedented threats to wellbeing loom if humanity's commons of seed for harvest becomes unalterably modified and proprietary. Enduring individual and societal flourishing in any age, as Ruskin observed when considering a sheaf of grain, requires the discipline of work and rectitude. Beneficial middle ways would temper entrepreneurship and technology through hard decisions protecting the common good.[37]

"To Do His Work as Well as Possible"

A common farewell expression I heard as a boy among our first-generation Volga German immigrants was "*ad´-yey*," but the word is not from their relict Hessian dialect nor apparently known among other Germans from southeastern Europe. I was told by a community elder that the term may have come to our ancestral villagers from the half-dozen *Grande Armée* prisoners of war who were sent there following Napoleon's catastrophic 1812 Russian campaign. We know that the forlorn survivors included some from French-speaking Lorraine who may well have contributed loan words like *adieu* that over time became the *ad´-yey* spoken by locals and their descendants in the Pacific Northwest. Tolstoy's *War and Peace* mixes fictional narrative with events of the Napoleonic wars and a scene near Smolensk relates the psychology of response to problems on the horizon. Peasants and nobles of Prince Andrei Bolkonsky's estate ponder their fates in the wake of invasion.

While Bolkonsky serves in the Russian army, his sister, Maria, seeks help from the family's serfs for their escape eastward. But her offer of grain from their reserve with prospect of safety in Moscow falls on deaf ears.

Both sides pillage crops en route to the capital that will soon fall to flame, and Tolstoy pauses to consider the general public's reaction: "The approach of the enemy did not make the citizens at all more serious; on the contrary, their levity seemed to increase as their position became more critical, as often happens. . . . In fact, at such a juncture, two voices are to be heard: one wisely preaches the necessity for estimating fairly the impending danger, and the means at hand for resisting it. The other . . . argues that the thought is too painful, since it is not given to man to escape the inevitable, and that it is better, therefore, to forget the danger and live merrily till it comes. Solitary men listen to the first, while the masses obey the second. . . ."[38] If contemporary mass media accurately reflects public sentiment, considerable levity and febrile complaint presently reigns. Leading politicians are preoccupied with matters of the moment while prominent religious leaders are obsessed with end times theology. Placating audiences threatens to supplant leadership in the face of climatic and terrestrial perils to peace and a renewable staff of life.

> *Wisdom cries aloud in the street,*   *at the entrance of the city gates*
> *in the market she raises her voice;*   *she speaks. . . .*
> *at the head of the noisy streets she cries out;*   —Proverbs 1:20-21

The characters who inhabit Tolstoy's stories have heroic capacity, even when cast as loners and losers. They walk familiar paths to work together in the fields and better understand the people and world around them. Among the most stirring moments in Tolstoy's *Anna Karenina* is when landlord Konstantin Levin, who sought the full life of mind, love, and labor, joins a scythe-wielding brigade in an afternoon of great satisfaction: "He . . . wished for nothing, but not to be left behind the peasants, and to do his work as well as possible."[39] In a day of conflict and indifference over progressive approaches to a more sustainable future, ancestral regard with considerations of Ruth and Boaz, muzhik harvesters, and Port William farmers offer restorative interpretations of land use and the human prospect.

> *Praise God for the harvest of farm and field,*
> *Praise God for the people who gather their yield,*
> *The long hours of labor, the skills of the team,*
> *The patience of science, the power of machine.*
> *Praise God for the harvest of conflict and love,*
> *For leaders and people to struggle to serve,*
> *To conquer oppression, earth's plenty increase,*
> *And gather God's harvest of justice and peace.*
>
> —Brian Wren, *"Praise God for the Harvest"* (1968)[40]

EPILOGUE

FOLLOWING A DECADE OF VISITS ABOUT AGRARIAN AESTHETICS across three continents with artists and authors, farmers, and museum curators and gallery directors, I found myself back home in the Pacific Northwest in the spring of 2024. My longtime friend Clifford Trafzer, Distinguished Professor of History at the University of California-Riverside, had hosted my recent travels exploring Southwestern art and agriculture so he joined me in June for a tour of historic sites and museums in Washington, Idaho, Oregon, and Montana. I had special interest in seeing conceptual artist Agnes Denes's remarkable new environmental installation *Wheatfield—An Inspiration* at Tinworks Art community space in Bozeman, Montana. Subtitled *The Seed is in the Ground*, Denes's provocative creation is located on a corner of the city's northeast where a suburban neighborhood meets a small field of ripening grain planted by community members who also received packets of Khorasan heritage wheat for their own plots. The surrounding cityscape invites consideration of farmland loss in the Gallatin Valley in the wake of unprecedented development and of humanity's wider impact on the planet. Denes and Tinworks sponsors explored these topics through public presentations on science and urban planning and with musical performances and poetry readings.

The 93-year-old Hungarian-born American artist-author is credited with originating the modern genre of environmental land art with her 1982 creation of *Wheatfield—A Confrontation* that featured

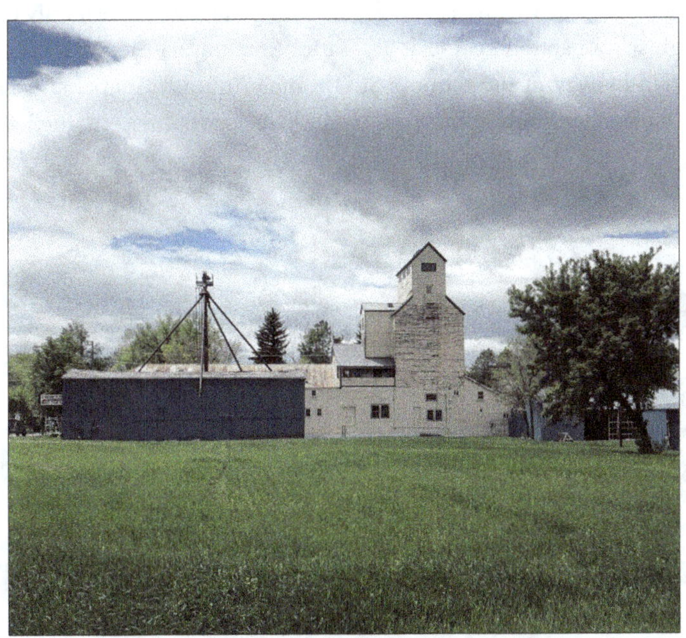

Agnes Denes, *Wheatfield—An Inspiration* **(2024)**
Tinworks Art Installation; Bozeman, Montana
Columbia Heritage Photograph

four reclaimed acres near Wall Street in lower Manhattan. For Milan Italy's 2015 "Feeding the Planet, Energy for Life" World Expo Denes designed a similar ten-acre installation in the city's fashionable Purto Nuova district. She has sought to extend the artistic experience from passive viewing to participation by stirring visitors to consider food sources, farming practices, and distribution systems—variables are often taken for granted until availability and supply chain interruptions threaten individual and community wellbeing. In the face of waste, hunger, and ecological threat, Denes uses art to advocate remedy through imagination, volunteerism, and collective effort. By involving participants in the physical experiences of harvest, local processing, and artisan baking, grain is celebrated as a vital symbol of generative potential amidst city traffic, suburban housing, and industrial development.[1]

On the return trip across the Rockies on I-90 we stopped at Old Mission Park near Coeur d'Alene, Idaho where Jesuit blackrobe Pierre-Jean DeSmet founded the Mission of the Sacred Heart among area Coeur d'Alene and other Indians in the 1840s. The ensemble of original frontier Baroque church—Idaho's oldest building, and adjacent period structures also includes a museum that features the impressive "Sacred Encounters" exhibit on area tribal and mission history. Coeur d'Alene families some of the earliest farms in the eastern Palouse Hills and local mission fields supplied wheat for milling into flour. One of the museum's religious treasures is an elaborate monstrance of polished silver and gold emblazoned with clusters of grain that was used to display consecrated Sacramental Bread during the Eucharist. The recently renovated Coeur d'Alene Casino Hotel farther south near Worley, Idaho displays a substantial wall of twenty-six spectacularly beaded panels mounted on convex wooden frames that tribal artisans designed to "celebrate the hills of the Palouse prairie." Coeur d'Alene women painstakingly fashioned hills of gold and green inhabited by birds and beasts while other panels feature cultivated landscapes with farm implements and fields of grain.

Back home in Richland, I drove on a sunny Saturday morning with the family to watch our granddaughter play a high school softball game at Tri-Cities Preparatory School in nearby Pasco. We approached the ballfield by way of the school's St. Thomas Aquinas Chapel and though we had not been to the school before we could tell by the sunlit exterior that it contained a remarkable series

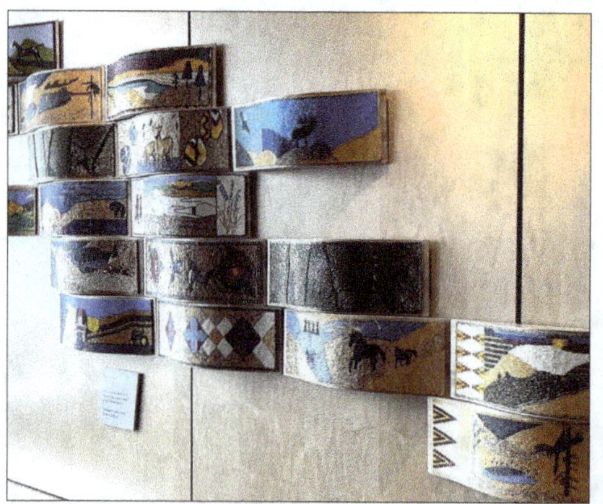
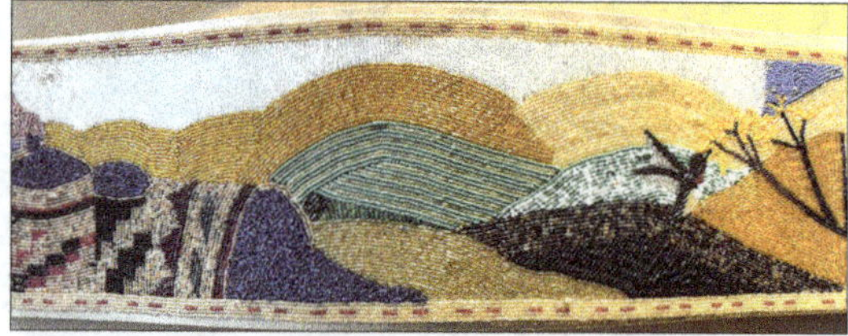
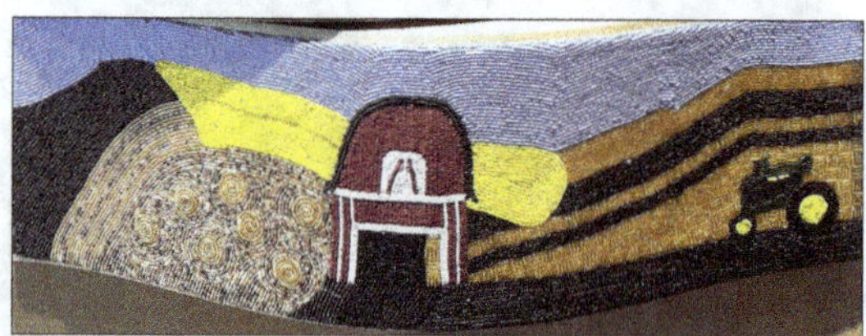

Qhesu'lumkw—Prairie (c. 2000)
Glass seed beads on fabric and wood panels
Coeur d'Alene Casino Hotel, Worley, Idaho
Debbie Wolfe Photograph

of stained-glass windows. During a break in the game I ventured over to the adjacent school's main entrance hoping the door might be open on the weekend and that someone would allow us to see the chapel interior. As had happened on many similar adventures, I was not disappointed. The door opened and in several steps I encountered several adults discussing plans for that evening's annual auction. I explained my interest in the windows and a member of the group, school president Lisa Jacobs, kindly guided me to the chapel and shared the story of its design. Despite substantial responsibilities that day, she patiently answered my questions and presented me with a copy of Nancy Roach's beautiful book, *Reflections of Faith: Tri-Cities Prep and the St. Thomas Aquinas Chapel* (2019). I learned that we had mutual friends in the Roach family as I had recently worked with members of the civic-minded clan on the local Franklin County Historical Museum's capital campaign. Moments later Lisa opened the door to the chapel to reveal the kaleidoscopic masterpieces brilliantly shining in the day's full sunlight.

The windows had been designed by Richland native Patrick Clark and built by him and Tri-City Prep art students over several years beginning in 2001. In the late 1960s Clark had moved with his family to Vienna where he learned about the history and production of stained-glass. Following graduation from Washington State University, he became a stained-glass artist and established workshops in Spokane and Rockaway Beach, New York. For the chapel windows Clark used both traditional German mouth-blown sheet glass as well as rare opalescent pieces salvaged from Louis Comfort Tiffany restoration collections in New York. Several of these masterpieces including the magnificent centerpiece *Eucharist Community* (2015) window, sponsored by thirty families and four area churches, contain images of grain and grapes representing communion bread and wine.[2]

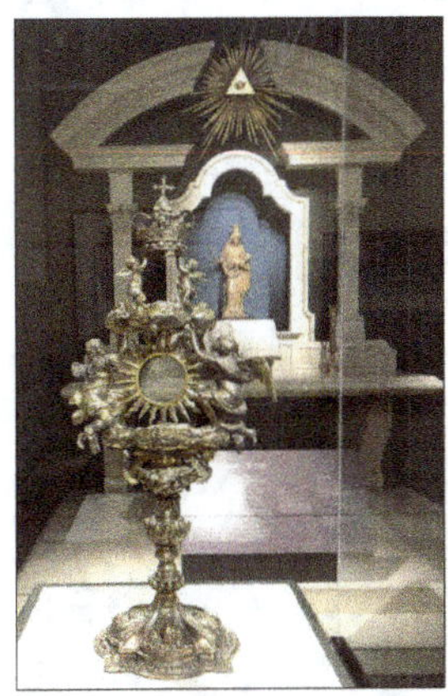

"Sacred Encounters"
Belgian Monstrance
(18th century)
Old Mission State Park
Cataldo, Idaho
Columbia Heritage
Collection Photograph

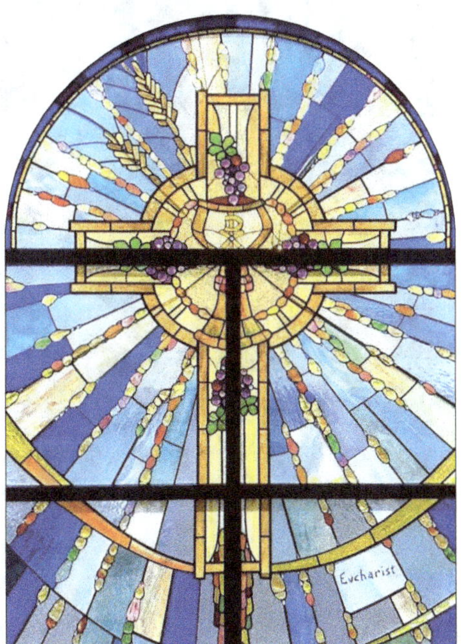

Patrick Clark, *Eucharist Community*
Stained-Glass Window (2015)
Bohemian jewel and gallery
glass grain stalks and grape clusters
St. Thomas Aquinas Chapel,
Tri-City Preparatory High School
Pasco, Washington
Columbia Heritage Collection Photograph

After my visit to the Pasco school I went to the nearby Richland Library to view several recently published photo essays on farm life. I was unexpectedly greeted by a substantial art exhibit recently assembled by local public school students and their teachers. The local district's K-5 Art Exhibition featured dozens of works in a colorful array of acrylic, watercolor, cut paper, and other media. The impressive display vividly attested to the district's curriculum goal to foster student participation in a range of aesthetic experience guided by instruction in the elements of art and principles of design. The state's learning standards for visual arts carry heavy emphasis on advertising and social media, but also contain explicit recommendations to "compare and contrast lines, colors, textures, and forms observed in urban and rural settings."[3] Area students clearly had found an outlet for their guided creativity through rural observations. Alongside paintings paying tribute to familiar scenes by van Gogh, Magritte, and Seurat were original renditions of bulbous orange and green squash and pumpkins, multicolored corn, and surrounding stands of yellow grain. Their art imparts a mood of gladness—that timeless attribute feeding wise souls since the dawn of existence. The fields and their harvest still enchant.

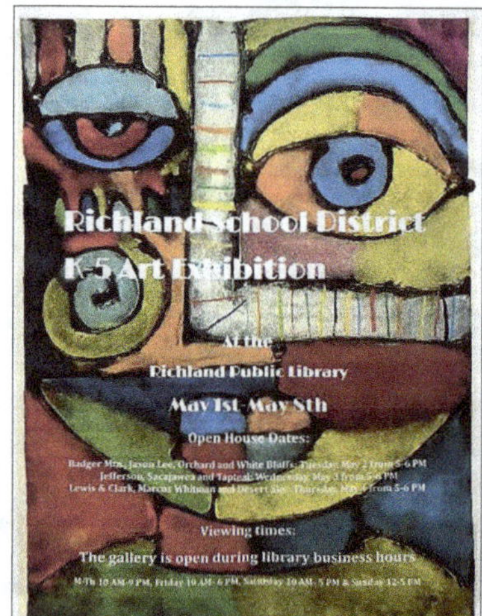

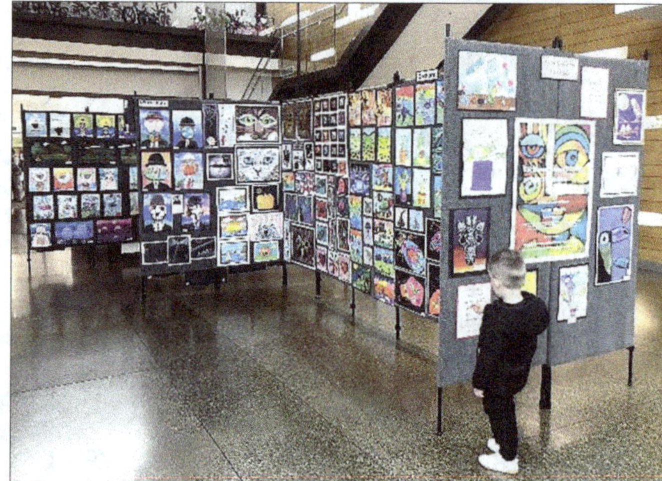

Richland School District Elementary Art Exhibition (2024)
Richland Public Library, Richland, Washington

APPENDIX A

The Columbia Heritage Collection of Agrarian Art

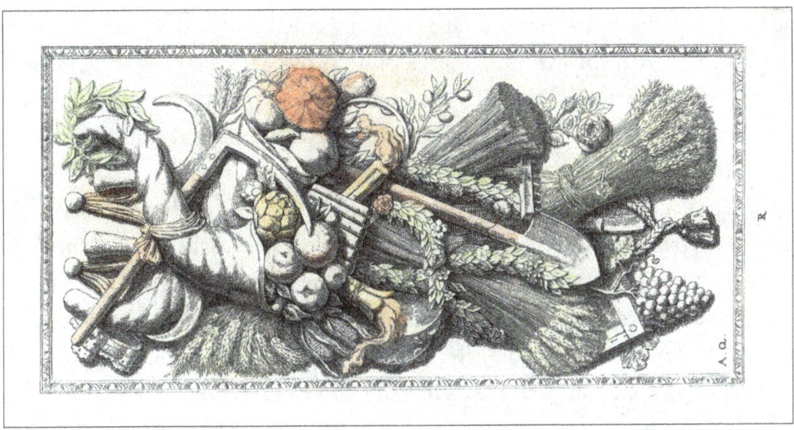

Hendrick Danckerts, after Artus Quellijn, Corncucopia with Grain Sheaves (1661)

The Columbia Heritage Collection of Agrarian Art and Literature consists of over 400 original paintings, limited edition prints, lithographs, and other media that depict harvest and related scenes of rural life from the sixteenth century to the present. The collection also includes books on agriculture and rural art history. The art collection includes approximately 160 woodcuts, engravings, and etchings, 230 lithographs and technical report color plates, several dozen magazine covers and period journal illustrations, 40 original works of art, and numerous photographs, and postcards. Nineteenth-century works are especially represented and reflect the proliferation of romantic painting and printmaking as the pace of industrialization fostered appreciation of receding agrarian experience in cultural imagination.

Works of art represent methods of hand-produced and limited editions that visually express important messages for artist and observer about personal, spiritual, social, and environmental values. Some artists and authors have been prominent in their fields, while others but little known beyond a particular region. Reproductions of their works as intaglio prints, lithographs, and photographs foster appreciation and understandings for those unable to visit museums and galleries that hold the original works. Printmaking can also involve a transformative process of creative commentary and critique as seen in the many depictions of Ruth and Boaz and other classical themes. The change in positioning figures, clothing styles, and equipment for harvesting all contribute to an unfolding dialogue among artists and viewers from ancient to modern times.

Among these paintings, prints, and photographs are many obtained through serendipitous gleanings online and from art dealers and antiquarian print shops. In addition to oil and watercolor paintings featured in this study, art prints represent a range of production methods. Woodcuts are made through a relief process in which grooves are carved on a soft wood surface bearing the artist's design so it remains standing in relief and is inked for the print. Wood engravings are similar but the spaces between the image's lines are left standing above the surface and the design itself prints in white. Intaglio techniques incise an image into a copper plate with an instrument to render a soft etching, or use a burin to create a sharper engraved print. Intaglio is also used for mezzotint by roughening the plate for a print of greater surface contrast.

Lithography traditionally produced an image because of the natural resistance of water and oil. Drawings made with a greasy material are applied on a finely ground limestone surface. The stone is dampened and rolled with ink which adheres only to the greasy areas and not to the wet stone. The inked image is then transferred to paper through a flatbed press. Modern lithography is a planographic process in which the picture is drawn and treated with inks and solutions on a stone or metal surface to make multiple black and white or color impressions.

I began studying period agrarian prints in 2012 to illustrate the book *Harvest Heritage: Agricultural Origins and Heirloom Crops of the Pacific Northwest* (Washington State University Press, 2013), which was produced in partnership with historian Alexander McGregor and photographer John Clement. Subsequent related volumes include *Harvest Horizons: Essays on Agrarian Themes in Northwestern Art and Literature*; *Hallowed Harvests: Agrarian Depiction from the Bible, Literature, and Art to Early Modern Times*; and *Harvest Hands: Reapers and Threshers in American and Modern European Art and Literature*.

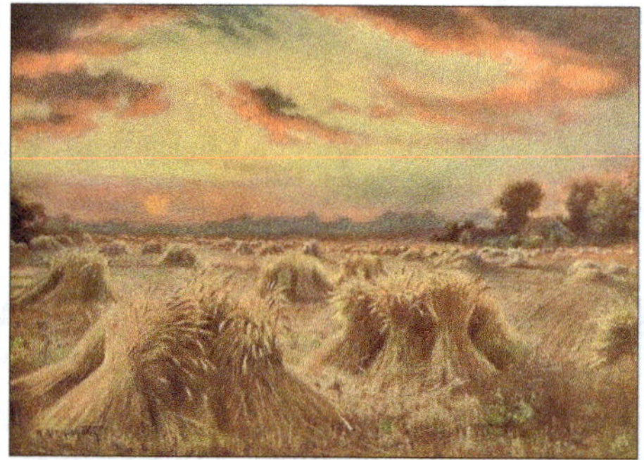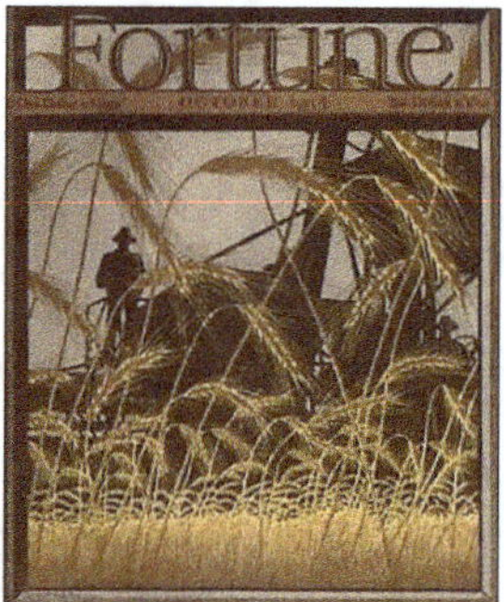

Robert Atkinson Fox, *After the Harvest* (c. 1910); lithograph on paperboard, 4 x 6 inches
Ernest Hamlin Baker, *Wheat Harvest, Fortune* (October 1935 cover), 10 ⅛ x 10 ⅝ inches

The idea of collecting agrarian art arose in the fall of 2013 after seeing a century-old postcard of Robert Atkinson Fox's luminous *After the Harvest* at the Washington State Fair in Puyallup. I did not buy it, but the scene stirred something deep within and kept appearing in my mind's eye. Several months later I made a random café stop near the state history museum in Tacoma. While awaiting lunch fare, I noticed a tattered array of magazines from the 1930s and '40s strewn about the tables including large-format copies of *Fortune*. I later learned they have been popular among collectors for their rich colors and high-quality paper stock. At my table was the October 1935 issue featuring a serigraph cover by Ernst Hamlin Baker of heads of grain in rich golds and browns silhouetted against a combine reel and header-tender. While paying my bill, I explained to the proprietor my special interest in the cover and for a few dollars he kindly parted with it. These happenstances got me to think seriously about agrarian art and I began acquiring affordable vintage prints related to country literature and agricultural history I had long enjoyed reading. Subsequent investigation turned up harvest scenes by Peter Helck and other prominent artists that had been featured in *Fortune, Country Life Magazine*, and other periodicals.

Little by little, I became fascinated with harvest themes in a broad art-historical context, and the availability of print forms for tactile chronological documentation. Prints of various kinds in Western art have constituted a prolific if fragile form ever since Europeans first appropriated papermaking and block-cutting from the Chinese in the fifteenth century. Masterful European woodcut prints first appeared in Italy and Germany from artists like Andrea Mantegna (c. 1431-1506) and Albrecht Dürer (1471-1528), followed later in the century there and in France with painstaking intaglio line engraving on iron and copper.

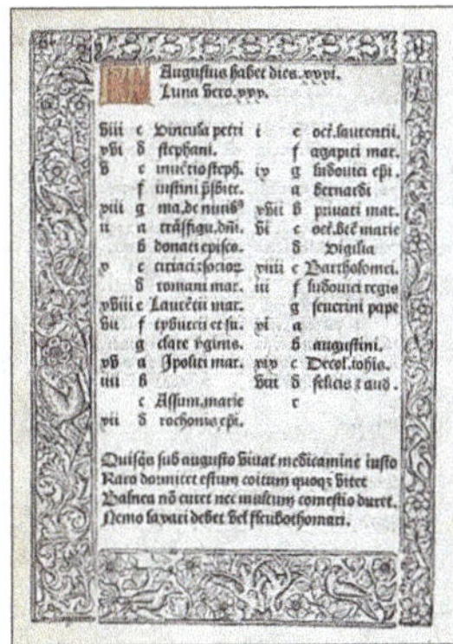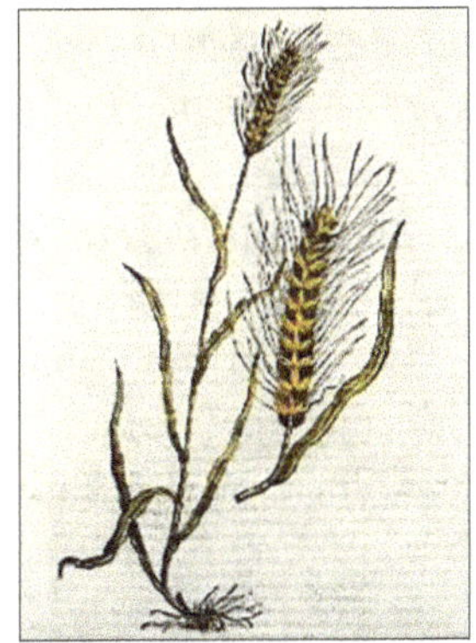

Left: Philippe Pigouchet, *Augustus* calendar of feast days, *Horae: ad usum Romanum* (1494)
Woodcut with illuminated metal engraving capitals in gold and red on vellum, 5 x 7 ⅖ inches.
Right: Wolfgang Mayerpeck and Giorgio Liberale, *Triticum* (Wheat)
Woodcut in P. A. Mattiolis, *Commentaria in Dioscoridem* (Venice, 1565), 13 ⅓ x 8 ⅞ inches

The earliest prints from engravings date from the 1440s, and the oldest ones in this collection are an illuminated July and August calendar leaf on vellum from a 1494 prayer-book by Paris master Phillippe Pigouchet and two 1565 colored botanical woodcuts by Wolfgang Mayerpeck and Giorgio Liberale. The first book printed in England had appeared only seventeen years earlier. Sixteenth-century Dutch printmakers like Hieronymus Cock (1518-1579) perfected shadowing techniques that provided depth to landscapes and other subjects, while engravings in the 1600s and 1700s are characterized by more ornamental design and the introduction of mezzotint for tonal shadings.

One discovery led to another with relevant acquisitions including important exemplars of various printing techniques including early grain woodcuts (1565) by Wolfgang Mayerpeck and Giorgio Liberale and the Dutch copperplate engraving *Augustus* (c. 1616) by Jan van de Velde II. Francesco Bassano's *Summer, with the Sacrifice of Isaac*, is remarkable for one of the earliest depictions of a realistic harvest scene and for being included as a print in the first-ever published art catalogue, Archduke Leopold Wilhelm's *Theatre Pictorium* (1660). This collection's Bassano engraving is from the series owned by the renowned Flemish artist and tapestry designer Baron Richard van Orley.

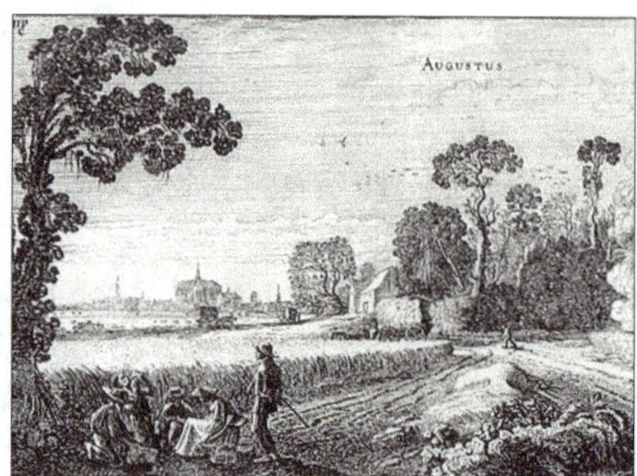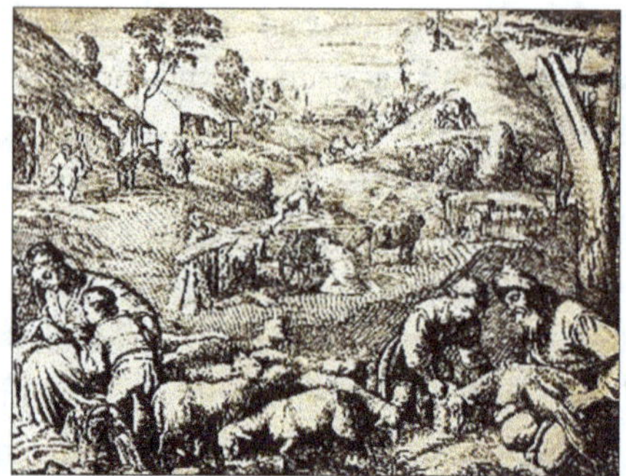

**Left: Jan van de Velde II, *Augustus*, from Twelve-Months (detail, c. 1616)
Copperplate engraving on wove paper, 11 ¾ x 17 ¼ inches
Right: Nicolaus van Hoy and Jan van Ossenbeek (engravers), after F. Bassano, *Summer* (detail, 1660)
Etching on laid paper, 8 ⅜ x 12 inches**

Other notable works in this collection include the Rembrandt-Watelet counterproof *Landscape with Hay Barn and Flock of Sheep* (1660/c. 1780), seventeenth- and eighteenth-century Labors of the Months harvest woodcuts, age-stained illustrations from early printings of the Book of Ruth, and steel engravings for James Thomson's *The Seasons* (1726). New York art critic Atherton Curtis characterized Rembrandt's scene one of the two or three finest landscapes the Dutch master ever created and a milestone in the history of art and printmaking. It shows a cottage, outbuildings, and copse of feathery foliage along Diemerdijk (Diemer Dike) Road near the artist's Amsterdam home. (This collection also includes the Albert Quantin-Firmin Delange 1880 heliograph plate of this landscape printed with other Rembrandt works in a Paris edition of 500 on Holland paper.)

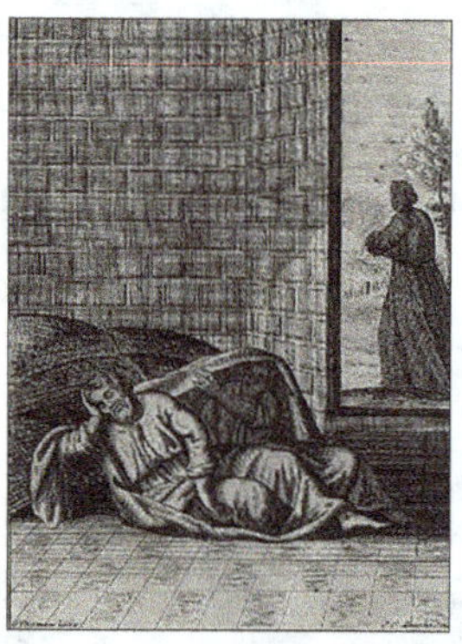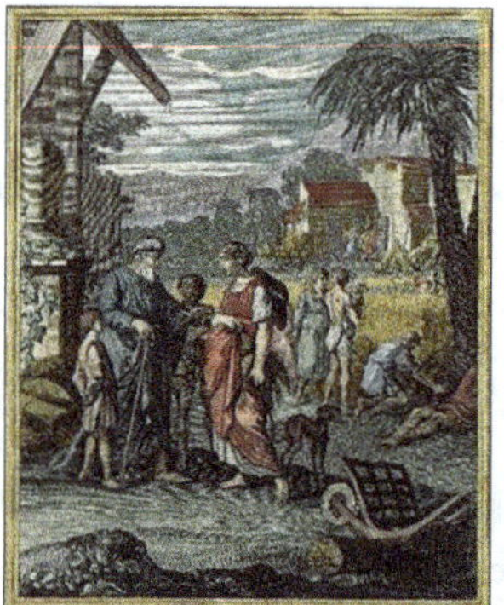

**Left: G. Freman, P[eter] Bouche (engraver), *Boaz Espouseth Ruth* (1688), laid paper, 7 ⅛ x 12 ¾ inches
Right: Caspar Luyken, *Ruth [and Boaz]—Ruth 2:8* (1708), laid paper, 7 ⅞ x 10 inches**

Peter Paul Rubens (1577-1640) also contributed significantly to printmaking through collaborations with engraver Shelte à Bolswert (c. 1586-1659) and publishers Martin van den Enden (1605-after 1654) and Aegidius (Gillis) Hendricx (active 1640-1677). Bolswert developed an engraving technique of delicately curved and crossed hatching that enhanced dimensionality and tonal variations of prints based on Rubens' panoramic masterpieces. The 1638 Bolswert-Hendricx "Small Landscapes" collection of twenty-one detailed engravings after Rubens included *Harvest Time* and

162

The Rainbow Landscape. Bolswert drew inspiration from various paintings and drawings by Rubens to create composite views of splendor and realism. Some of the working drawings were retouched by Rubens with pen and ink and the prints are considered among the most spectacular produced during the Dutch Golden Age.

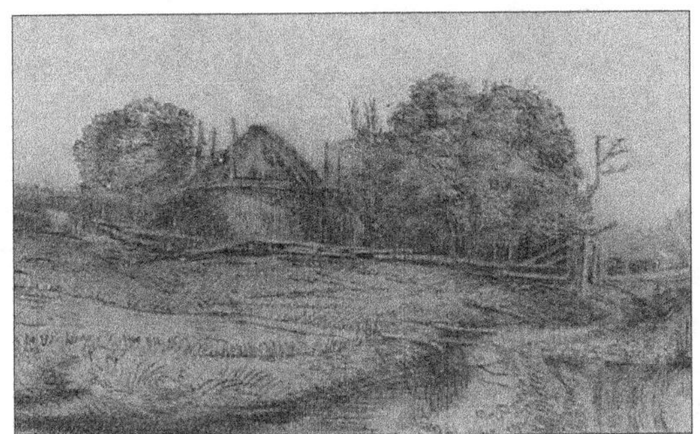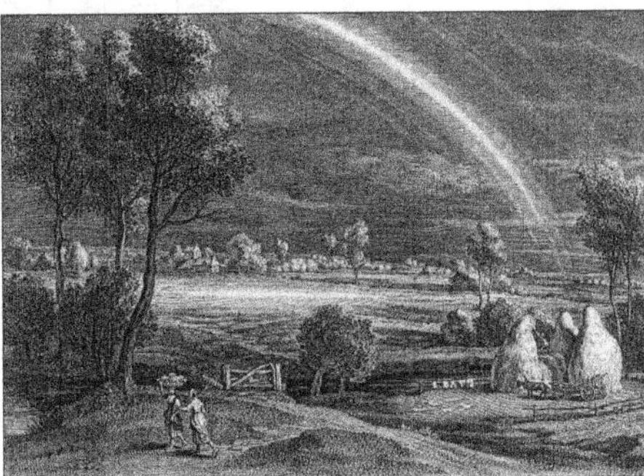

Left: Rembrandt van Rijn, engraver; *Landscape with Hay Barn and Sheep* **(detail, 1650)**
Reverse copperplate counterproof on laid paper (c. 1780), 4 ⅛ x 7 ½ inches
Right: Shelte à Bolswert, after Peter Paul Rubens, *The Rainbow Landscape* **(1638)**
12 ⅗ x 17 ½ inches, ink on laid paper

Eighteenth-century English engraver William Wynne Ryland (c. 1738-1783) developed stipple etchings as a variant of French engraving methods by using dots as well as lines for rich tonal variations. Italian Francisco Bartolozzi (1727-1815), who learned the process in England, became the most famous practitioner of stipple and was probably the first to use color with the process. The advent of lithography in the eighteenth century led to detailed color reproductions of fine art, and nineteenth-century printmaking innovations included application of steel coating to copperplate to make more impressions possible.

Detailed illustrations for Denis Diderot's landmark *Encyclopédie* by A. J. Defehrt (c. 1723-1774) and other engravers attest to new uses of art for labor efficiency. Defehrt's "The Threshing Barn" interior view not only depicts the traditional practice of grain flailing and storage but divides it into a numbered series of "figure" stages to represent the division of labor in novel didactic clarity. Other illustrations show cutaway views of workshops, tasks, and tool assemblies as with various types of sickles, scythes, and rakes. Art has come to serve an incipient transfer of power from field laborer and craftsman to cultural elite who analyze worker methods in ways that promote emerging industrial processes. The specialization fostered an enduring division within humanity into those who *know* and those who *do.*

London master printer John Boydell (1719-1804) became one of the era's most influential publishers who procured the services of Bartolozzi and such leading British artists as Benjamin West (1738-1920) and Richard Westall (1765-1836) for a series of collected works. His magisterial edition of *Liber Veritatis* (1774-1777), a precursor to the modern coffee table book, contained two hundred drawings of works by influential French landscapist Claude Lorrain that had come to be owned by William Cavendish, Duke of Devonshire. Claude's scenes, known for figurative details and depth of field, were reproduced for by engraver Richard Earlom (1743-1822) as distinctive mixed-method colored mezzotint for washes and etching for pen lines. The series soon became a standard for aspiring artists to study and significantly influenced the art of J. M. W. Turner, John Constable, and Samuel Palmer.

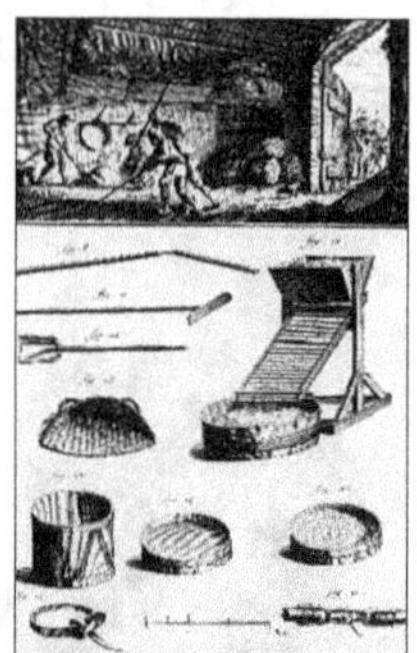 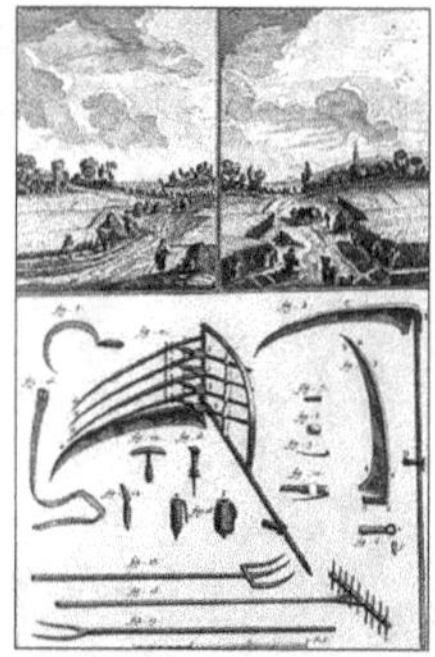

J. A. Defehrt, *Agriculture*, D. Diderot and A. Le Breton, *Encylopédie ou Dictionnaire Raissoné* (1762)
Engraving on laid paper, 10 ½ x 16 ¾ inches

British engraver Thomas Bewick (1753-1828) revived the popularity of printmaking in the nineteenth century by introducing dense end-grain boxwood blocks for carved relief instead of more costly and labor-intensive metal plates. The splendid handiwork of Bewick and his Scottish apprentice John Anderson is represented by country scenes from an early printing (1802) of Bloomfield's book. During the nineteenth century many high-quality woodcuts were used as illustrations including agrarian scenes featured here that appeared in *The Graphic* (London), *The Illustrated London News*, *L'Universe Illustré* and *Le Petit Journal* (Paris), *Gleason's Pictorial* (Boston), *Harper's Weekly* (New York), and other newspapers and periodicals. *The Illustrated London News*, the world's first fully illustrated weekly newspaper, notably featured stories and pictures of harvest-related endeavors in summer and fall issues. *The Graphic*, published by engraver William Luson Thomas (1830-1900), was printed on larger and better paper to enhance the artistic quality of its engravings. Dozens of these British illustrations were collected by Vincent van Gogh as subjects for his drawings and paintings.

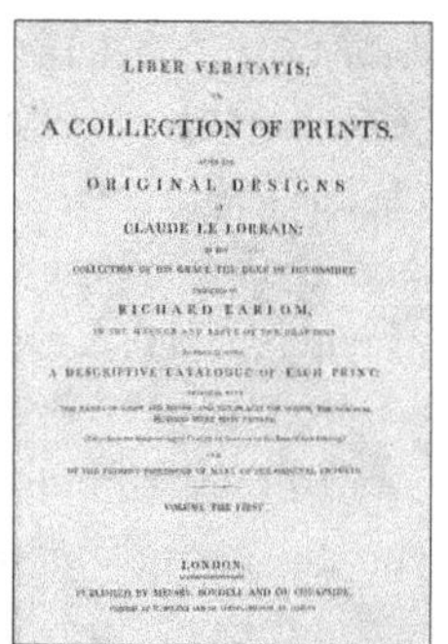 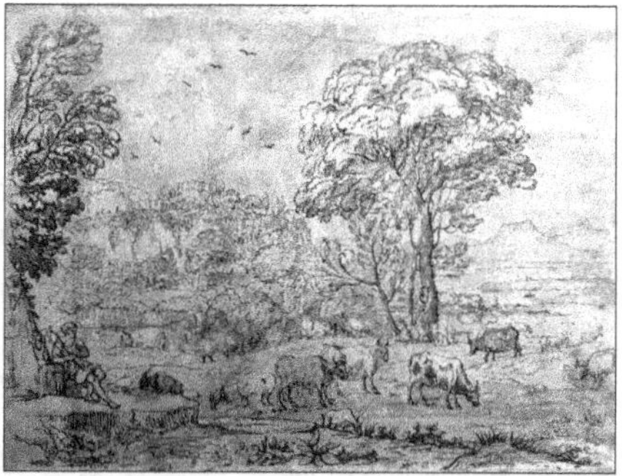

Richard Earlom, after Claude Lorrain, Title Page and Landscape with Livestock print
Etching and mezzotint on paper, 8 x 10 ⅛ inches
John Boydell, *Liber Veritatis* (1777)

Youthful *L'ete* (Paris, 1824) by Louis-Léopold Boilly, who may have coined the term *trompe l'oeil*, is one of the earliest commercial lithographs, and *Southfleet [St. Nicholas] Church* (London, c. 1865) by John Colling is among the first chromolithographs. (Colling's published studies of Early Decorative Medieval architectural style feature detailed representations of grain motifs like *Carved Vesica— Wheat and Tares*, are also included in this collection.) The John Constable (1776-1837)-David Lucas

(1802-1881) *Grainfields near Brighton* (London, 1855) is from a mezzotint series (*English Landscape Scenery*) considered the finest ever produced.

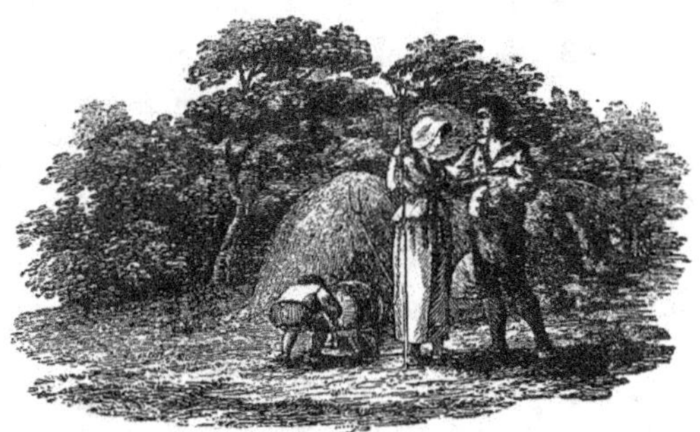

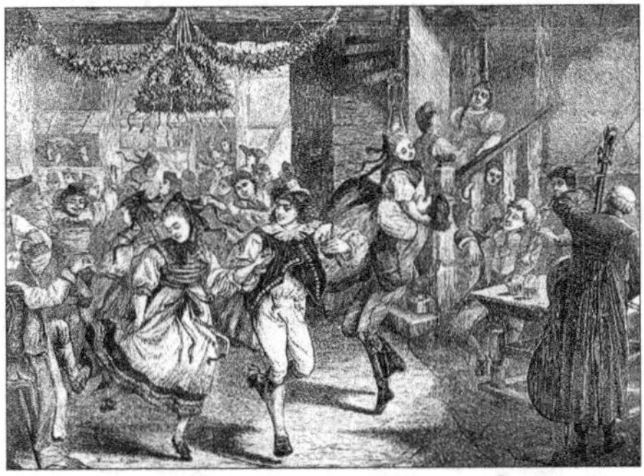

Left: Thomas Bewick (attributed), from *The Farmer's Boy* (London, 1802), 2 ¼ x 3 ¾ inches
Right: L[ionel] C. Henley, *Harvest Festival in Germany*, *The Graphic* (July 2, 1870), 11 ¾ x 8 ⅞ inches

The grand manner art of the French Barbizon masters Jean-François Millet (1814-1875), Jules Breton (1827-1906), and Léon-Augustin Lhermitte (1844-1925) spurred emergence of a new European Realism through the loose association of Naturalism. Peasant farmers appear in these works as central figures after centuries of exile to the distant margins of landscape art. Millet is best known for his iconic 1850s *Gleaners*, *Angelus*, and *Sower* series that earned him such popular titles as "evangelist of the fields" and "apostle of peasants." These early works were inspired by biblical themes as well as the artist's regard for realistic narrative content above color and light as seen in his *Les travaux de champs* ("Labor of the Fields," 1853) that included evocative "everyman" portrayals of work with scythe, rake, flail, and other farm tools. The series of ten woodcuts by engraver Jacques Adrien Lavieille (1818-1862) was published in 1855. Breton was raised in the rural Artois village of Courrières and studied classical painting at the Academy of Fine Arts in Ghent and later at the Paris Ecole des Beaux-Arts where he trained with several Realist painters. He is considered a leading interpreter of rural life which he depicted both in paintings and illustrated literary works like *Le Champs et la Mer* (1875) and *Jeanne* (1880).

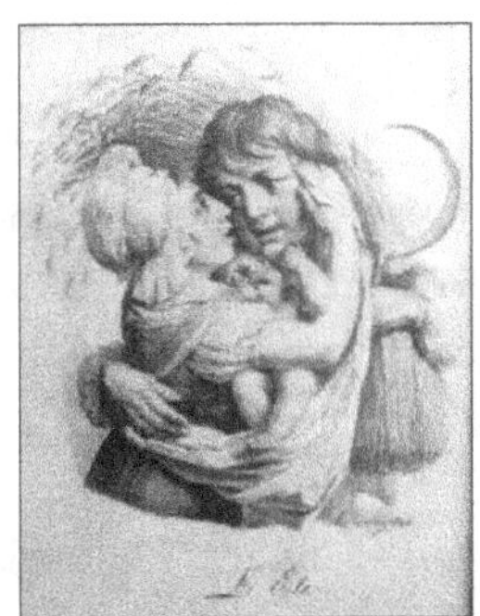

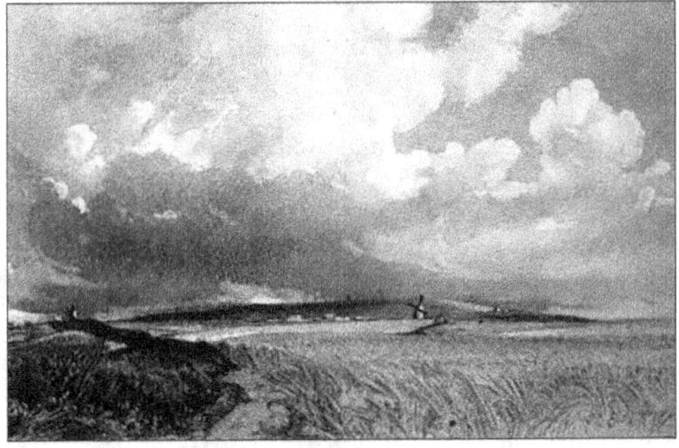

Left: Louis-Léopold Boilly, *Summer* (1824); lithograph on paper, 4 ¼ x 6 ⅛ inches
Right: John Constable, artist (1824); David Lucas, engraver (1846), *Grainfields near Brighton*
***Various Subjects of Landscape, Characteristic of English Scenery* (London, 1855)**

Lhermitte painted numerous canvases titled *The Gleaners* and *The Harvesters* that were inspired by Millet's views but with expressive details that depict recognizable individuals and their endless toil. His numerous Brittany summer scenes earned him the nickname "Singer of Wheat." (Several of Lhermitte's oil and pastel paintings, like *Paying the Harvesters* (1882) and *The Harvest* (1883), feature the stoic peasant hero known as Le Pére Casimir who is believed to have been modeled on a Breton worker named Casimir Dehan.) Major works by these Barbizon masters were reproduced for the covers of popular French periodicals and as high-quality hand-colored photogravures by Philadelphia publisher George Barrie. Author-publisher André Theuriet (1833-1907) commissioned Lhermitte to execute several dozen charcoal compositions engraved by Clément Bellenger (1851-1898) for a magisterial limited edition quarto volume, *La Vie Rustique* (1888). The Columbia Heritage Collection copy is from the library of British art historian Cornelius Verheyden de Lancey.

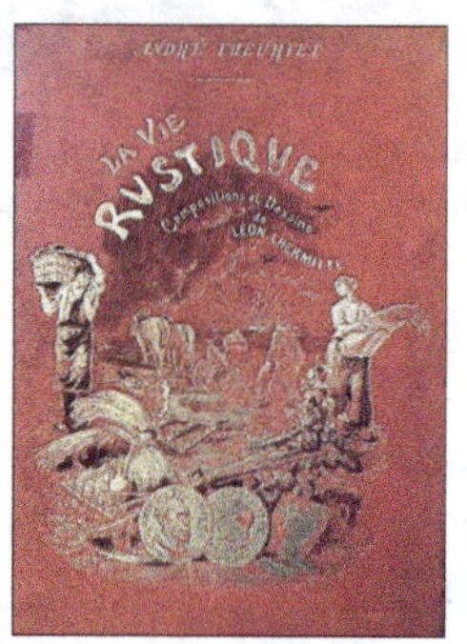 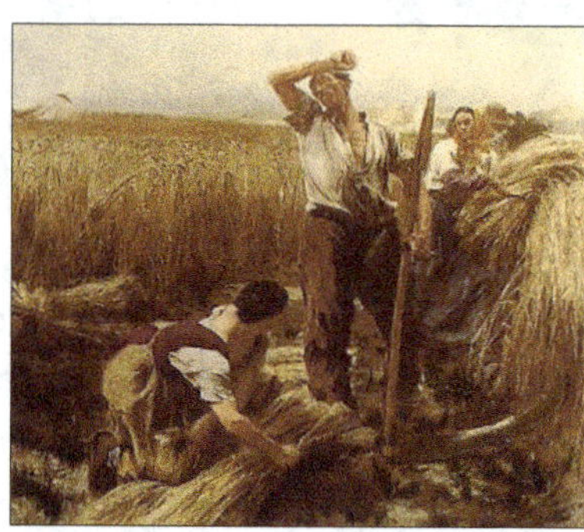 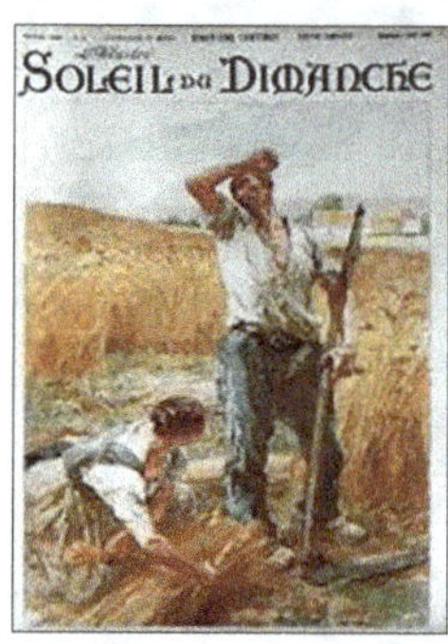

André Theuriet, *La Vie Rustique* red cloth and gilt cover (1888)
Center: Léon Lhermitte, *The Harvest* (1883), hand-colored photogravure, 8 ½ x 7 ½ inches
Chefs d'Oeuvre de L'Exhibition Universelle **(Philadelphia: [George] Barrie, 1889)**
Right: Léon Lhermitte, *The Harvest, L'Illustré Soleil du Dimache* (August 8, 1889), 9 ½ x 12 inches

The European Etching Revival was a nineteenth-century reinvigoration of etching as a printmaking artform that emphasized surface tones in the spirit of Dürer and Rembrandt. By the mid-1800s lithography had made possible larger print runs and use of color inks that caused a decline of traditional etching. Steel-face electroplating of copper plates had been patented in 1857 and made them more durable so a larger number of quality impressions could be produced. (Steel engraving was also popular at this time but done on solid iron plates and more suited for mezzotints.) Publishers, magazines, and organizations significantly contributed to etching's popularization which led to high quality fixed editions of original prints commonly featuring landscapes, portraits, and genre scenes. Artists Francis Seymour Haden (1818-1910), his brother-in-law James Whistler (1834-1903), and British art critic Philip Gilbert Hamerton (1834-1894) were among the movement's prime movers and among the first to exploit the benefits of steel-face printmaking. Haden's *About Etching* (1866) and Hamerton's *Etching and Etchers* (1868) became standard works, and in 1870 the latter founded an influential monthly periodical devoted to the genre, *The Portfolio*.

This survey collection also includes a complete set of Eugène Graff's grain variety color prints (66) from Henri de Vilmorin's *Les Meilleurs Blés* (Paris, 1880) with supplemental black-and-white plates (27) published in 1909 by Vilmorin's son, Philippe; and James Wilson & Son's twenty-nine steel engravings of harvest equipment (1884). Exquisite lithographs of grains raised in North American are from James Hall, *A Natural History of New York* (1851) as well as in foreign publications and

USDA reports like Mark Carleton's *The Basis for the Improvement of American Wheats* (1900, from the library of Phillipe Vilmorin) that featured notable work by women artists like Deborah Passmore.

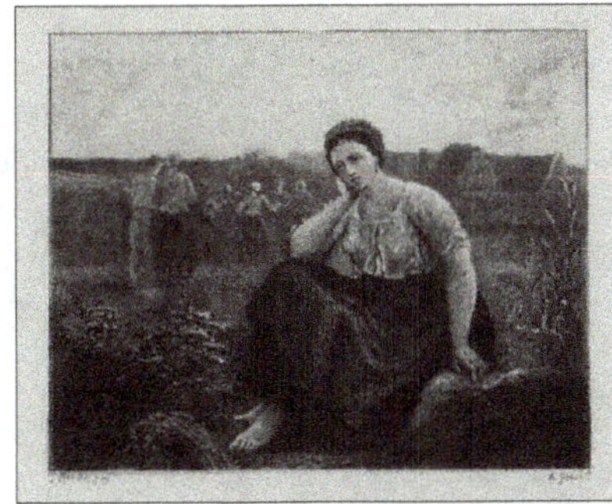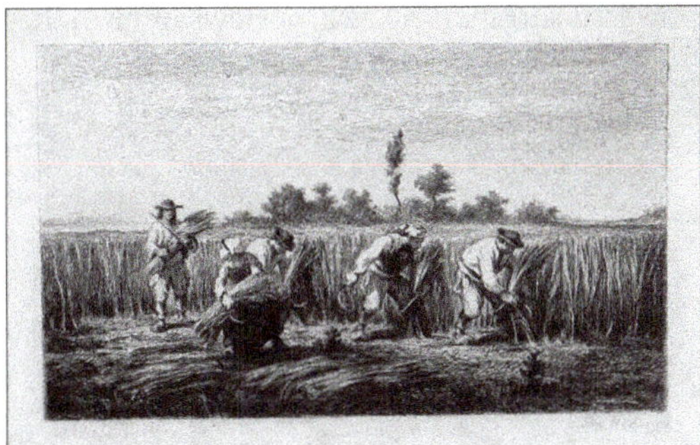

Left: A[médée] Greux, after J. Breton, *Peasant Girl*; 4 ½ x 5 ¼ inches
Right: P. E[dmond] Hèdouin, after A. Leleux, *Reaping*; 4 x 6 inches
Phillip G. Hamerton, *The Sylvan Year* (London, 1876)

When the existence of paintings is unknown, as with William H. Coffin's evocative *Moonlight Harvest* (1897), such works can only be seen through rare period reproductions. *The International Studio's* 1911 chromolithograph of Jean Charles Cazin's *Les Meules*, Job Nixon's signed drypoint proof, *Harvest Field, Ely: Evening* (1926), and Harry Wickey's catalogue print *Wind Over Wheat* (1936) are among several works not listed at any other public institution. Among several watercolors of harvest scenes are watercolors by British agrarian landscapists Raymond Turner Barker and Jane Dryden Alexander.

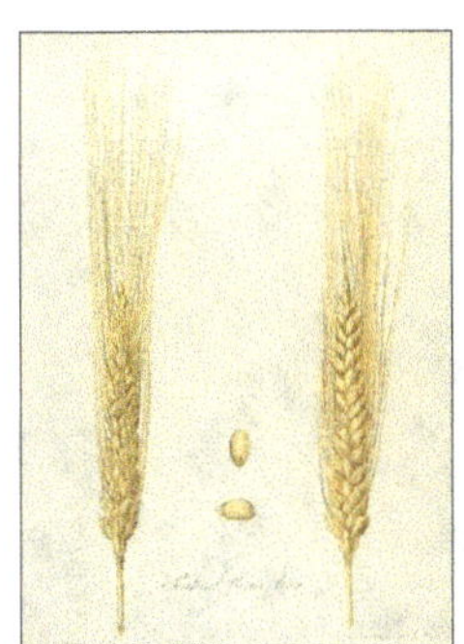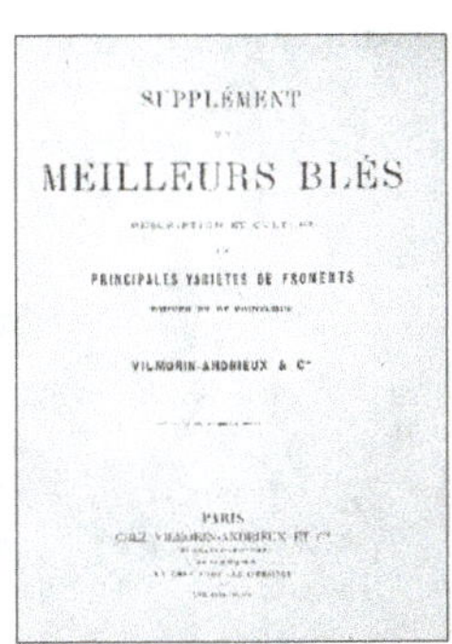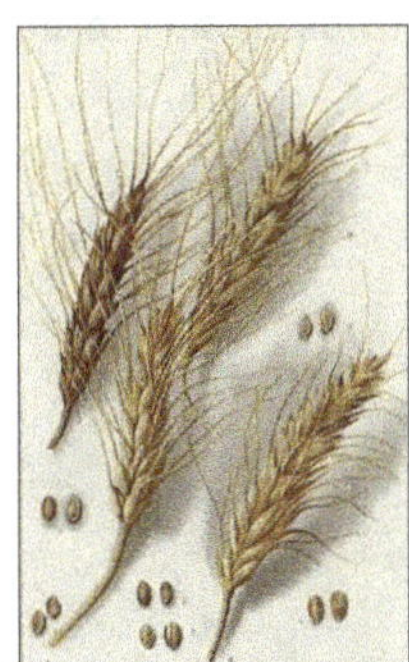

Left: Eugène Graff, *Red March Wheat*, 8 ¾ x 12 ½ inches; *Les Meilleurs Blés* (1880)
Center: Vilmorin *Supplément aux Meilleurs Blés* Title Page (Paris, 1909)
Right: Deborah Passmore, *Drought-Resistant Wheats* (Washington, D.C., 1900), 5 ½ x 9 inches

Throughout most of the nineteenth-century rural inhabitants witnessed developments in agricultural mechanization transforming old ways of harvesting that were photographically documented. Some of the earliest photographs ever made were the 1840s-50s print-negagtive calotypes of British farm scenes made by English inventors William Henry Fox Talbot (1800-1877) and his protégé Hugh Owen (1808-1897). Later rural American photographers artfully captured panoramic farm scenes including Washington's R. Raymond Hutchison and Joshua Elmer, W. A. Raymond of Oregon, Wyoming's Joseph E. Stimson, and Nebraskan Solomon D. Butcher. Their work typically featured wide-format sepia tone pictures of crews who gathered grain from the fields and tended massive steam-powered threshers as well as views of farm life throughout the year.

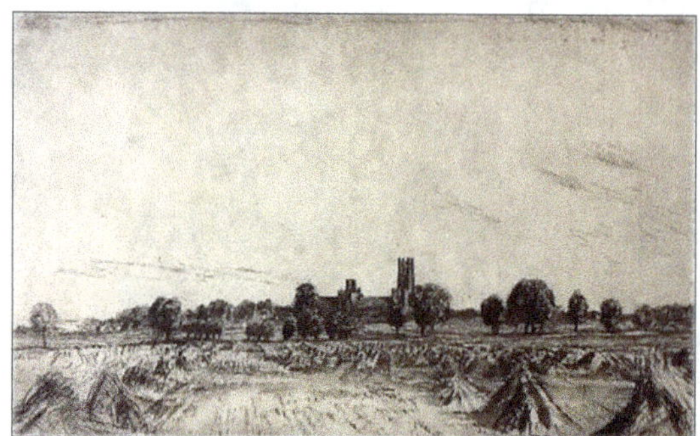

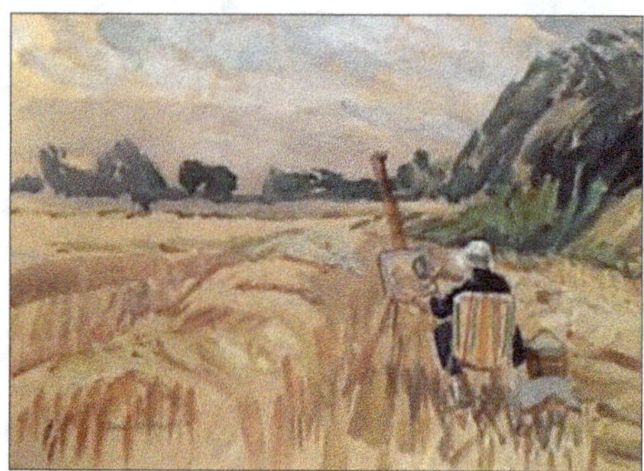

Left: Job Nixon, *Harvest Field, Ely: Evening* (1/6, 1926); signed proof on wove paper, 10 ⅞ x 5 ½ inches
Right: Jean Dryden Alexander, *Painting at Thelveton* (1991), gouache on paper, 10 ½ x 7 inches

The Works Progress and Farm Security Administrations (WPA/FSA) originated in the 1930s as two of President Franklin Roosevelt's signature New Deal programs to promote rural recovery in the wake of the Great Depression. Prominent photographers associated with the Farm Security Administration from 1935 to 1943 included Arthur Rothstein, Dorothea Lange, and Ben Shahn. The stark and stunning visual record amassed by the group from 1938 to 1942 yielded a prodigious collection of some quarter-million negatives ranging in size from 35 mm to 8 x 10 inches. Approximately 170,000 FSA images, digitized in the 1990s, survive as a national treasure and are now housed under controlled conditions at the Library of Congress.

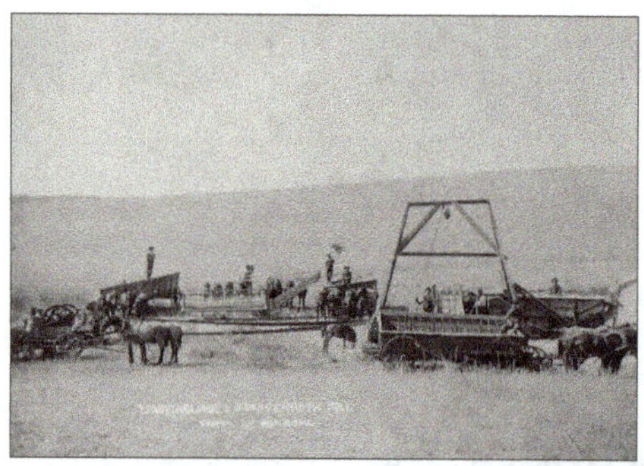

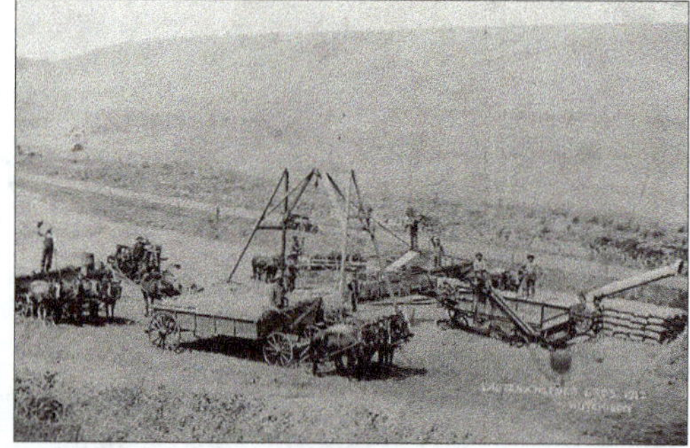

Left: R. Raymond Hutchison, "Laudenschlager & Paffenroth" Threshing Outfit (1911)
Right: R. Raymond Hutchison, "Lautenschlager Bros." Threshing Outfit (1912)

Lithuanian-born Shahn (1898-1969) was already an accomplished National Academy of Design artist and printmaker when also hired as one of the first FSA photographers in 1935. He used his pictures not only to advance the agency's moral mission to inform the wider population to support rural economic and social reform, but also as models for various forms of agrarian art including many harvest paintings and lithographs including *Bountiful Harvest* (1944), *Beatitudes* (1952), and *Wheat Field—Ecclesiastes* (1958). It was also used for Shahn's illustration of the third chapter of Ecclesiastes ("To everything there is a season. . . .") in a collection of photo-lithographs rendered with handwritten and illuminated text by the artist for *Ecclesiastes Or, The Preacher* (Paris: The Trianon Press, 1967).

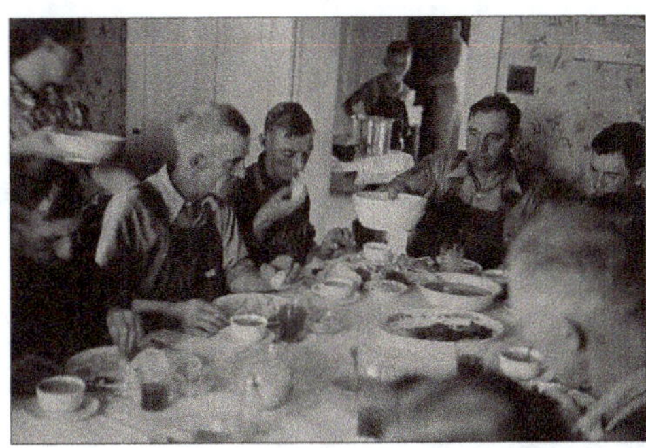 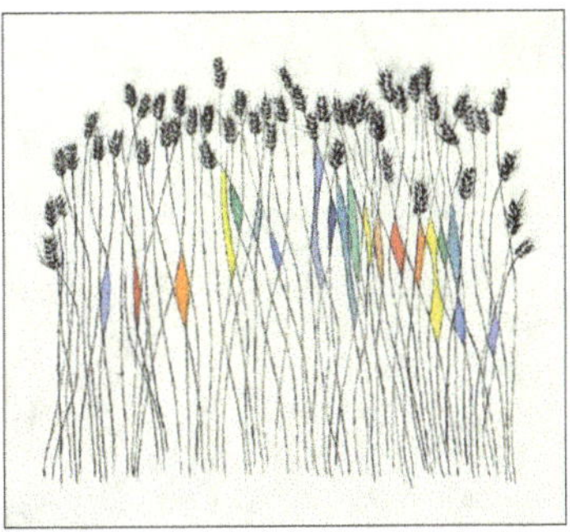

Ben Shahn, *Harvest Dinner* (1938), digitized silver gelatin print
FSA Historical Section Collection, Prints and Photographs Division, Library of Congress
Ben Shahn, *Wheatfield* (1958), *Ecclesiastes Or, The Preacher* (Paris: Trianon Press, 1967)

The advent of modern digital technologies has made possible a wide range of printmaking innovations including photography. The longtime collaboration of Richland, Washington, landscape photographer John Clement and historian-entrepreneur Alexander McGregor of Colfax-based The McGregor Company has yielded numerous books on regional history and other illustrated works. Published since the 1988, color images for their "Northwest Drylands" calendars show the influence of two prominent American watercolor artists whose works Clement has closely studied—Winslow Homer and Andrew Wyeth. Clement's unpeopled landscapes, which earned him induction into the Professional Photographers of America International Hall of Fame, typically feature evidence of humanity's waning presence—dilapidated barns and fences, retired farm machinery, and fields of maturing grain. His ideas about the "saturating luminosity" of dawn and dusk suggest affinity with the nineteenth-century American Luminists and pioneers of color photography whose detailed agrarian views beneath soft, hazy skies engender feelings of melancholy and meditation.

Washington State University's Manuscript, Archives, and Special Collections Division at Holland/

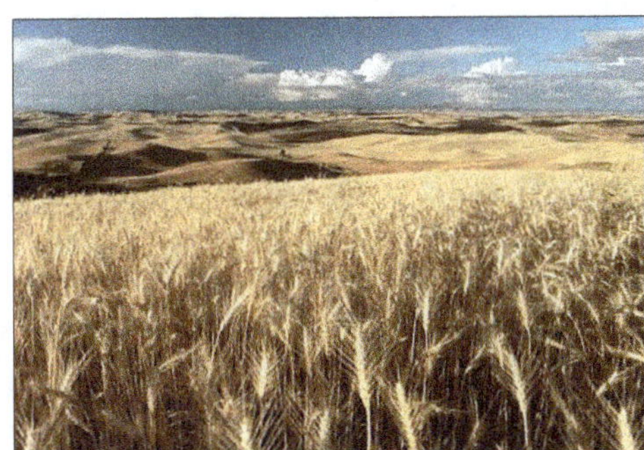 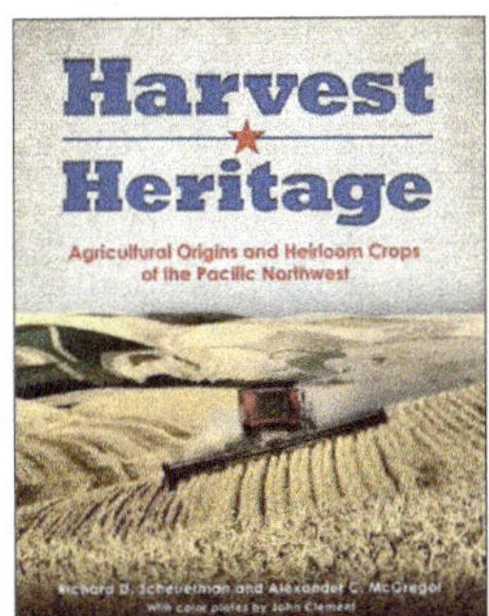

John Clement-Alexander McGregor Calendar (1994) and Book (2013) Covers

Terrell Libraries in Pullman will serve as the primary repository of the collection while the Franklin County Historical Society Museum in Pasco, Washington, serves as a custodial center for exhibitions and related educational activities. The purpose of the collection is to strengthen both institutions as cultural destinations for members of the general public and scholars who wish to explore the vital role of agrarian experience throughout history.

Richard D. Scheuerman
Richland, Washington

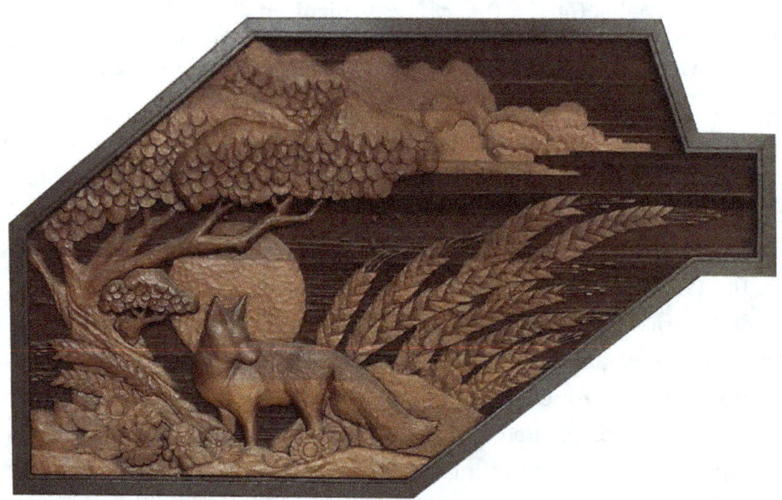

Stylized Coyote and Grains Cedar Sculpture (c. 1975)
Red Lion Hotel; Pasco, Washington
Columbia Heritage Collection Photograph

ARRANGEMENT OF THE COLLECTION

I.   **EARLY WOODCUTS**—(A) Pages and Plates, (B) Art Book Illustrations

II. **OLD MASTER AND MODERN ENGRAVINGS AND ETCHINGS**—(A) Plates, (B) Art and Other Book Illustrations

III. **LITHOGRAPHS AND OTHER PRINTS**—(A) Popular Editions and Offprints, (B) Newspapers, Magazines, and Books: 1. British, 2. Other European, 3. Harper's, Scribner's and Other Nineteenth-Century American Publications, 4. Collier's, Fortune, and Other Modern American Publications, 5. Farm Collector Magazine Harvest Machinery Advertising Chromolithograph Reproductions, (C) Technical Reports

IV. **PAINTINGS, LIMITED EDITION PRINTS, AND OTHER ORIGINAL FINE ART**

V.   **FOLK ART**

VI. **POSTCARDS AND NOTECARDS**—(A) Postcards: 1. International Harvester Company, 2. Pacific Northwest, 3 Other United States and Canada, 4. Foreign, 5. Thanksgiving Cards with Harvest Motifs, (B) Notecards

VII. **PHOTOGRAPHS AND PHOTOGRAPHIC REPRODUCTIONS**—(A) Photographic Reproductions, (B) Keystone and Other Stereoviews, (C) R. R. Hutchison Harvest Photographs, (D) Clement-McGregor Calendars and Publications, (E) Richard Scheuerman Digital Images

VIII. **BOOKS WITH ART PLATES**

TERMS

counterproof: an image made from a freshly printed sheet with the ink still wet to create a proof that faces the same way as originally etched onto a plate.

engraving: the impression from the surface of a metal plate that has been cut with a tool called a burin. The plate is inked, placed into a press, and a sheet of paper placed over it with pressure applied to make the print.

etching: a print made from a copper or zinc plate coated with acid-resistant varnish on which lines are drawn through the coating. The plate is then immersed in acid which bites into the drawn lines. The plate is then inked and wiped so ink remains only in the etched area.

intaglio: a printmaking process using an engraved image from a plate or block.

laid paper: paper with fine ridges on the surface originally produced during manufacture.

mezzotint: an intaglio form of engraving in tone rather than line.

photogravure: an intaglio print process used to produce high-quality reproductions from photographs.

vellum: prepared animal skin or membrane used for writing and prints.

woodcut: a sharp tool carves a design into a woodblock's surface leaving only the design. The surface is inked and paper placed over to print the image.

wove paper: paper with a uniform surface and no texture or watermark.

When reproduction of a work is chronologically referenced in this catalog, the entry is placed in the category of its later rendering, with the name of artist listed first followed by the etcher or engraver. Thus, David Lucas's 1855 engraving of John Constable's original c. 1824 painting *Cornfields near Brighton* is dated (c. 1824/1855). Unless otherwise noted, print sizes indicate overall dimensions including border.

I. EARLY WOODCUTS

A. Pages and Plates

Philippe Pigouchet, *Julius* and *Augustus* calendar of feast day pages (2), from *Horae: ad usum Romanum* (*Book of Hours: Roman Service*), for Paris bookseller Simon Vostre (October 23, 1494), woodcut botanical borders with illuminated metal engraving capitals in gold and red on vellum, 5 x 7 ⅖ inches. From the Alexandre Rosenberg Collection, New York

Wolfgang Mayerpeck and Giorgio Liberale, *Triticum* (Wheat) and *Hordeum* (Barley), from Pietro Andrea Mattiolis, *Commentaria in Dioscoridem* (Venice, 1565), laid paper, 13 ⅕ x 8 ⅞ inches

Sebastian Münster, Scythe and Peasant Scenes (2 views), from *Cosmographia* (Basel, 1578), laid paper, 8 ¼ x 12 ¼ inches

Sebastian Münster, Peasants Flailing Grain, from *Cosmographia* (Basel, 1592), laid paper, 8 x 12 ⅞ inches

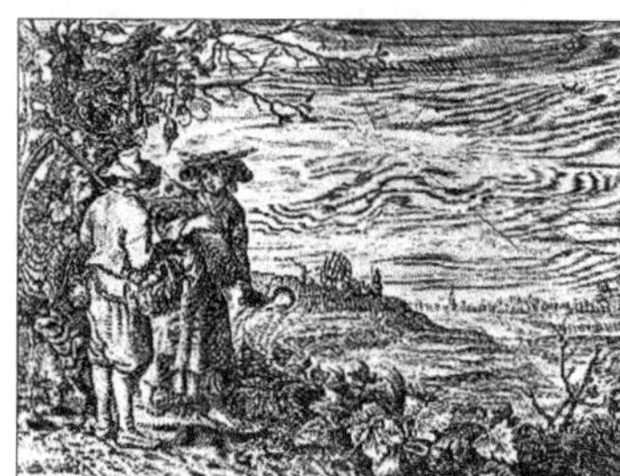
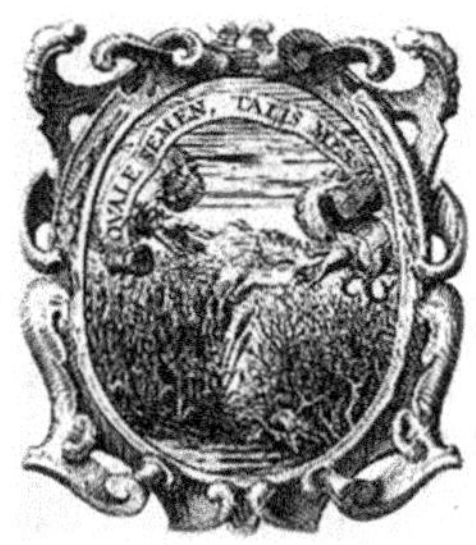

A. B.

A. *Summer*, from J. P. van de Vinne, *Houwelick* (Middelburg, Netherlands, 1625), laid paper, 5 ½ x 4 ¼ inches

B. A. Julien Hayneuve and Michael Cuvelier, *Quale semen, talis messis* ("The quality of seed, so the harvest"), from *Scala Salutis E Solidis Veritatibus Constructa* (Cologne, 1650), laid paper, 2 ¼ 2 ⅝ inches

Matthäus Merian the Elder, *Der Gerste* [The Barley], in Prince Ludwig zu Anhalt-Köthen, *Der Fruchtbringenden Geselschaft Nahmen/Vorhaben/Gemälde und Wörder* Vol. I (Frankfurt, a. M., 1646) laid paper, 7 ⅜ x 5 ¾ inches

Nicolaus van Hoy and Jan van Ossenbeek (engravers), after F. Bassano, *Summer* (1660), etching on laid paper, 8 ⅜ x 12 inches (from the collection of 18th century Flemish artist and tapestry designer Baron Richard van Orley)

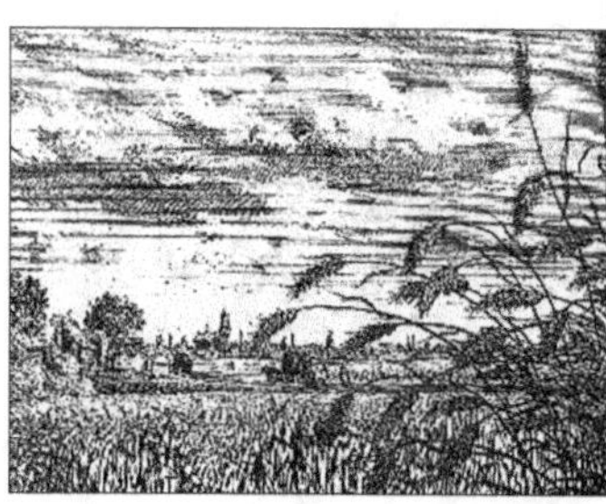
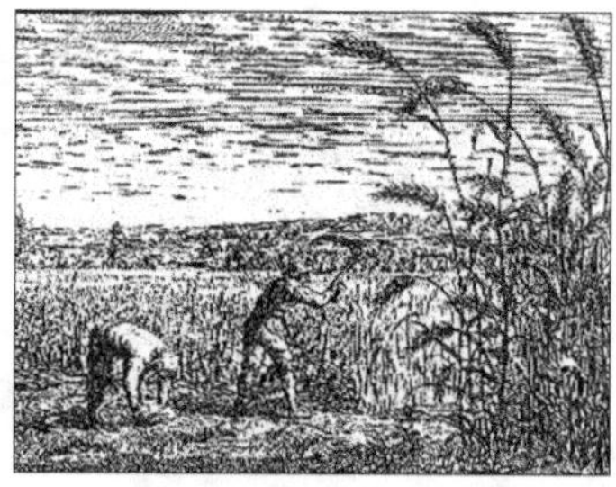

Vincent van der Vinne, Grainfield scenes (3), from Adriaane Spinniker, *Vervolg der Leerzame Zinnnebelden* (1757), laid paper, 4 ½ x 3 ½ inches

Hendrick Danckerts, after Artus Quellijn, Corncucopia with Grain Sheaves, from Jacob van Campen, *Afbeelding van't Stadt Huys van Amsterdam* (Amsterdam, 1661), laid paper, 5 ¾ x 11 ¼ inches

Attributed to Bartholomew Breemberg, *Joseph Distributing Grain to the Egyptians—Genesis 41* (1700), laid paper, 25 x 11 ⅛ inches

Cornelis Dusart, J. L. C. Zentner (engraver), Harvest Scene (c. 1700), laid paper, 8 x 5 ⅞ inches

Triticum Froment, from Joseph P. Tournefort, *Institutiones rei Herbariae* (Paris, 1700), laid paper, 7 ¼ x 9 ¼ inches

Romeyn de Hoogie, Grain Shield Framing Ruth and Boaz (1710), laid paper, 5 ¾ x 4 ¾ inches

Moisson, from Louis Liger, *La Nouvelle Maison Rustique* (1762), laid paper, 6 ½ x 10 inches

Moses Griffith, Grignion (engraver), *Women at the QUERN and the LUAGHAD with a view of TALYSKIR*, from Thomas Pennant, *A Tour of Scotland* (1772), laid paper, 8 ¾ x 5 ⅝ inches

B. Art Book Illustrations

Adolf Bartels, *Der Bauer in der deutschen Vergangenheit* [The Farmer in German History], (Leipzig, 1900), 100 15th-18th century woodcuts and other illustrations on wove paper

Paolo Paschetto, *Psalm XXIII*, 5 ¾ x 6 ½ inches; Muriel Jackson, *Harvesters, Italy*, 5 ¾ x 4 ⅝ inches; M. Desligniè, *Vanneurs-Corses*, 7 ⅜ x 6 inches; Emil Orlik, *Woman Carrying a Sheaf*, 6 x 7 ¾ inches; from Malcolm Salaman, *The New Woodcut* (London: The Studio Limited, 1930)

II. OLD MASTER AND MODERN ENGRAVINGS AND ETCHINGS

A. Plates

Jan van de Velde II, August (c. 1616), copperplate engraving on laid paper, 11 ¾ x 17 ¼ inches

Shelte à Bolswert (engraver), after Peter Paul Rubens, *Return from Harvest (Landscape with Rainbow)*, from Aegidius (Gillis) Hendricx, *The Small Landscapes* (Antwerp, 1638), laid paper mounted on hardboard, 12 ⅗ x 17 ½ inches

Alain Manesson Mallet, St. Denis [Cathedral, Paris] harvest scene, from *Description de l'Universe* (Paris, 1683), copperplate engraving on laid paper, 5 ½ x 8 ¼ inches

G. Freman, P[eter] Bouche (engraver), Boaz Espouseth Ruth, from Richard Blome, *History of the Holy Bible* (London, 1688), laid paper, 7 ⅛ x 12 ¾ inches

Caspar Luyken, Ruth [and Boaz]—Ruth 2:8, from *Historia Celebriores Veteris Testamenti* (Nürnberg, 1708), laid paper, 7 ⅞ x 10 inches

After Mattias Scheits, Boaz Talking with Ruth. Ruth 2:8-9, from *Tafereelen des Ouden en Nieuwen Testaments . . . Tableaux du Vieux & Nouveaux Testament* (Amsterdam: Joshua and Reinier Ottens, 1754), laid paper, 7 x 8 ⅝ inches

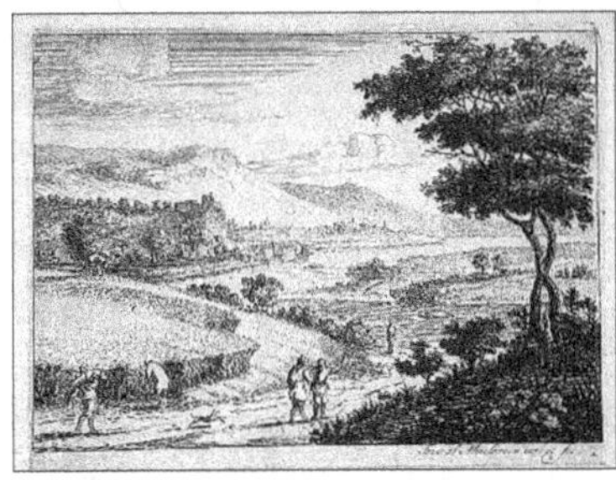

Jan van Almeloveen, *The Road (Landscape with Harvesters)* (1683), laid paper, 6 x 7 ⅗ inches, from the Dutch Cannenburch Castle collection of Baron Frederik Carel van Isendoorn à Blois

Richard Blome, *Agriculture or Husbandry, Gentleman's Recreation* (1686), 9 x 14 ½ inches

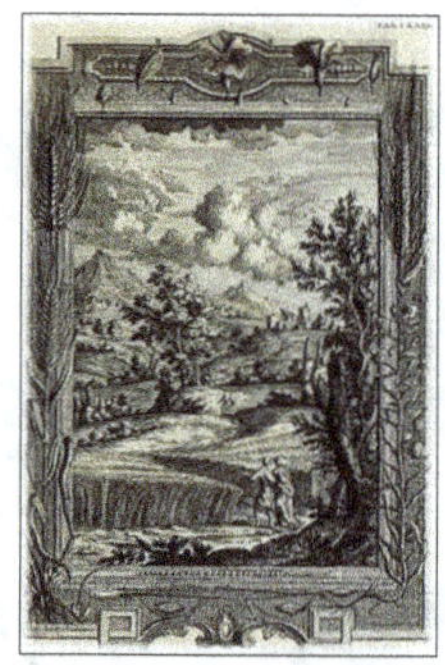 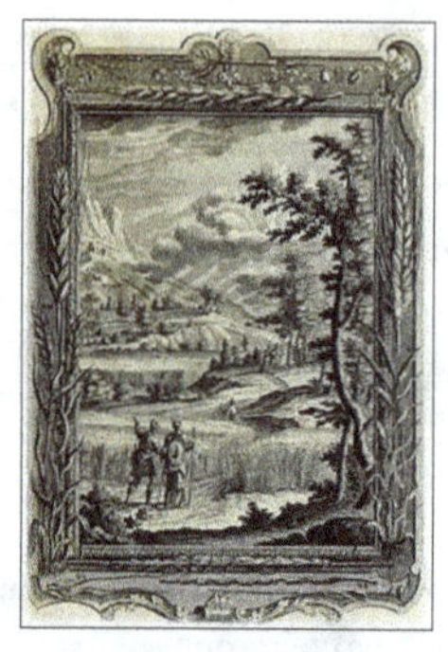 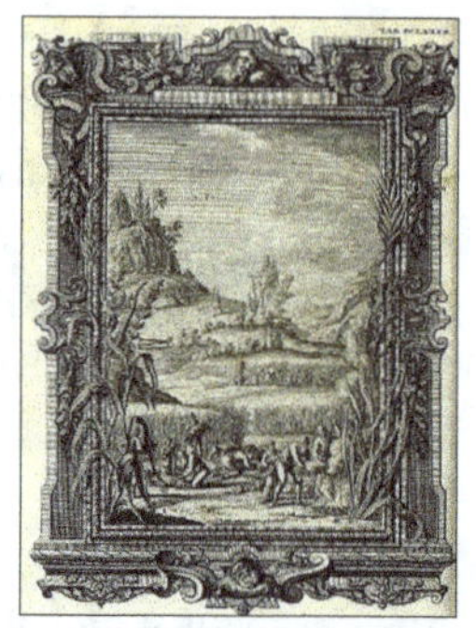

A.   B.   C.

A. Jacob Fredrich, *Flax and Barley (Exodus 9: 31)*

B. M. Tyroff, *Emmer and Wheat (Exodus 9:32)*

C. G. D. Heüman, *Gideon's Threshing Floor*, and I. G. Pintz, *Tares in the Wheat (Matthew 13:25)*, from Johann Scheuchzer, *Physica Sacra* (1731-1735), laid paper, 10 ¼ x 16 inches

Gerard Hoet, *Ruth Gleaneth in the Fields of Boaz. Ruth 2:3*, from Pieter de Hondt, *Taferelen der voornaamste Geschiedenissen van het oude en nieuwe Testament* (Gravenhage [The Hague], 1728), laid paper, 7 x 10 ¼ inches

Jan Luyken, *Bounty of the Land* (Zechariah 8:12), from *Beschouwing der Wereld* (Amsterdam, 1735), laid paper, 3 ⅛ x 3 ½ inches

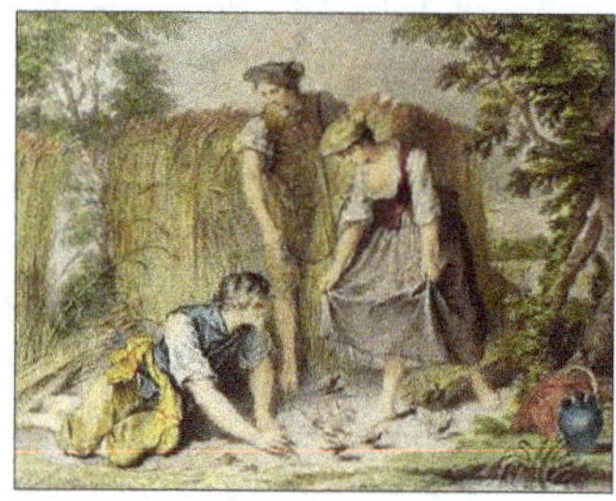

Nicolas Lancret, after J. G. Hertel, *Summer* (c. 1735), mezzotint on laid paper, 14 x 16 ½ inches

P(ierre) L. Surugue, after David Teniers the Younger, *L'Ete (Summer)*, (Paris?, 1749), engraving on paper, 6 ½ x 9 inches

M(ichel) Picquenot, after David Teniers the Younger, *Du Cabinet de Mr. Poullain (Village Fête)* (Paris?, 1780), engraving on laid paper, 7 ½ x 6 inches

F[rans] Bleyswijck, Harvest Scene (detail), copperplate engraving on laid paper, Georges-Louis Leclerc, Comte de Buffon, *Histoire Naturelle* (Amsterdam, 1775), 1 ½ x 3 inches

J. A. Defehrt, *Agriculture* (2 plates), from Denis Diderot and André Le Breton, *Encylopédie ou Dictionnaire Raissoné* (1762), laid paper, 10 ½ x 16 ¾ inches

F[rans] Bleyswijck, copperplate engraver, Harvest Scene, Georges-Louis Leclerc, Comte de Buffon, *De Natuurlyke Historie* (Amsterdam, 1775), laid paper, 1 ½ x 3 inches

Rembrandt van Rijn, engraver, *Landscape with Hay Barn and Flock of Sheep* (1650), Claude Henri Watelet, printer, copperplate counterproof on laid paper (c. 1780), 4 ⅛ x 7 ½ inches; Firmin Delange-Albert Quantin, Paris printer, heliograph on Holland paper (1880)

Thomas Stothard, *Gleaning and Harvest Home*, James Thomson, The Seasons (London, 1794), steel engravings on laid paper, 2 ⅞ x 3 ⅝ inches

Manner of Threshing in the East, Augustine Calmet, *Fragments of the Holy Scriptures* (1800), 5 x 8 inches

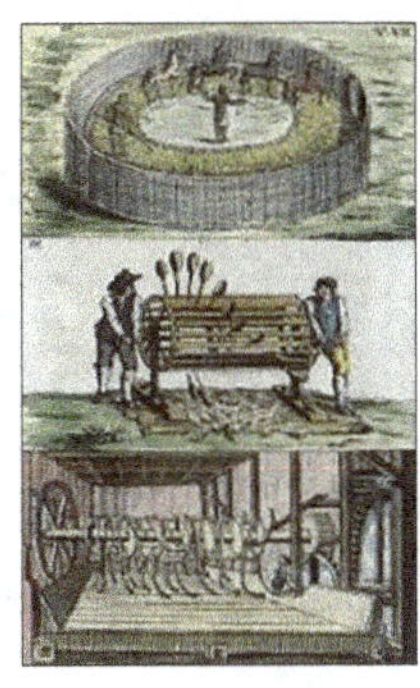 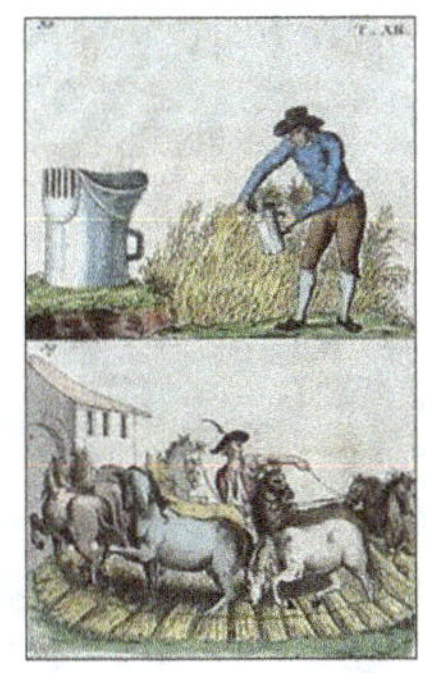

A.   B.

Vue de la Halle au Ble [*View of the Hall of Grain*], J. B. De Saint-Victor, *Tableau Historique et Pittoresque de Paris* (Paris, 1808), laid paper, 7 ½ by 10 inches

A. & B. Attributed to Jacob Schmutzer, "Methods of Threshing and Flailing Grain," Gottlieb Tobias Wilhelm, *Unterhaltungen aus der Naturgeschichte* (Augsburg, 1810), laid paper, 3 ¾ x 6 ½ inches

R[obert] Hills, etcher; J[oseph] C. Stadler, engraver, "Brussels, from Lacken," Robert Hills, *Sketches in Flanders and Holland* (London, 1816), 13 ⅔ x 10 ⅗ inches

Jacopo Bassano (artist), E. Achille Reveil (engraver), *The Harvest* (c. 1575), *Musée de Peinture et de Sculpture*, Vol. XII (Paris, 1831)

W. H. Bartlett, "The Corn Market, Colchester, Essex, 1831," Virtue's *Picturesque Beauties of Great Britain* (London, 1834)

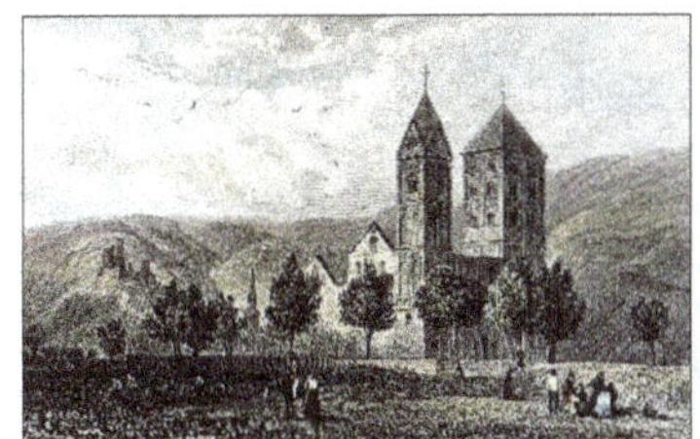 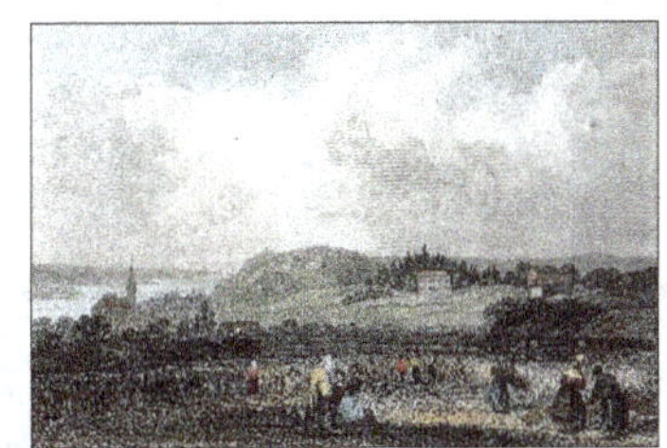

A.   B.

A. William Tomblesons, "St. John's Church, near Niederlahnstein" and

B. V. Bing, "Village of Oosterveek near Arnhem," N. G. van Kampen, *The History and Topography of Holland and Belgium* (London, 1837), 6 x 3 ⅞ inches

T. Atalm, E. Benjamin (engraver), *Perth* [Harvest], (London, 1837), 7 x 4 ½ inches

J(ohn) M(assey) Wright (artist), G. Presburg (engraver), *The Rigs o' Barley* (London, 1839), 7 ⅞ x 10 ¼ inches

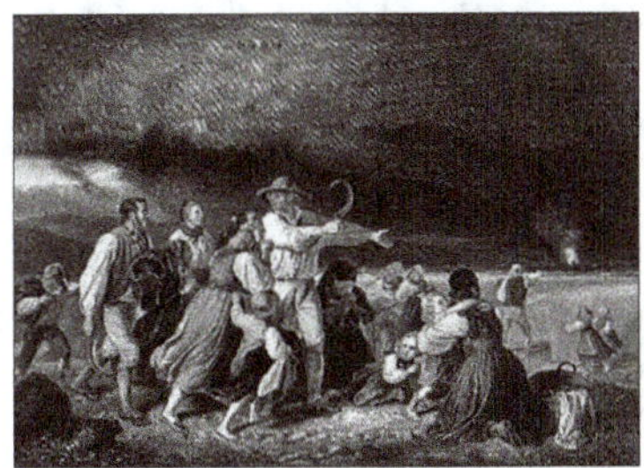

Jacob Becker, W(illiam?) French (engraver), *Die Schnitter* [*The Reapers*] (Dresden, 1840), 6 ½ x 4 ¾ inches

Frederich Gauermann, L. Koch (engraver), *Die Ernte* (also *Harvesting in the Foothills of the Alps*) (c. 1840/1880), 9 ½ x 6 ¾ inches

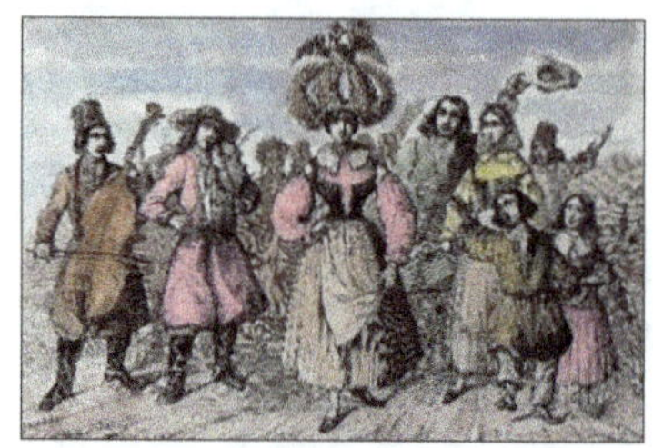

A.

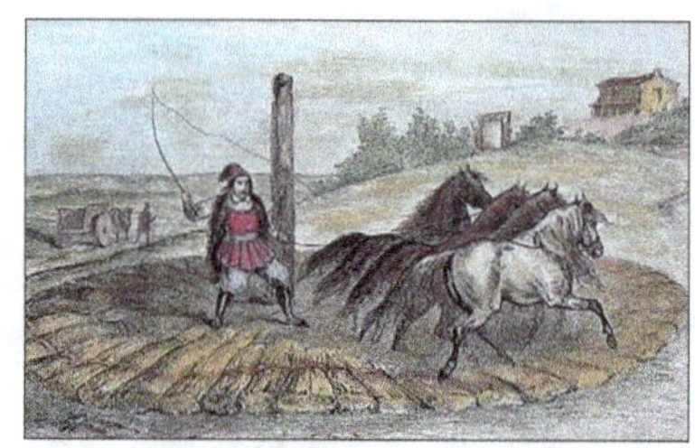

B.

[A. C.] Lemaître, engraver, after [Charles] Vernier, *Russia—Trader Calculating Grain Sale; St. Petersburg—Peasants at the Grain Market* (1838)

A. *Poland—Festival of the Harvesters* (1840)

B. *Threshing Wheat—Sardinia* (1847), Paris: *L'Univers Pittoresque*, 5 ¾ x 4 ⅛ inches

Charles Tschaggeny (artist), J. Cousen (engraver), *The Harvest-Field* (c. 1850) *The Art-Journal* (London, 1857), 11 ⅞ x 8 ⅝ inches

William H. Bartlett, Milo Osbourne (engraver), *Harvest Scene in New Hampshire* (New York, c. 1855), 7 ⅛ x 4 ⅜

(William?) French (engraver), after (Franz?) Pittner, *Ernte in der Romagna* (c. 1850), 10 ¾ x 7 ¾ inches

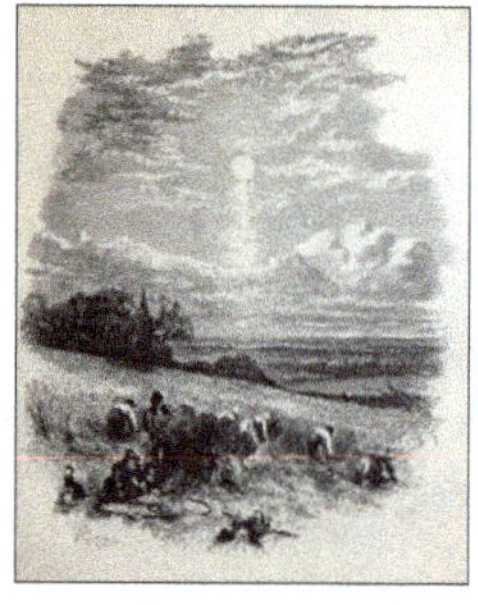

A.

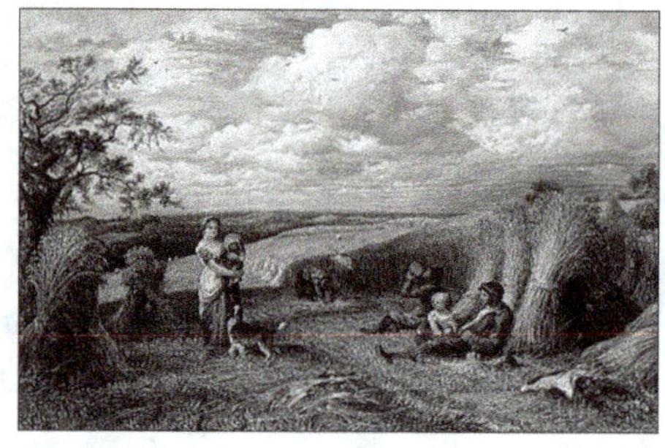

B.

A. Birket Foster, *Summer* Harvest Scenes (2), Robert Bloomfield, *The Farmer's Boy* (New York, 1858), 3 ½ x 4 ½ inches

B. John Linnell, *Harvest Rest*, from *The Art-Journal* (London, 1862), 9¾ x 6 ½ inches

Henri F. Schopin, C. Cook (engraver), Luke Limner (border), *Boaz and Ruth* (London: Blackie & Son, 1862), 7 ⅜ x 10 inches

Cornelius Varley, artist; J. W. Lowry, engraver, *Bell's Improved Reaping Machine* (London: Blackie & Son, c. 1865), 9 ¾ x 6 ½ inches

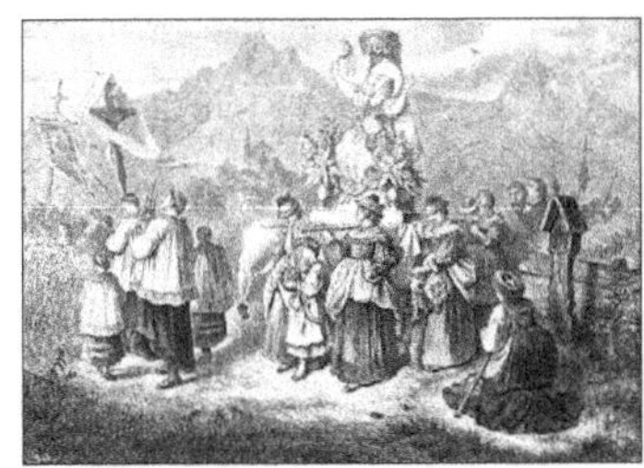

A.

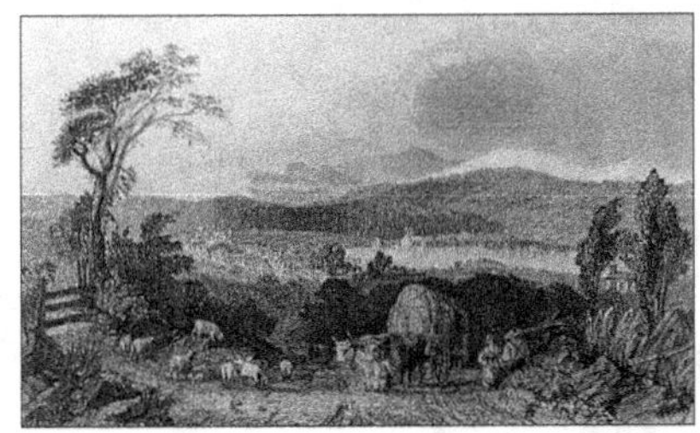

B.

A. Heinrich? Streller, *Erntebittgang* (Leipzig: Ernst Keil Nachfolger, 1868), 11 ⅛ x 7 ¾ inches

B. William H. Bartlett, *A Harvest Scene in New Hampshire* (c. 1838), *Pictorial History of the World's Great Nations*, Vol. III (1882), 4 ¼ x 7 ¼ inches

John Constable, *Cornfields near Brighton* (c. 1824), David Lucas (engraver, 1855), 8 x 12 ¾ inches

Karl Hermann Schmolze (1823-1861), *Harvest Nooning* (c. 1860), 3 ⅝ x 4 inches

Emil Hochdanz, W. Votteler (engraver), German Harvest Home Scene (1865), ¾ x 5 ½ inches

B. Vereschaguine, A Marie (engraver), *Moissonneur* [Harvester], M. Edouard Charton, Du Monde (Leipzig, 1869), 4 ⅝ x 3 ⅛ inches

Jules Breton, Amadee Greux (engraver), *Harvest Field Worker* (c. 1870), 3 ⅝ x 3 inches

Emmanuel Leutze (1852), *Mrs. Schuyler Firing Her Corn Fields on the Approach of the British* (New York: Johnson, Wilson & Co., 1874 copyright), 7 ¼ x 10 ¾ inches

Adolphe Pierre Leleux, P. E[dmond] A. Hedouin (engraver), *Reaping* (1874), 4 ⅞ x 3 inches

T. Gérard, G. C. Finden (engraver), *Returning Home* (New York: D. C. Appleton & Co., 1873/1878), 10 ⅛ x 6 ⅝ inches

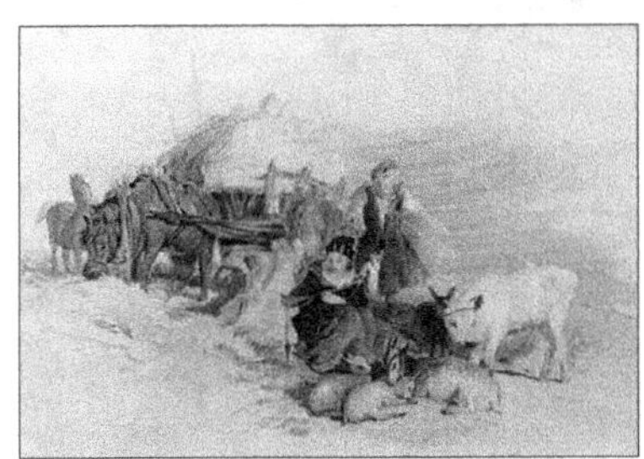

Edwin Henry Landseer, C. G. Lewis (engraver), *Harvest in the Highlands* (New York: D. Appleton & Co., 1831/1878), 13 x 9 ½ inches

Gustav Doré, A. F. Pannemaker (engraver), *Boaz and Ruth* (color and black & white versions,1880?), 7 ¾ x 9 ¾ inches

Richard Beavis, C. Cousen (engraver), *Threshing Corn*, from *Picturesque Palestine* (color and black & white versions, 1881), 9 ⅞ x 7 inches

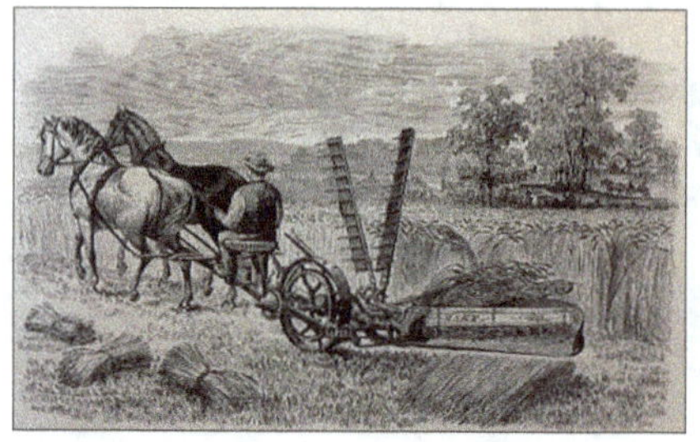

James Wilson & Son, *Art Designs in Harvesting Machinery* (1884), 29 steel engravings on paper, 8 ¾ x 12 ¾ inches: *Champion Sheaf Carrier, Sheaf Carrier at Work, Sheaf Carrier Attachment, Clover and Flax Attachment, Straight Line Pitman, Champion New Mower (3), Champion New Mower Passing a Tree, Champion New Mower, "The Flyer," Champion Combined No. 4 Self-Raking Reaper and Mower, Champion No. 1 Single Reaper, Champion Combined Chopper and Mower, Grain Wheel Lever, Champion Light Binder (3), The Champion Knotter, Relief Rake, Champion Automatic Weight Trip, Master Wheel Gear, Head Raker Head Raker Attachment, Canvas Cleaner, Champion Ball Joint Mower, Driving Power, Factory Scenes (3)*

Albert Henry Fullwood, *An Australian Harvest* (1888 reproduction), 7 x 6 ½ inches

Rudolf Lehmann, *The Reaper and the Flowers* (1893, steel engraving), 10 x 7 inches

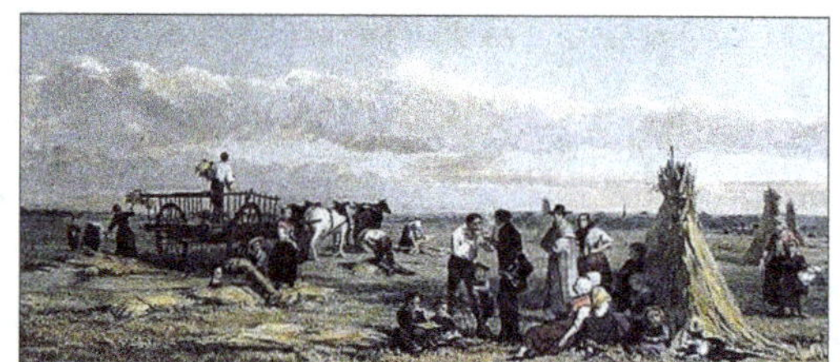

A.

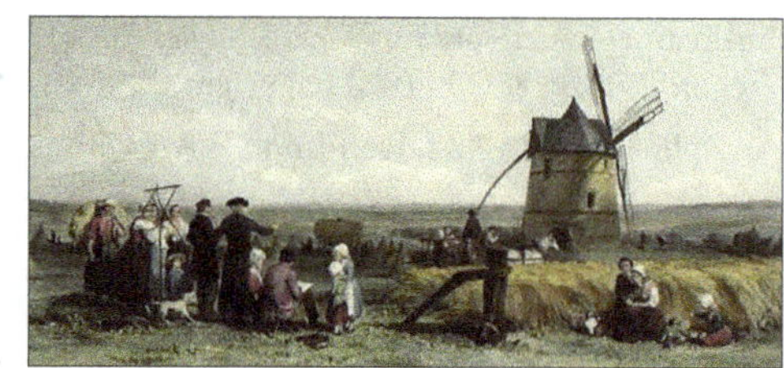

B.

A. John Absolon, *The Field of Agincourt* (E. Radcliffe, engraver)

B. *The Field of Cressy* (H. Robinson, engraver; 1857/c. 1890), 18 x 34 inches

Edward Gay, *Ripe and Ready (Pelham Bridge, Westchester County, New York)*, c. 1890, 7 ¼ x 4 ¾

Emil Limmer, Harvest Scenes Collage, 1890 (Germany, 1891), 6 ¾ x 9 ¾ inches

Schnitterfest in Kleinrussland [*Reaper Procession in Ukraine*], c. 1895, 9 ½ x 6 ⅛ inches

Julius Agghazy, Gustav Morelli (engraver), *Kranztragende Schnitter* [*Reapers Procession*], *Die Öesterreichische-Ungarische Monarchie in Wort und Bild* (Vienna, 1900), 7 ⅜ x 5 inches

Jean-François Millet, *Gleaning* [*The Gleaners*], 1857 (London: The Studio, 1902 etching), 6 ⅞ x 5 ⅛ inches

Leopold Kalckreuth, The Reapers, Charles Holm, *Modern Etching and Engraving* (London: The Studio, 1902), 5 ¾ x 6 ¾ inches

Vicat Cole, Clough Bromley (engraver), *A Surrey Landscape, from Modern Etching and Engraving* (London, 1914), 10 ½ x 6 ⅛ inches

Léon Lhermitte, *Paying the Harvesters* (1882) (New York: The Mentor Association, 1918 reproduction), 8 ⅞ x 6 ¼ inches

Henry J. Stuart Brown, *Harvest Field, Ely: Evening* (1926), drypoint on paper, 5 ½ x 10 ⅞ inches

C. R. W. Nevinson, *Sussex Downs*, 5⅞ x 9 ¼ inches

André Smith, Reaper's Rest, *Fine Prints of the Year*, New York (c. 1930), 5 ⅜ x 4 ¼ inches

T. Shibbe?, *Harvest on the Sea* (1930), 13 ⅜ x 18 ½ inches

B. Art and Other Book Illustrations

Myles Birket Foster, *Building the Haystack* (color), *The Reapers*, *The Hayfield* (London: *Pictures of English Landscape*, 1863), 5 ¼ x 7 inches

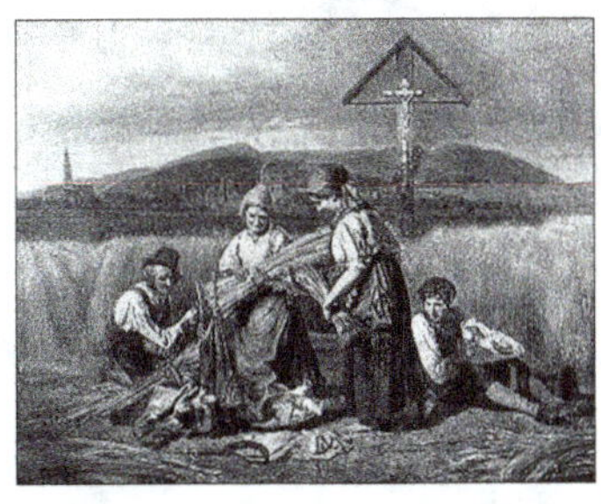

A.

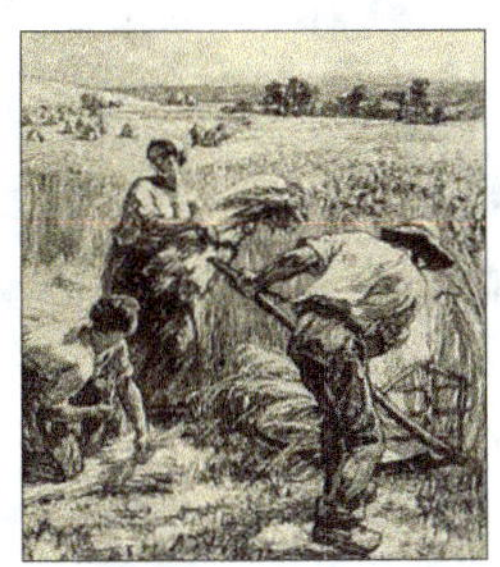

B.

A. Ferdinand Waldemüller, Friedrich Oldermann (engraver), *Die Ernte* (1847/1887), 7 ½ x 9 inches

B. Léon Lhermitte, Clément Bellinger (engraver), from André Theuriet, *La Vie Rustique: Compositions et Dessins de Léon Lhermitte* (Paris, 1888)

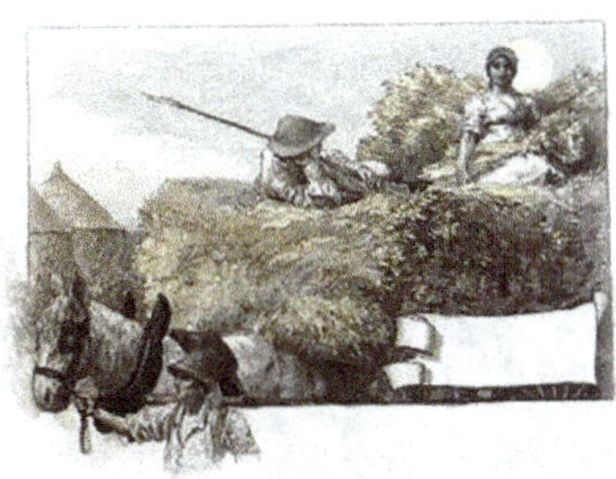

Marie & West [?], untitled harvest color lithographs (2) to accompany poetry from Helen J. Wood ("But the harvest wain goes rumbling on/And the air is soft and dim. . . ."), *Through the Woodland and Meadow & Other Poems* (Boston: E. P. Dutton, 1891), 7 x 10 and 8 x 6 inches

Léon Lhermitte, in André Theuriet, *Rustic Life in France* (Boston, 1896), 26 engravings

Lèopold Lesign, *La Porte de Mars à Reims, 1901*, in *La Revue de l'Art ancient et modern*, Vol. XI (1902), laid paper, 8 ¾ x 10 ½ inches

Harold Sutton Palmer, *Wet Harvest Time near Dalmally, Argyllshire*, in A. R. Hope Mancrieff, *Bonnie Scotland* (London, 1904), 6 x 8 inches

Otto Ubbelohde, grain sheaves engravings, in *Die Erntezeit*, (Munich, 1921)

Frank Ormrod, harvest woodcuts (4), in Fred Kitchen, *Life on the Land* (London, 1941), 4 ¾ x 6 ¾ inches

Vincent van Gogh, in L. Goldscheider and W. Uhde, *Vincent van Gogh*, Grainfield series paintings (London: Harrison & Sons [color plates]; New York: The Oxford University Press, 1945), various large format sizes

Charles Tunnicliffe, *Threshing, Cutting the Wheat Farm Workers Raking Hay, The Old and the New*, from *Both Sides of the Road: A Book about Farming* (London: Collins, 1949), 10 ¾ x 8 inches

III. LITHOGRAPHS AND OTHER PRINTS

A. Popular Editions and Offprints

Louis-Léopold Boilly, *L'ete* (1824), 4 ¼ x 6 ⅛ inches

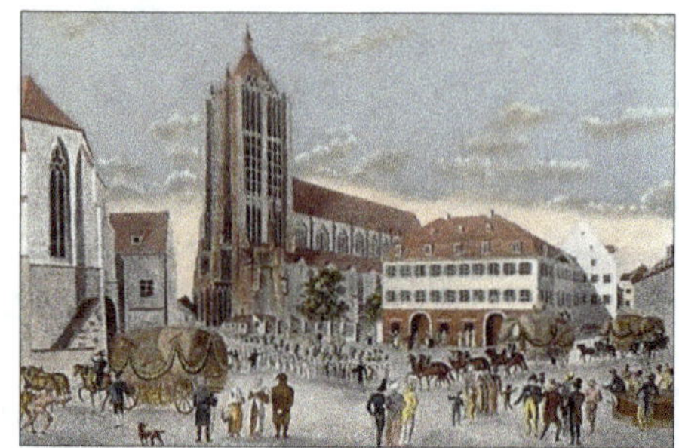

A.

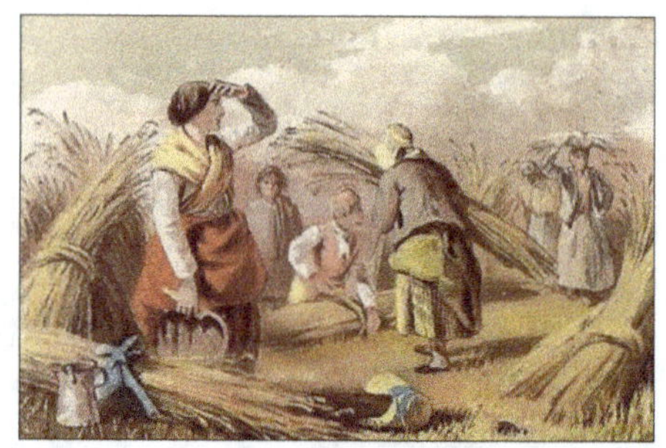

B.

A. Johann Hans, *Celebration of the Harvest Festival in Ulm*, (1817/c. 1900), 9 ½ x 13 ¼ inches

B. William Lee, Edwin Buckman (lithographer), *Poetry of the Year* (London, 1867), 5 ¼ x 3 ½ inches

James K. Colling, *Southfleet [St. Nicholas] Church* (London, c. 1870), 7 ¼ x 11 inches

Frederick A. Bridgman, *Pyrenees Peasants Returning from the Harvest Field* (1872), George W. Sheldon, *American Painters* (New York: D. Appleton, 1881)

Charles Stanley Reinhart, *Primitive Thrashing* (Spain, 1883)

After Ferdinand Waldemüller, *The Harvest* (1847), 7 ½ x 9 inches (1887)

Thomas Rowlandson, *The Harvest Home* (1821/c. 1860), 8 ½ x 4 ⅞ inches

J. C. Hook, *A Devonshire Harvest Cart* (London: J. S. Virtue & Co., 1887), 9 ⅞ x 7 ½ inches

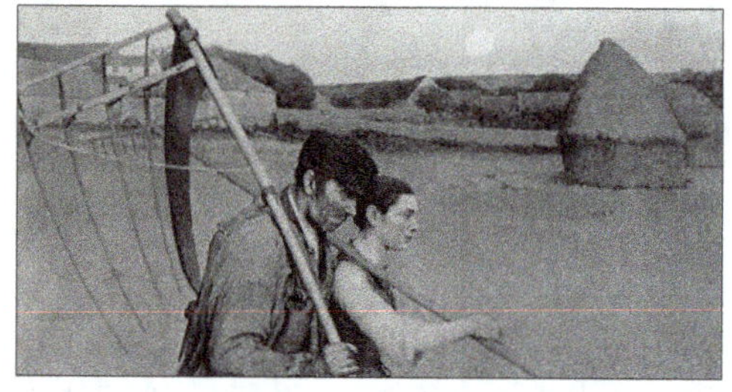

A.

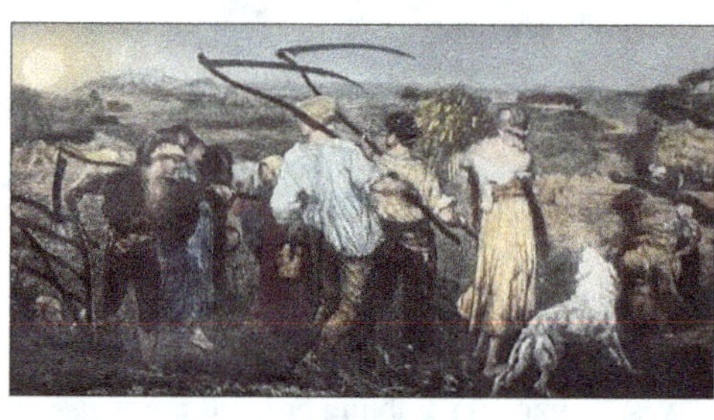

B.

A. Guy Rose, *The End of the Day* (New York: D. Appleton, 1891/1893), 12 ⅞ x 5 ¼ inches

B. George F. Wetherbee, *The Harvest Moon* (1881), 10 ½ x 27 ½ inches (mezzotint, c. 1900)

Léon Lhermitte, *The Harvest* (1883/undated lithograph reproduction), 8 ½ x 7 ½ inches

Jules Breton, *The Evening Call* (1889, Philadelphia: [George] Barrie), hand-colored photogravure from *Chefs d'Oeuvre de L'Exhibition Universelle*, 1889), 10 x 6 ¼ inches

F[rederick] A. Bridgman, *Pyrenees Peasants Returning from the Harvest-Field* (1872/1889), 4 ⅝ x 7 inches

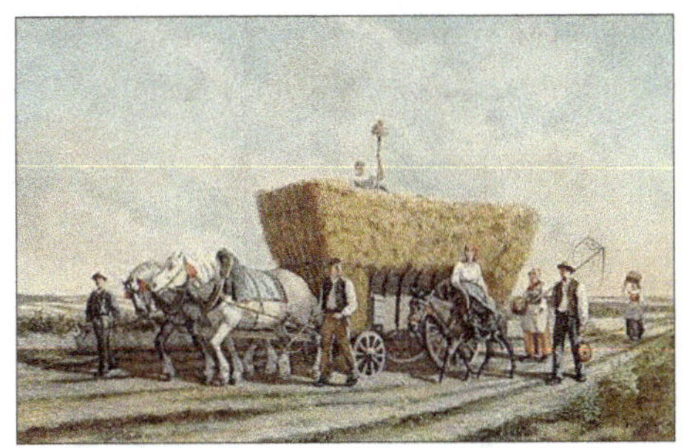 A.

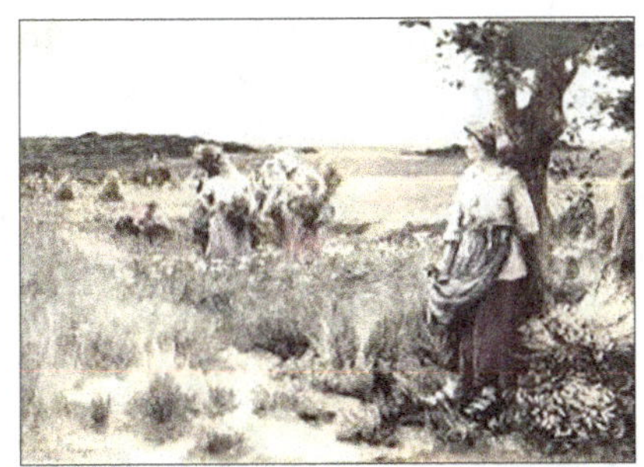 B.

A. Jacques Veyrassat, *Last Load of Wheat* (1892/1893), 11 ⅞ x 8 ⅛ inches

B. Georges Laugee, *Matinée d"Eté* [Summer Morning], (1899), 7 ⅜ x 5 ⅜ inches

Paul-Albert Baudoüin, *The Corn-Field—Carters: A Decorative Frieze* (1892/1893), photolithograph, London: Sprague & Company, 9 x 31 inches

Florence A. Saltmer, *Field of the Cloth of Gold* (c. 1895), 6 ½ x 9 inches

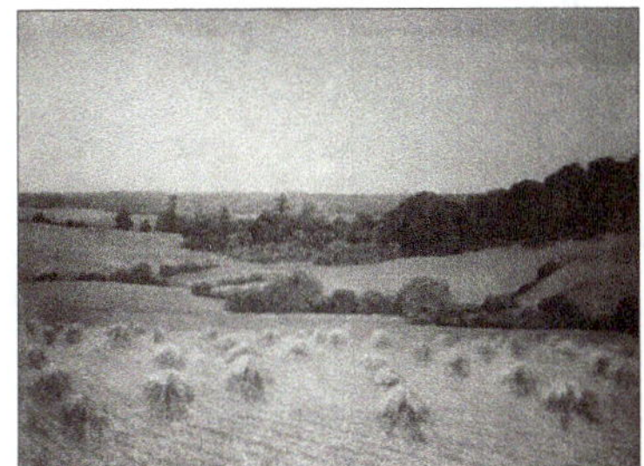

William H. Coffin, *Moonlight in Harvest* (1897), 10 ½ x 14 ⅜ inches

Wilfred Ball, *Harvesting*, 5 ¾ x 3 ½ inches (c. 1900 reproduction)

Myles Birket Foster, Gleaner Group and Witley Church Harvest Scenes (1906), 5 ½ x 3 ⅝ inches

Ellen Wikinson, *Field with Skamlingsbanken in the Distance*, 4 ½ x 3 ½ inches, and *A Danish Farmstead* (1909 reproductions), 5 ½ x 3 ½ inches

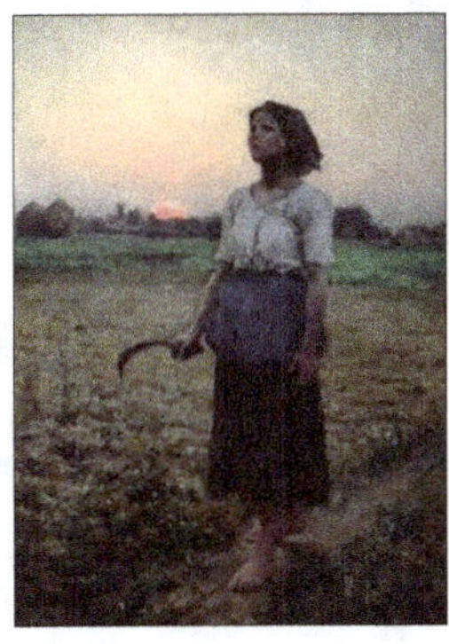 A.

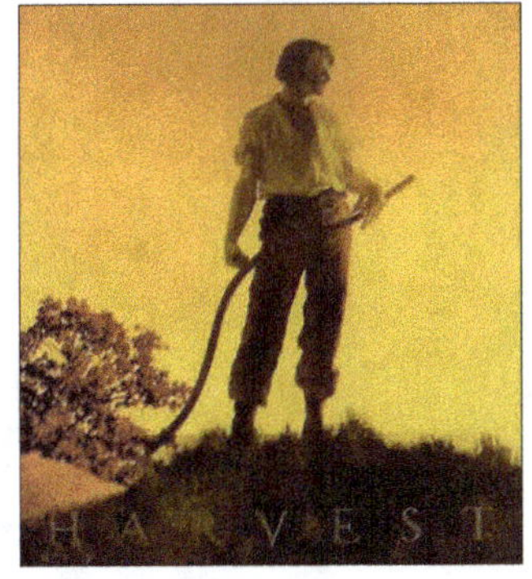 B.

A. Jules Breton, *Song of the Lark* (1884, c. 1910 reprint), 8 x 12 inches

B. Maxfield Parrish, *Harvest, Collier's* (September 24, 1905, c. 1910 and 1915 calendar reproductions), 5 ¾ x 7 ⅝ inches

Niccolo dell Abbate, *Harvest* (c. 1571/Berlin, c. 1910 reproduction), 6 ½ x 5 inches

William Taggart, *Harvest at Broomieknow* (1897), *The International Studio* (1909), 8 ⅛ x 5 ½ inches

Josef Jungwirth, *Beim Erntemahl* (c. 1910), 7 ½ x 5 ⅛ inches

C[harles] Murand, *The Seasons* series (c. 1910), 4 color prints, 6 x 7 ½ inches

O. E. Bossert, *Harvest*, *The International Studio* 12 (1910), 5 ¾ x 4 ½ inches

Edward Davies, *Harvest Time, Evening* (c. 1895), *The International Studio* (1911), 8 x 11 ½ inches

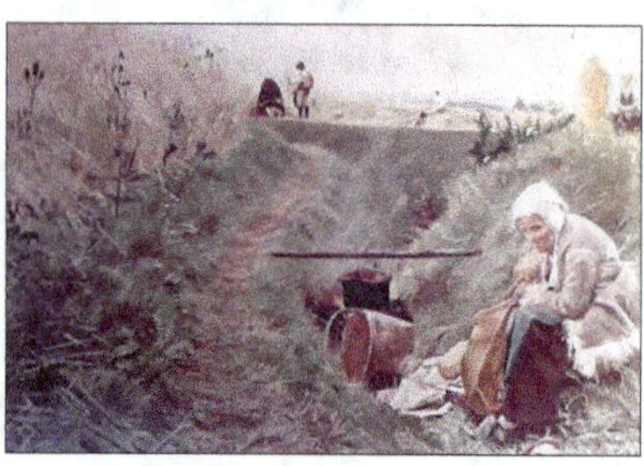

Anders Zorn, *Our Daily Bread* (1886), *The International Studio* (1911), 8 x 11 ½ inches

Jean Charles Cazin, *Les Meules* (c. 1885), *The International Studio* (1911), 8 x 11 ½ inches

E. A. Waterlow, *The Harvest Moon* (1914 color reproduction), 7 ¼ x 5 inches

Jules Breton, *The Gleaner* (1887/c. 1920 reproduction), 6 ⅜ x 11 ¾ inches

F. Mason, *All Among the Barley* (c. 1920 reproduction), 7 ⅛ x 4 ⅝ inches

Peter De Wint, *A Harvest Scene* (c. 1830/c. 1920 reproduction), 8 ⅞ x 6 ½ inches

H[?]. F[rank?]. English, *Western Wheat Field* (c. 1920), 9 x 6 ½ inches

Jean-François Millet, *Les Glaneuses* (1857/c. 1925 reproduction), 6 ½ x 4 ¾ inches

Edizioni Falteri, *Summer* (c. 1850/c. 1930 reproduction), 7 ¼ x 7 ¾ inches

Frank English, *Harvest Time and Blossom Time* (c. 1930 color reproductions), 9 ⅞ x 7 ⅞ inches

Wilhelm Steinhausen, *Landschaft mit Garben* [Landscape with Sheaves] (1930 color reproduction), 5 x 8 inches

Francesco Bartolozzi (engraver), after a painting by William Hamilton, *August* (c. 1790/c. 1930 Sidney Lucas reproduction), 10 ½ x 7 inches; *Agosto* (1761/c. 1950), 11 ¾ x 10 inches

The [Canadian] Prairie in Harvest Time (c. 1935 color reproduction), 6 ¼ x 5 inches

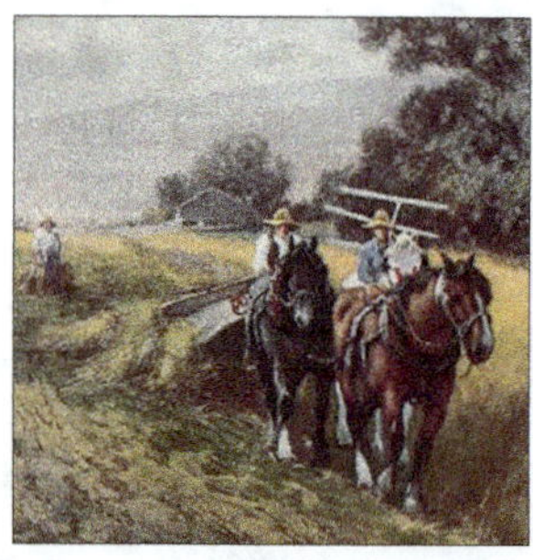

"F. M. T.", Harvest Reaper scene for Hampen Market, Soap Lake, Washington calendar (1935), 5 ¼ x 9 ¼ inches

Adolf Dehn, *Minnesota in August* (1938/1939), 21 ½ x 18 inches

John Dukes McKee, untitled harvest scene (c. 1950), 17 x 12 inches

Allen Mold, *Threshing* (1949), 6 x 5 inches

N. C. Wyeth, *The Scythers* (1908/c. 1970), 6 x 8 ¾ inches

N. C. Wyeth, *Mowing* (1907/c. 1970), 6 x 8 ¾ inches

Robert Motherwell, *Summer Light Series* Portfolio: *Harvest, with Blue Bottom*; *Harvest, with Leaf*; *Harvest, with Two White Stripes* (1973 color reproductions), 7 ¼ x 8 ¼ inches

Fannie Palmer, *The Farmers Home-Harvest* (1864); *The Wheat Harvest*; *The Harvest Moon*, 16 ¼ x 23 ½ inches (Currier & Ives, c. 1970 calendar reprints)

Vincent Van Gogh, *The Harvest* [also *Noon, Rest from Work*] (1891/c. 1975), 15 ½ x 11 ⅝ inches

Marc Chagall, *Woman Reaping* (c. 1951), *The Peasant* (1926), from Werner Haftmann, *Marc Chagall: Gouches, Drawings, Watercolors* (New York, 1984), 8 ⅞ x 8 ¾ inches

Peter De Wint, *A Cornfield* (c. 1815); Alexei Venetsianov, *The Reapers* (c. 1828) and *Harvest Time, Summer* (1827), art gallery portfolio reprints, various sizes, c. 1980

Samuel Palmer, *Cornfield by Moonlight* (1830, c. 1980 reprint), 9 ¼ x 6 inches

Leopold Graf von Kalckreuth, *Summer* (1890, c. 1985 reprint)

Robert McGinnis, *Harvest Home* (1985), 22 x 39 inches

Kazimir Malevich, *Going to the Harvest: Marfa and Vanka* (c. 1928/c. 1990 reproduction), 8 ½ x 11 ½ inches

Maxfield Parrish, *Young Gleaner* (c. 1990 reprint), 5 ⅛ x 8 inches

Tom Weeks, Art About Agriculture Scyther poster (Corvallis: Oregon State University, 1993), 17 x 24 inches

Tom Weeks, Art About Agriculture Combine poster (Corvallis: Oregon State University, 1996), 17 x 24 inches

S. T. Gill, *Summer* and *September*, Bob Raftopoulos, ed., *S. T. Gill's Rural Australia, 1818-1880* (1989 reprint), 7 x 10 ½ inches

Der Erste Erntewagen, color copperplate (1817) and *Ernte bei Praunheim*, color painting reproductions (Frankfurt, Germany, c. 2000), 7 x 5 ⅜ inches and 7 x 3 ⅞ inches

William Morris Hunt, *Peasant Girl* (1852); Berthe Morisot, *In the Wheatfield at Gennevilliers* (1875); Hieronymus Bosch, *The Haywain* (1515); Thomas Gainsborough, *Mr. and Mrs. Andrews* (c. 1750); Jules Breton, *The Song of the Lark* (1884); New York: International Masters Publishers (color reproductions, 2008), 9 ¾ x 13 ⅜ inches

Ivan Shiskin, *Harvest* (c. 1850/2010 Kazan State Art Museum, Russia), 15 ⅜ x 11 ⅜ inches

L. A. Ring, *Harvest* (1885/c. 2010 National Gallery of Denmark), 9 x 11 ¼ inches

Keep Within Compass, Carrington & Bowles, London (1784/2010)

Giulio Casanova, Decorative Grain Frieze Design (1917/c. 2015)

Norman Rockwell, Reaper Holding Bird scene, *The Saturday Evening Post* (August 18, 1923/c. 2015 reproduction), 10 ¼ x 14 ¾ inches

Olga Volchkova, *Saint Wheat*, signed limited edition print (2015/c. 2020), 12 x 14 inches

John Rogers Cox, *Gray and Gold* (1942/c. 2020), The Cleveland Museum of Art/Bentley Global Arts, 11 x 14 inches

Vincent van Gogh, *Field with Stacks of Grain* (1890/c. 2022), Robert Bayer Photo Reproduction, 11 x 17 inches

Leon Hushcha, *Blue Skies Over Golden Wheat Fields* (2013), The [Minneapolis] Russian Museum of Art limited edition print (2013), 9 ½ x 12 ½ inches

B. Newspapers, Magazines, and Books

1. *The Illustrated London News*, *Sphere*, and Other British Publications

F. P. Stephanoff, W. Finden (engraver), *The Harvest Home*, from *Finden's Tableaux for 1841* (London: Black & Armstrong, 1840), 7 ¾ x 9 ½ inches

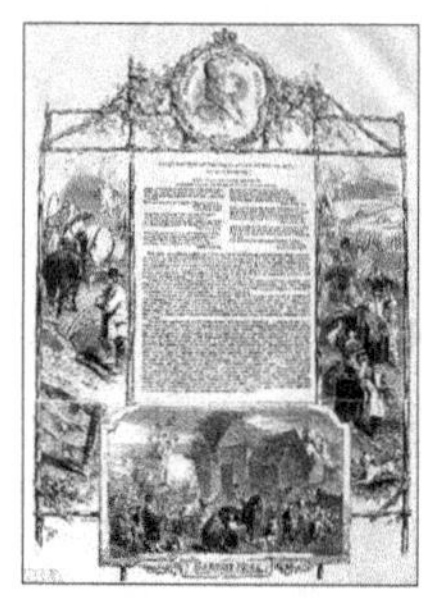 A.   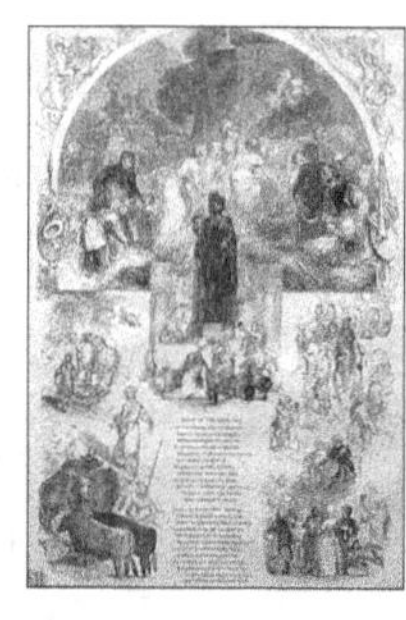 B.   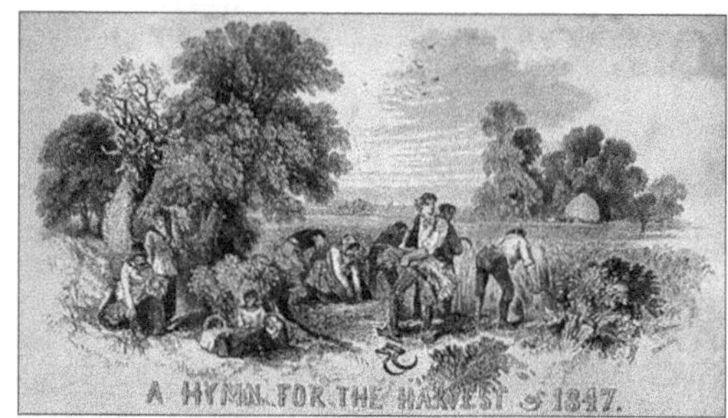 C.

A. Frederick J. Smyth, *Great Meeting of the Royal Agricultural Society at Southhampton with Harvest Home, The Illustrated London News* (July 27, 1844), 9 ¼ x 13 ¾ inches

B. William Harvey, G[eorge] Dalziel (engraver), Harvest, *The Illustrated London News* (September 26, 1846), 1 ½ x 14 ¼ inches

C. Frederick J. Smyth, *A Hymn for the Harvest of 1847 (Composed for the Thanksgiving-Day), The Illustrated London News* (August 9, 1847), 9 x 5 ¼ inches

William Harvey, *Autumn, The Illustrated London News* (September 25, 1847)

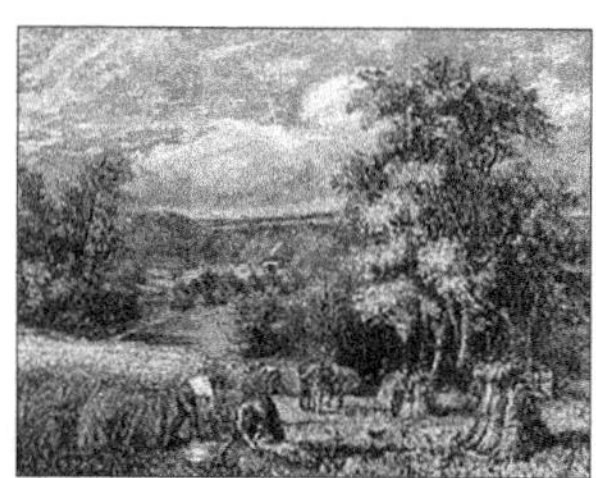

H[enry] Lupton, *The Harvest Field, The Illustrated London News* (February 24, 1849), 9 ¼ x 6 inches

An Irish Harvest Home, The Illustrated London News (September 15, 1849)

D. Jacque, *Harvest-Home Custom in France, The Illustrated London News* (September 2, 1854), 9 x 7 ⅛ inches

F. Underhill, *The Wheatsheaf, The Illustrated London News* (February 24, 1855), 8 ¾ x 6 ¾ inches

McCormick's Reaping Machine, John Morton, ed., *Cyclopedia of Agriculture* (1855), 8 ⅛ x 5 ¾ inches

The Harvest Home, Peterson's Magazine (1857), 5 x 6 ¾ inches

Harvest Operations—Reaping (A.) *Carrying, Stacking* (B.) *Gleaning, The Illustrated London News* (September 4, 1858), 9 x 14 ¼ inches

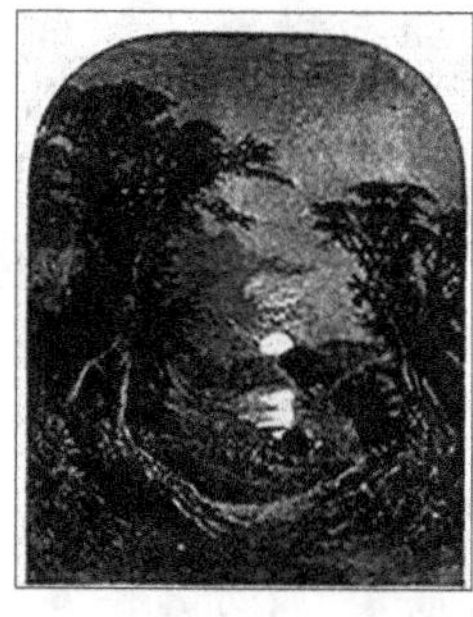 A.   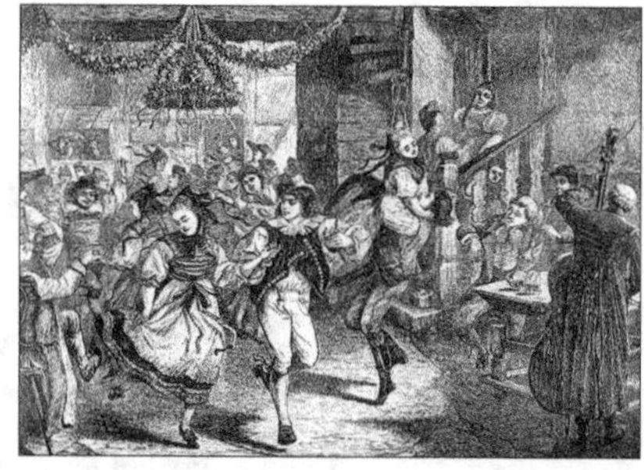 B.

A. Herbert Shepherd, *The Reapers' Song to the Harvest Moon, The Illustrated London News* (September 4, 1858), 9 ¼ x 11 ¾ inches

B. L[ionel] C. Henley, *Harvest Home in Germany, Every Saturday, An Illustrated Journal of Choice Reading* (March 20, 1870) and *An Ernteball or Harvest Festival in Germany, The Graphic* (July 2, 1870), 11 ¾ x 8 ⅞ inches

Horse-Treading and Steam-Threshing—The Old-Fashioned Process in Chili and *The New Threshing Machine in Chile, The Illustrated London News* (September 25, 1869), 9 ½ x 4 ¼ inches

Harvest Trophy at the Crystal Pavilion, The Illustrated London Times (October 14, 1871)

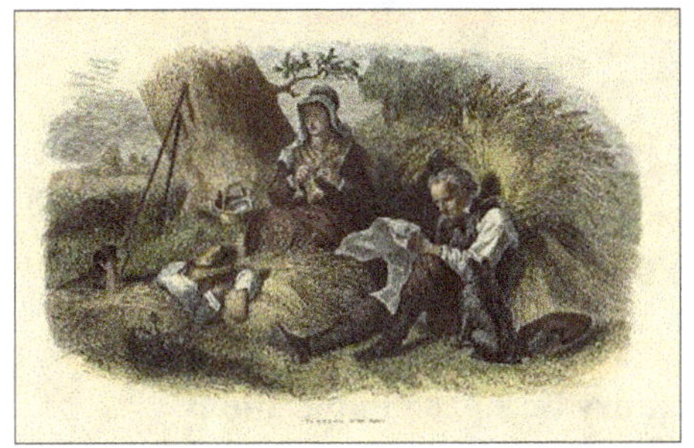
A.

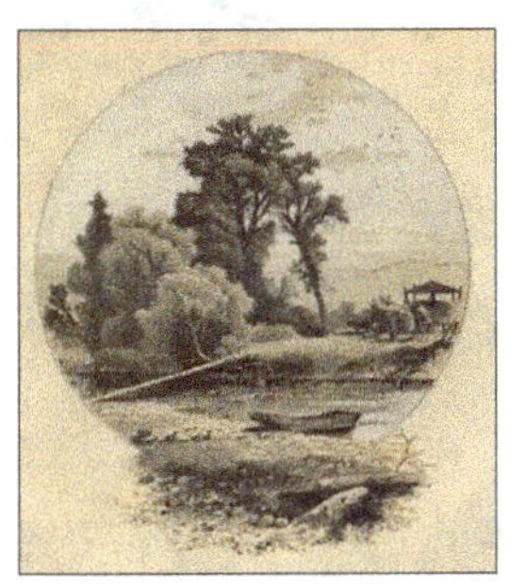
B.

A. F. O. C. Darley, Henry Linton (engraver), *The Nooning, The Aldine, An American Journal* (January 1872), 8 x 13 inches

B. John D. Woodward (engraver), *The Four Seasons—Summer*, from Samuel W. Duffield, *The Aldine: The Art Journal of America* (June 1875)

Land und Hauswirthschaft ["Farm Economy" Harvesting Equipment], *Bilder-Atlas: Ikonographische Encyklopädie der Wissenschaft und Künste* (Leipzig: F. A. Brockhaus, 1875), 10 ¼ x 13 ½ inches

Harvest Festival Service in the Church of St. Edmund, The Illustrated London News (October 30, 1875)

Slotter Ol, or *Harvest Home in Sweden, Leslie's Monthly* (August 1876), 6 ⅜ x 4 inches

James K. Colling, *Carved Vesica—Wheat and Tares, Art Foliage for Sculpture and Decoration* (London: By the author, 1878)

H[orace W.] Petherick, *Harvest Scenes, The Illustrated London News* (September 20, 1879)

Festival at Grimston Park, Yorkshire [Harvest Queen, Harvestman, Car of Ceres, etc.], *The Illustrated London News* (September 12, 1885), 9 x 13 inches

Harvest in Norway, The Illustrated London News (September 28, 1889), 12 ⅜ x 8 ⅞ inches

J. Walter Wilson, *A Day Among the Corn, The Illustrated London News* (September 26, 1891)

Mary Walker, *Preparing for the Harvest Thanksgiving, The Illustrated London News* (October 3, 1891), 9 x 11 ¾ inches

Gilbert James, *Ruth the Moabitess*, "Ruth in the Field of Boaz" and "At the Feet of Boaz," *The Sketch* (February 3 and 24, 1897)

F. W. Burton, *The Merry Harvest Time, The Illustrated London News* (August 27, 1898), 9 x 12 ¾ inches

Archibald Hartruck, *A Harvest Festival in the Cotswolds, The Sphere* (London, September 21, 1901)

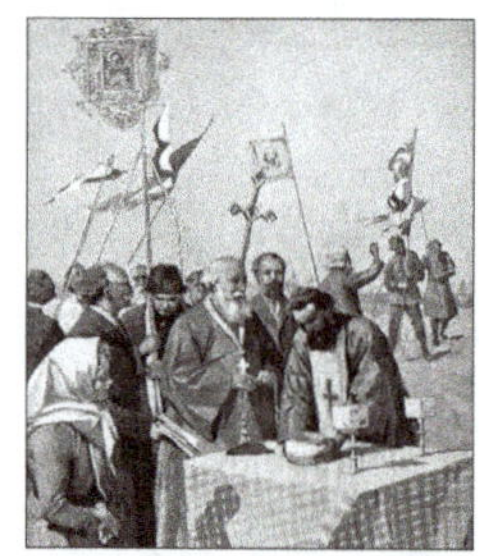 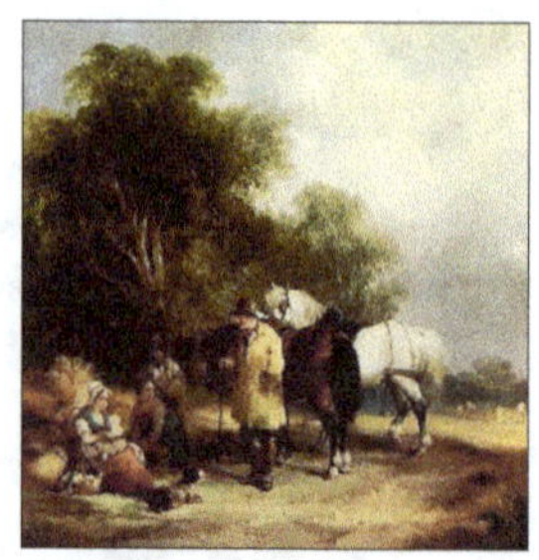

A. B.

A. Victor Prout, *The Russian Harvest, The Sphere* (September 14, 1901)

B. William J. Shayer, *Harvest Time* (c. 1850), *The Connoisseur* Vol. LXVI (January/April 1926), 6 ⅞ 8 inches

Sydney Higham, *Corn-Growing in Canada: Harvesters Entraining at Winnipeg Railway Station, En Route to the Wheat-Fields* (1904), *The Graphic*, July 29, 1905, 8 ¼ x 11 7/8 inches

2. *Le Petit Journal, Jungen,* and Other Nineteenth-Century European Publications

La Derniére Gerbe—Retour de la Moisson [The Last Sheaf—Return of the Harvesters], *L'Univers Illustré* (1867), 9 ¼ x 8 inches

J. Nash, *La Ligue Agraire en Irlande—Le Capitaine Boycott et sa Familile Paisant Leur Moisson Sous la Protection de Soldats* [The Agrarian Agitation in Ireland—Captain Boycott and His Family Gathering Their Harvest under Protection by Soldiers], *L'Universe Illustré* (Paris, 1880), 8 ¾ x 11 ¾ inches

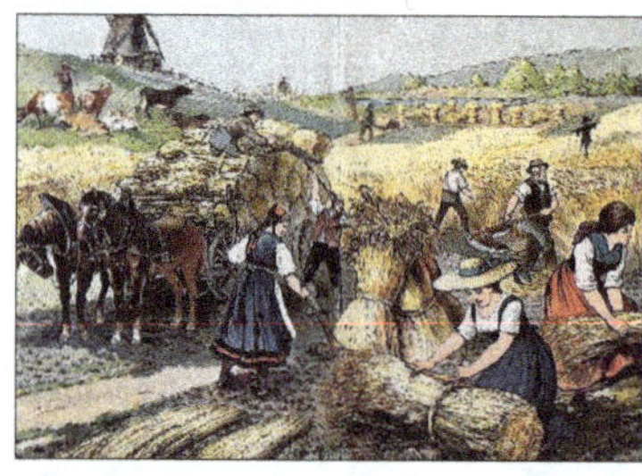 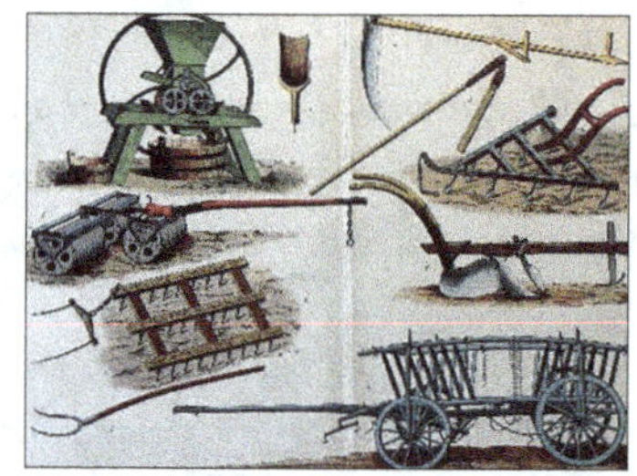

Das Feld zur Zeit der Ernte [The Field at Harvesttime] *and Landwirtschaftliche Geräte* [Farm Implements], from Eduard Walther, *Bilder aum Anschauungsundericht* (Eslingen: Baden-Württemberg, 1881)

G. Koch, *Das Rapsausreiten* [threshing scene with horses] (1882), 9 ¼ x 6 ⅝ inches

Albert Kappis, E. Hofmann (engraver), *Dreschmachine auf einem Bauerhofe* [Threshing Machine in the Farmyard], *Die Neue Gartenlaube* (1885), 14 ⅛ x 9 ½ inches

R[obert F.] Brendamour, Paul Böhm (engraver), *In der Pukta* [harvesters resting] (c. 1885), 15 ⅜ x 8 ½ inches

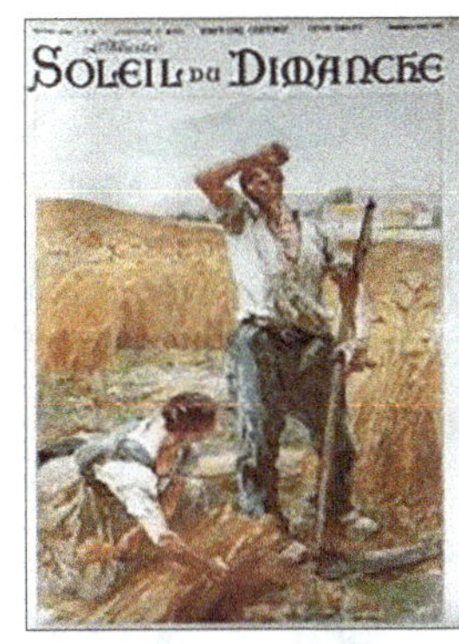

A.

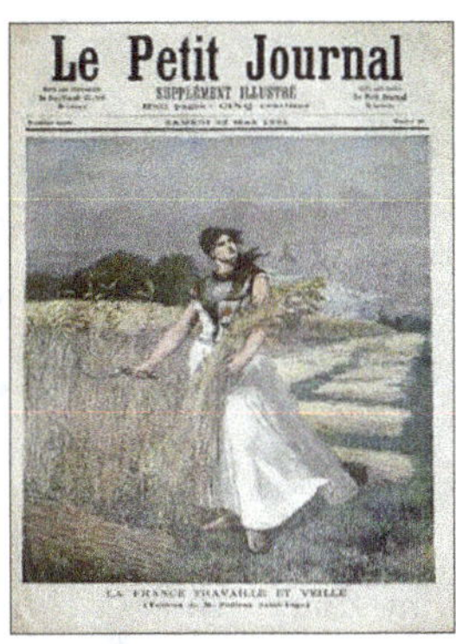

B.

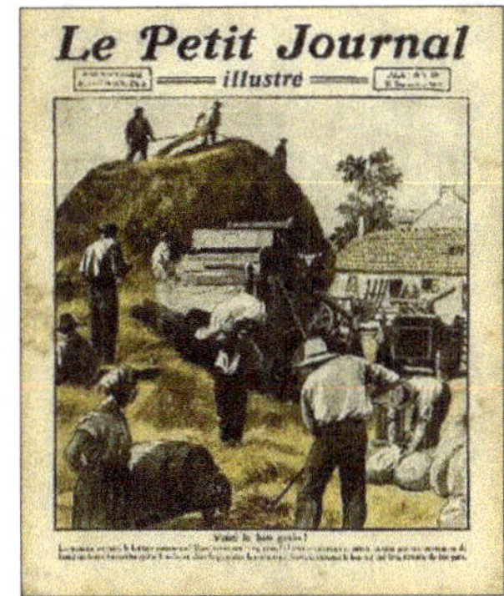

C.

A. L[éon] Lhermitte, *La Moisson* [*The Harvest*], *L'Illustré Soleil du Dimache* (August 8, 1889), 9 ½ x 12 inches

B. M. Poilleux Saint-Ange, *La France Travaille et Veille*, *Le Petit Journal* (May 23, 1891), 10 ½ x 12 ¼ inches

C. *Voici le bon Grain!* [Here is the Grain!], *Le Petit Journal* (September 21, 1924), 10 ⅜ x 10 ⅞ inches

Albert Rigolot, *La Batteuse, Loiret* [The Threshing Machine, Loiret] (c. 1893), 7 x 4 ¾ inches

A. Reinhold Eichner, *Die Schnittner* [*The Scyther*], and (B.) Walther Georgi, *Zur Erntezeit* [*The Harvest Time*], *Jugend* 40 (Munich, 1901), 7 ¼ x 9 ⅝

F. Menter, R. Mahn (engraver), *Klosterskörd* [*Cloister Harvest*] (c. 1910), 9 ½ x 12 ¾ inches

R. M. Eaglins (?), *Love in the Grain Field*, *Jugend* (May 30, 1919), 9 ½ x 12 ⅜

M. Köche-Wichmann, *Beim Korndreschen* [*At the Thresher*], (1924), 11 ⅛ x 8 ¾ inches

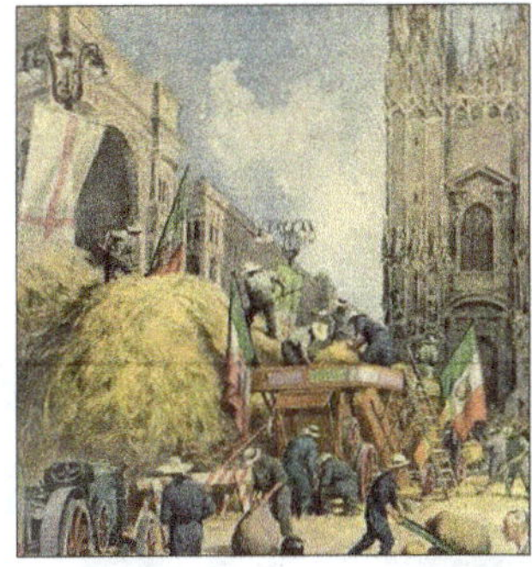

W[alter] Molino, *[Threshing] Spectacle in the Piazza del Duomo, Milan*; *La Domencia del Corriere* (July 5, 1942)

3. *Harper's*, *Scribner's*, and Other Nineteenth-Century American Publications

Farming Tools of the Present Time, as Exhibited by Robert B. Bradley & Col, New Haven, Conn(ecticut), *Agriculture in the United States*, (1868), 5 ¾ x 9 inches

Summer [Harvesters Resting scene], *Harper's Weekly* (July 17, 1869), 13 ¾ x 9 ¼ inches

Champion Combined Mower & Self-Raking Reaper, *American Agriculturalist* (May 1878), 5 ½ x 7 ⅜

Irish Harvest Scene, in Kilkenny, Ireland, *Gleason's Pictorial* (December 11, 1852), 9 ¼ x 5 ⅝ inches

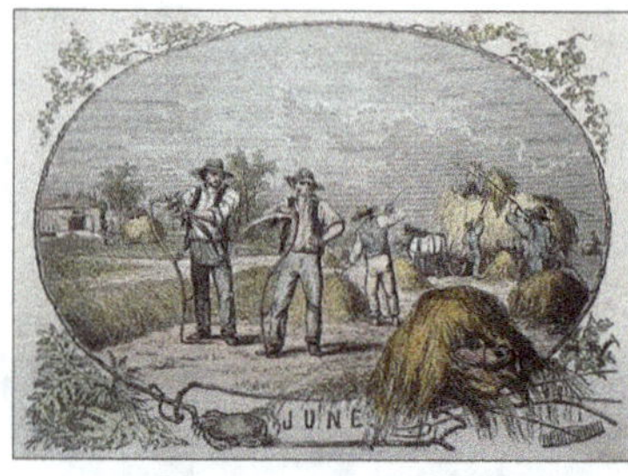
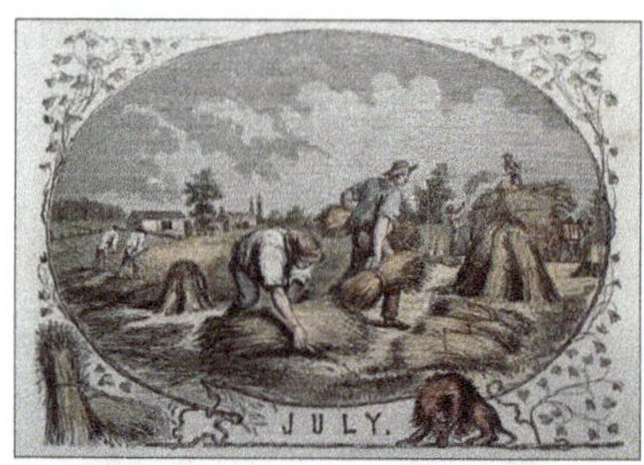

June and *July*, *Gleason's Pictorial* (June 17 and July 15, 1854), 7 x 9 ½ inches

Reaping in the Olden Time and *Reaping in our Time*, *Harper's Weekly* (August 1, 1857), 9 ¼ x 5 ¼ inches

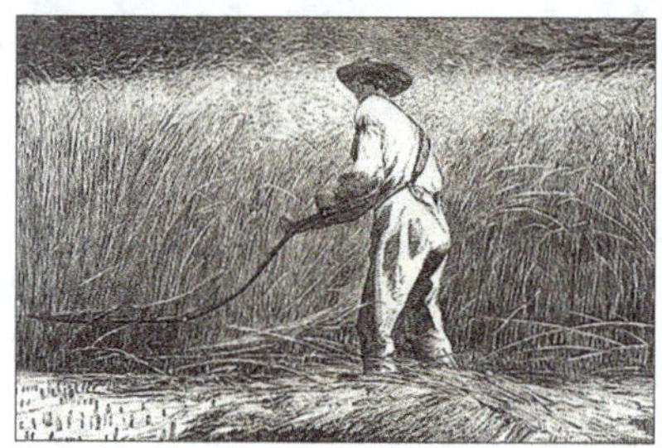
A.

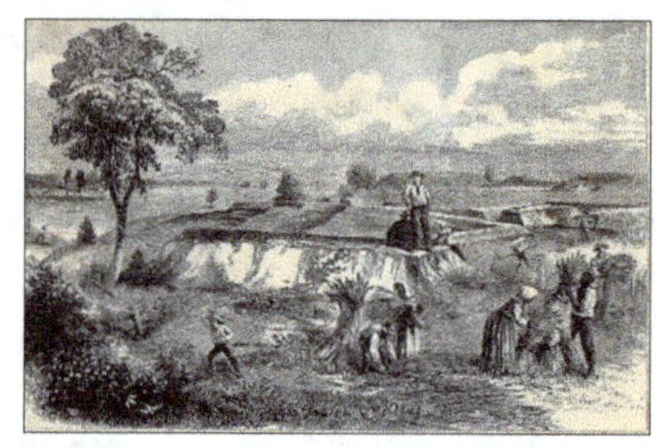
B.

A. Winslow Homer, *The Veteran in a New Field* (1865), *Leslie's Illustrated Newspaper* (July 13, 1867)

B. William L. Sheppard, *Harvest on Historic Fields—A Scene in the South*, *Harper's Weekly* (July 27, 1867), 13 ¾ x 9 ¼ inches

Farming Tools of the Present Time, as Exhibited by Robert B. Bradley & Col, New Haven, Conn(ecticut), *Agriculture in the United States*, (1868), 5 ¾ x 9 inches

Summer [Harvesters Resting scene], Harper's Weekly (July 17, 1869), 13 ¾ 9 ¼ inches

Champion Combined Mower & Self-Raking Reaper, *American Agriculturalist* (May 1878), 5 ½ x 7 ⅜

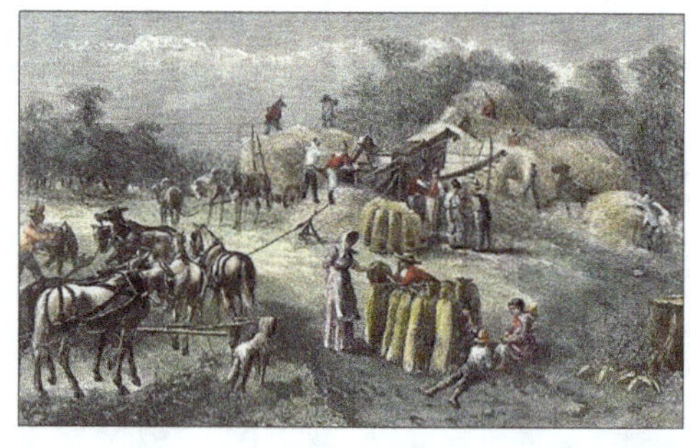
A.

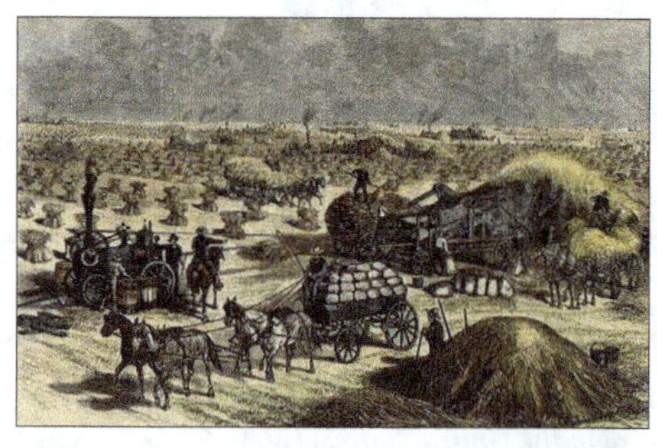
B.

A. William H. Rogers (after O. C. Steinberger), *Harvest Time in the West*, *Harper's Weekly* (September 21, 1878)

B. Albert Berghaus, *Threshing by Steam on the Dalrymple Farm*, *Frank Leslie's Illustrated Newspaper* (October 19, 1878)

George? Gregory, Hill (engraver), *Harvesting in Scotland*, *Harper's Bazaar* (February 16, 1878), 19 ¾ x 13 ½ inches

O. D. Steinberger, *A Harvest Scene in the West—Threshing Grain in the Field*, *Harper's Weekly*, (September 21, 1878), 13 ½ x 9 inches

Thaddeus Welch, *Gleaning* (St. Paul's Bay, Quebec), *Harper's New Monthly Magazine* (August 1883)

William Robinson Leigh, *Dakota Grainlands, Reaping with Right-Hand Binders, Shocking a Sea of Wheat, Steam Threshers at Work,* and others, from *Scribner's Magazine* (November 1897)

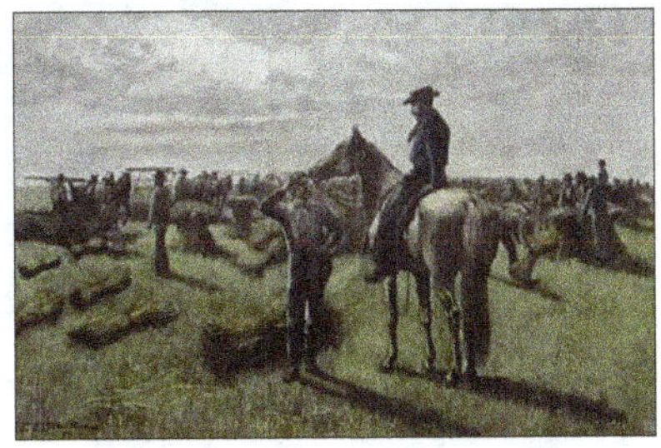

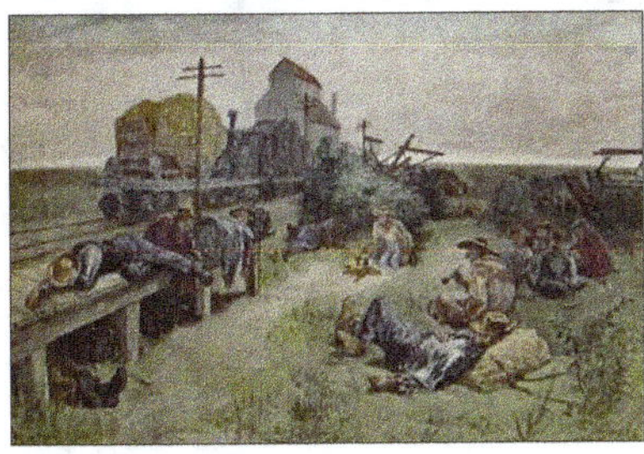

A. B.

A. R. F. Zogbaum, *Wheat-Harvesting in Dakota, Harper's Weekly* (July 30, 1887), 13 ⅝ x 9 ⅛ inches

B. William Rogers, *Harvest Hands on Their Way to the Wheat Fields of the Northwest, Harper's Weekly* (December 13, 1890), 12 ⅞ x 9 ⅛ inches

R. Hurtt, *The Minneapolis Harvest Festival, Harper's Weekly* (October 3, 1891), 9 ¼ x 13 ½ inches

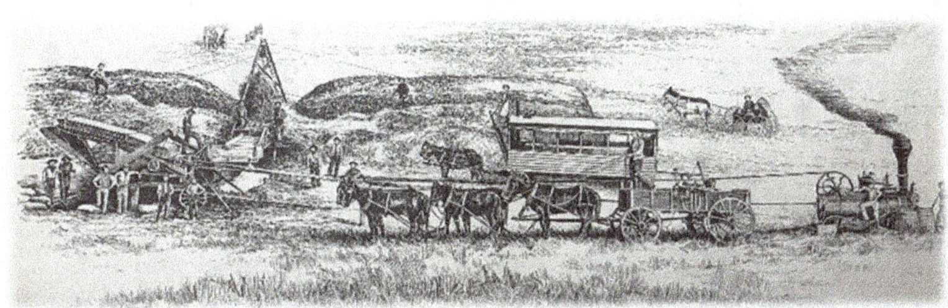

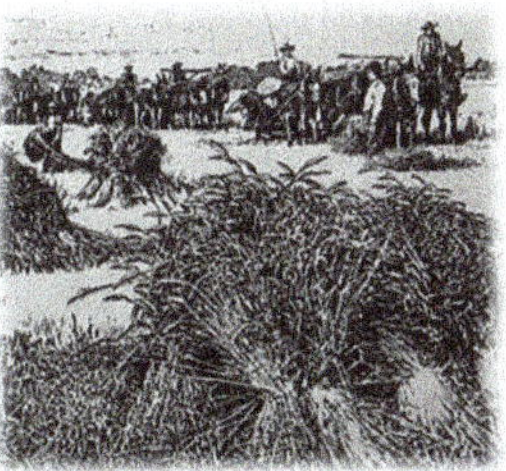

C. E. Burr, *The Harvest in Washington—Gathering and Threshing Wheat on the Burrell Farms in "The Palouse Country"* (October 10, 1891), 14 x 9 inches

The True Agent of Prosperity [Farmer carrying wheat sheaf], *Puck* (August 11, 1897), 8 x 9 ⅜ inches

H. Vogel, in Marguerite Merington, "A Rustic Calendar" (4 illustrations), *Scribner's Magazine* (August 1897), 4 ⅞ x 7 ½ inches

4. *Collier's, Fortune,* and Other Modern American Publications

Maurice Leloir, *The Harvest Festival* (1882), A. LaLauze (engraver), in C. Cook, *Art and Artists of Our Time,* 1888), 8 ½ x 10 ¼ inches

Harvest Scene cover photograph, *Leslie's Weekly* (September 4, 1902), 11 ¼ x 9 inches

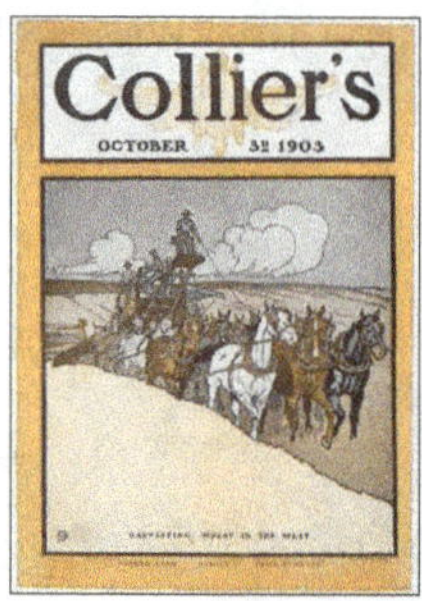 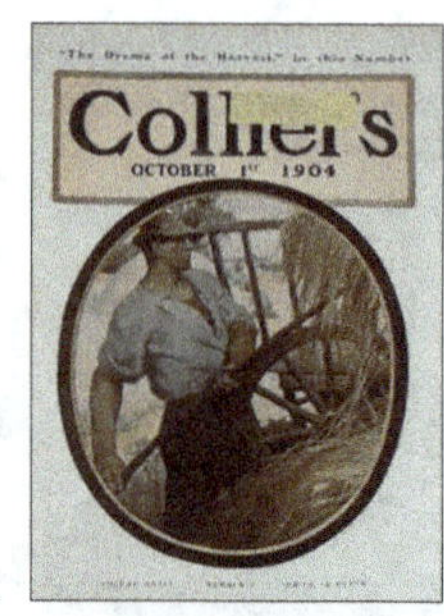 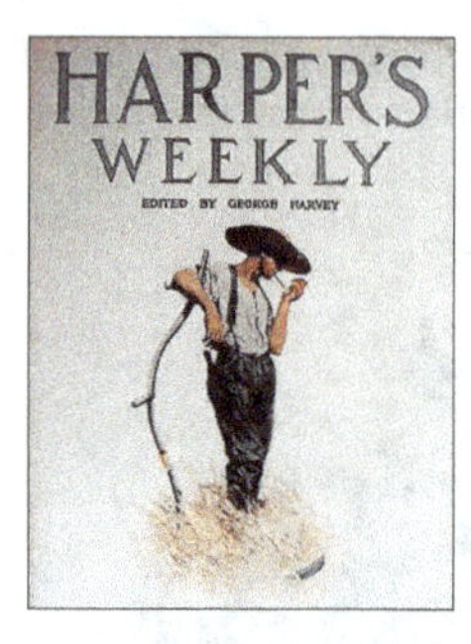 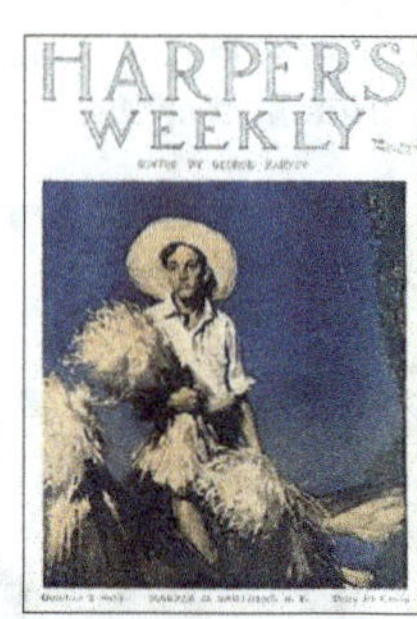

A.   B.   A.   B.

A. Edward Penfield, *Harvesting Wheat in the West, Collier's* (October 3, 1903)

B. Denman Fink, "The Drama of the Harvest" (Harvester Pitching Stalks cover), *Collier's* (October 1, 1904), 8 ⅞ x 10 inch oval

C. W. H. Dunton, Scyther Smoking Pipe cover, *Harper's Weekly* (August 7, 1909), 5 ¼ x 9 ⅞ inches (image)

D. Harry Townsend, Boy with Sheaf cover, *Harper's Weekly* (October 2, 1909), 9 ½ x 10 ¾ inches (image)

"Harvest Number" (Woman Raking Grain cover), *The Youth's Companion* (August 26, 1909), 8 ¾ inch circle

Harrison Cady, *The Harvest Home Supper at Tinkham's Corners, The Country Gentleman* (October 3, 1914), 9 x 9 ¼ inches

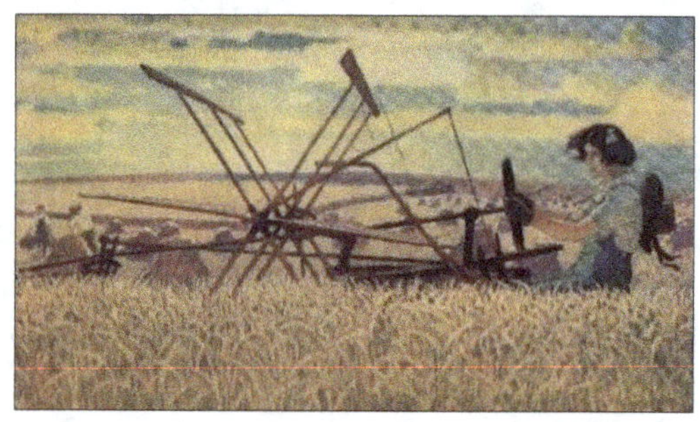

N. C. Wyeth, *From the Battlefields of France to the Wheat Fields of America, The Ladies' Home Journal* (October 1919), 14 ⅜ x 8 ¾ inches

A Rainbow was Shattered and the Fragments Fell in the Grain Market of Samarkand, National Geographic (November 1919), color photograph, 8 x 5 ½ inches

Harold Copping, *"And Certain of the Pharisees said unto them. . . ."* [Jesus and disciples gathering grain on the Sabbath], *The Ladies Home Journal* (March 1923), 9 x 11 ½ inches

Clare Leighton, "British Yeomanry: Four Woodcuts" (*Turning the Plough, Threshing, Milking, Lambing*), *The Forum* (September 1926), 4 ⅜ x 6 ⅜ inches

L. A. Shipton, *New Deal Harvest, The Literary Digest* Cover (September 23, 1933)

Peter Helck, *Combine Harvester and Railroad, Fortune* (September 1930), 10 ¼ x 10 ¼ inches

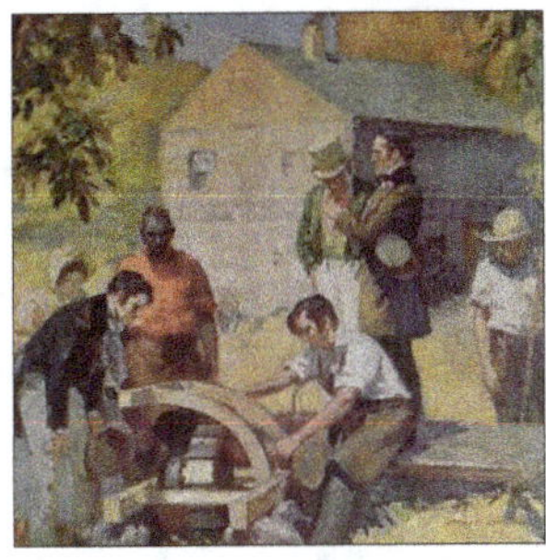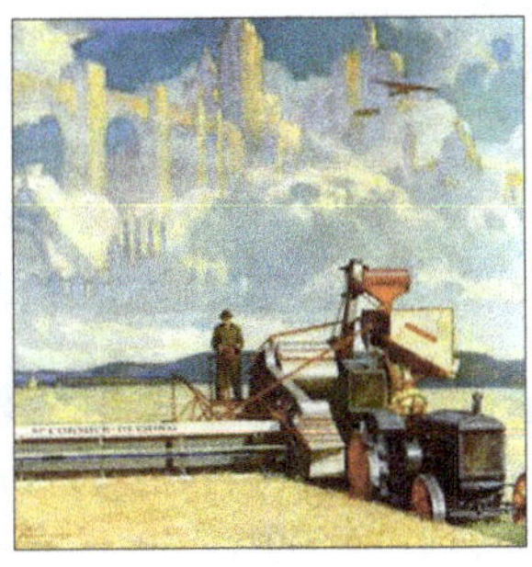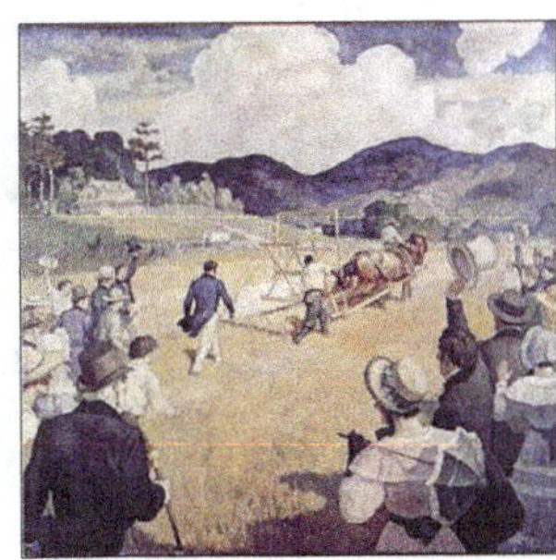

Phillip Lyford, *The Reaper Takes Form* (March 1931); *The McCormick-Deering Harvester-Thresher* (May 1931); N. C. Wyeth, *The World's First Reaper* (April 1931), *The Country Gentleman* (McCormick-International Harvester 1831-1931 Centennial Series), 9 ⅜ x 9 ⅜ inches

F. V. Carpenter, Tractor and Reaper scene, *Successful Farming* cover (July 1931), 8 x 8 inches (image)

L. Shipton, Three Pull-Combines with Tractors cover, *The Literary Digest* (September 1933 cover), 8 ⅝ x 9 ⅜ inches (image)

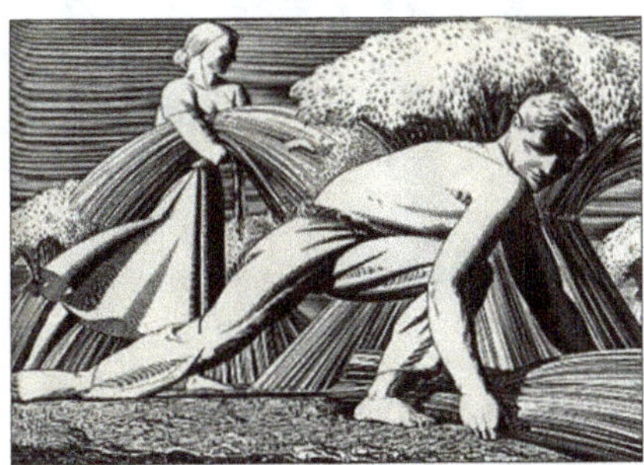

Rockwell Kent, *Harvest Time* (September 1934), *Fortune Magazine*, 11⅛ x 7 ¾ inches

Grant Wood, *"When Tillage Begins, Other Arts Follow"* Barn triptych, *Fortune Magazine* (January 1935), 4 x 10 ¾ inches

S. R. Badmin, *The Threshing Machine Steam Engine,* "A Farm in Illinois," *Fortune* (August 1935)

Ernest Hamlin Baker, *Wheat Harvest, Fortune Magazine* (October 1935 cover), 10 ⅛ x 10 ⅝ inches (image)

B. Joe Jones, *Threshing, Fortune Magazine* (November 1937), 9 x 6 ¾ inches

Harry Wickey, *Wind Over Wheat* and *Sultry August Afternoon* (1936), in *Harry Wickey Print Club Invitation* (1937), 4 x 5 ½ inches (original lithographs listed as 14 ¾ x 10 ¼ inches)

"Canada the Siren," *Fortune* (September 1938): Allen Saalburg *Three Line Elevators*

Gertrude A. Kay, Sheaf and Wagon scene, *Country Gentleman* cover (September 1940), 8 ¼ x 10 ¾ inches (image)

Saul Tepper, *The Harvest Dinner, Country Gentleman* (1941), 9 x 7 ⅝ inches

[Peter] Helck, *Farmers Drive Steel Horses* [caption for hillside harvesting scene], *The Saturday Evening Post* (August 16, 1941), 9 ⅜ x 6 ¾ inches

Dale Nichols, *Wheat Harvest, Country Gentleman* cover (June 1942), 10 ⅝ x 11 ⅛ inches (image)

Field Notes from Farm Country," *Fortune* (June 1943): John Atherton, *The War on the Farm*

[John] Atherton, *Harvest, The Saturday Evening Post* cover (September 25, 1943), 10 ¾ x 13 ¾ inches

N. C. Wyeth, *On the Hay Load, Country Gentleman* cover (June 1944), 10 ¾ x 12 ¾ inches (image)

B. Wheel Tractor and Reaper scene, *Country Gentleman* (July 1945 cover), 10 x 10 ¾ inches (image)

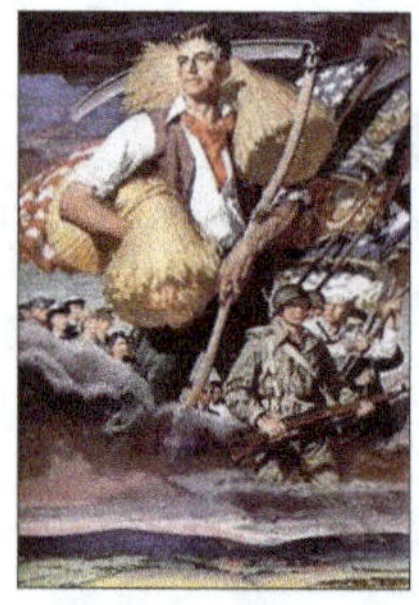
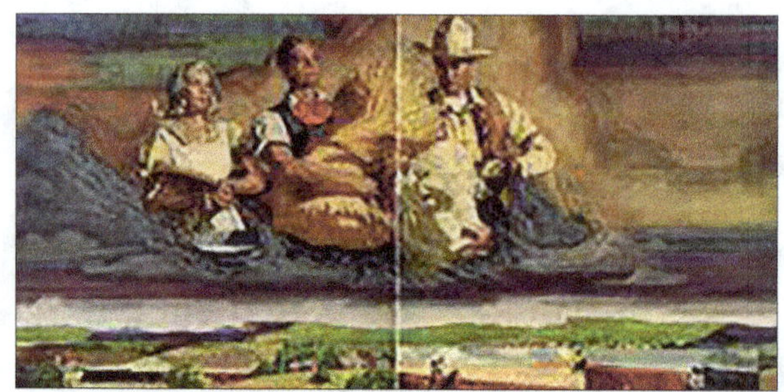

A.   B.

A. Dean Cornwell, *The Land of Plenty*, *The Saturday Evening Post* (June 1944), 6 ¾ x 12 ¼ inches

B. Frank Reilly, *Good Providers*, *Holiday Magazine* (September 1946), 21 x 20 inches (bi-fold)

Thomas Hart Benton, *Threshing Wheat* (c. 1938), *Life Magazine* (September 16, 1946), 10 ¼ x 6 ¾ inches

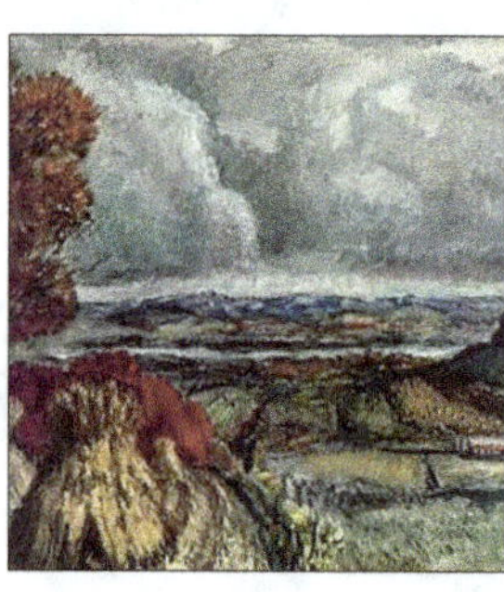
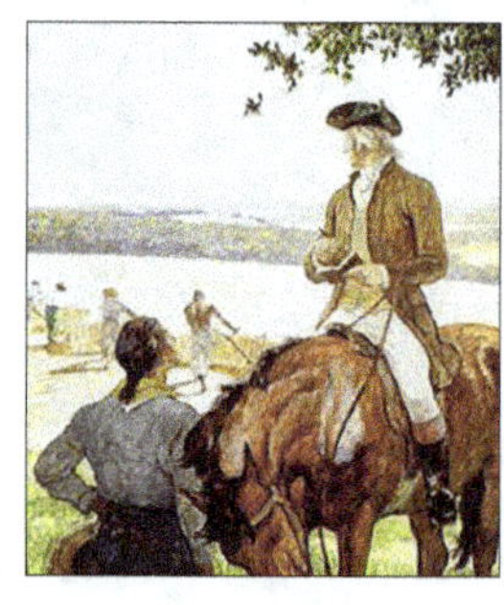
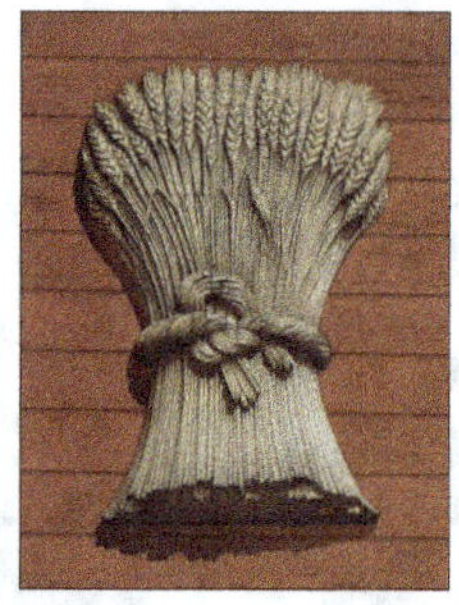

A.   B.   B.

A. John Steuart Curry, *Valley of the Wisconsin*, *Country Gentleman* cover (October 1946 ["one of (his) last and finest landscapes. . . ."]), 10 x 11 inches (image)

B. N. C. Wyeth, *First Farmer of the Land*, *Country Gentleman* (February 1946), 10 ⅝ x 12 ⅞ inches

C. "America's symbol of plenty, the sheaf of wheat. . ." [19th century baker's shop sign], National Gallery of Art, and "The Harvest Table," *House Beautiful* (November 1950), 9 ½ x 12 ½ inches

"Family Harvest" Allis-Chalmers Tractor and All-Crop Harvester advertisement, *The Saturday Evening Post* (June 28, 1947), 9 ⅜ x 6 ¾ inches

Edmund Lewandowski, *Boom on the Farm*, *Fortune* (October 1947), 10 ½ x 13 inches

Bread for the World [caption for McCormick-Deering combines], *Country Gentleman* (June 1948), 10 ½ x 6 ⅛ inches

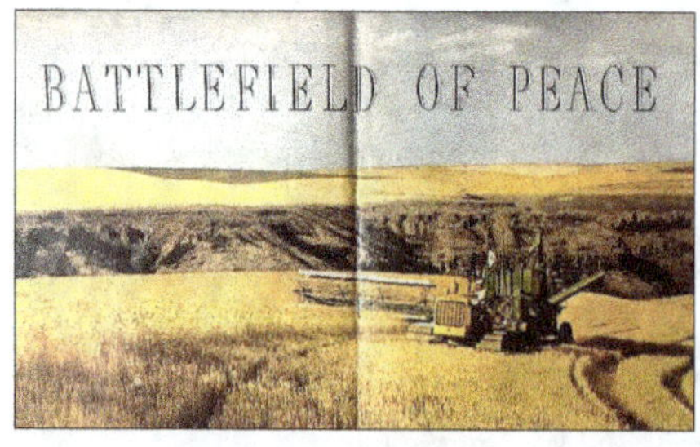

Battlefield of Peace [caption for Caterpillar tractors], *Saturday Evening Post* (June 19, 1948), 21 x 13 ¾ inches [Picture taken above Stewart Canyon on the Walter Faerber Farm near Uniontown, Washington.]

Stanley Meltzoff, *The Genealogy of Wheat*, *Scientific American* 189:1 cover (July 1953)

Best Wheat in a Lifetime, Cutting a Wheat Field, and other Kansas harvest photographs, *LIFE Magazine* cover story (July 14, 1958)

John Deere, Oliver, and Massey-Ferguson combine advertisements (7), *Washington Farmer* (June 1959), various sizes

Allis-Chalmers Gleaners, *Washington Farmer* (May 1963), 10 ½ x 14 ¼ inches, two-page and single-page spreads

Massey-Ferguson Models 300 (9 ⅜ x 8 ½ inches, two-page spreads), Super 92 (single page), and Super 35 (single page) color advertisements (c. 1963)

Ilonka Karasz, Harvest Scene, *The New Yorker* cover (September 16, 1967)

Enid Kotschnig, Grain Heads, *Scientific American* 244:1 cover (January 1981), 7 ¼ x 7 ½ inches (image)

Joseph Farris, Grain Stalks and Blue Vase Still Life, *The New Yorker* cover (October 18, 1982)

Reaper engraving reproductions, *Small Farmer's Journal* (Winter 1988 and Spring 1989), 8 ¾ x 5 ½ inches

5. Farm Collector Magazine Harvest Machinery Advertising Chromolithograph Reproductions

The Advance Banner Boy (c. 1900/January 2013) 16 x 10 ¾ inches

The Marsh Harvester *(c. 1878/March 2013)* 16 x 10 ¾ inches

The Milwaukee [Harvester Company] Leads (c. 1890/November 2021) 16 x 10 ¾ inches

Sara Quinn, We'll Still Be Here When This Is Over [Ukraine War], Tumbleweird 7:4 (April 2022 cover), 10 ¼ x 14 ½ inches

John Broadley, "Empathy & the Economy" blackline illustration (scyther-baker-buyer), New York Review of Books LXIX:19 (December 8, 2022), 5 ¾ x 7 ½ inches

C. Technical Reports

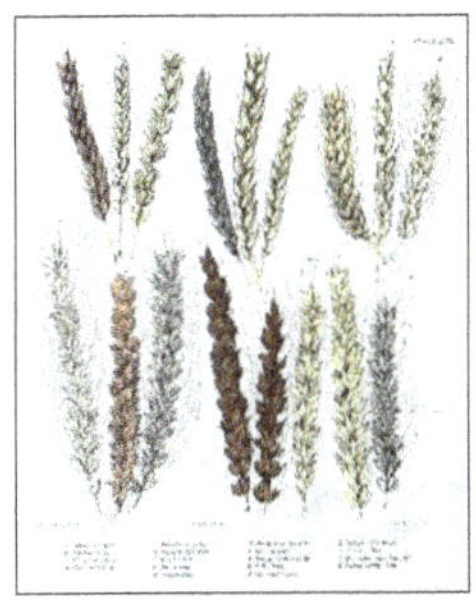 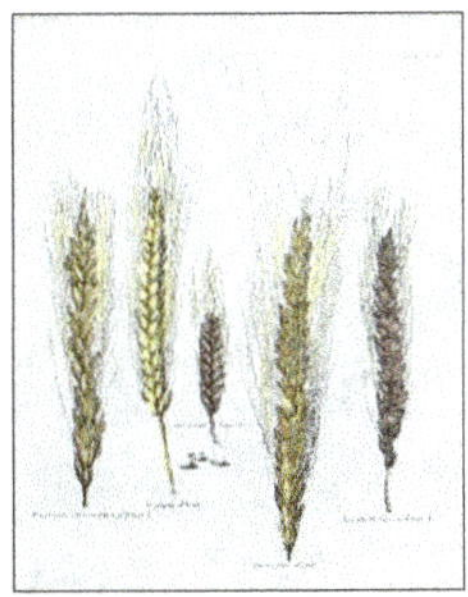

Ebenezer Emmons, New York Wheats, James Hall, *A Natural History of New York*, Division V:4 (3 color prints, 1848), 8 ⅝ x 11 ½ inches

Eugène Graff, from Henri de Vilmorin, Les Meilleurs Blés: Description et Culture des Principales Variétés de Froments (Paris, 1880), 66 color plates, 12 ¼ x 8 ¾ inches

Flandre, Victoria blanc, Chiddam à épi blanc, Hunter, Trump, Blanc de Hongrie, Roseau, Chili, Chiddam de mars, Richelle, Zélande, Talavera, Mareuil, Crépi, Noé, Touzelle, Saumur de mars, A épi carré, Hicklin, Tunstall, A duvet, Odessa, Victoria d'automne, Hallet, Saumur d'automne, Red Chaff Dantzick, Chiddam à épi rouge, Rousselin, Rouge d'Ecosse, Spalding, Prince Albert, Rouge inversible, Rouge de Hongrie, Rouge de Saint-Laud, Browick, Rouge de Provence, Mars rouge, Sicile, Hérisson sans barbes, Blé seigle, Californie, Shireff, Mars barbu, Victoria de mars, D'automne rouge, Mars rouge, Japon, Hérisson, Poulard blanche, Pétanielle, Lausanne, Poulard d'Australie, Pétanielle noire de Nice, Miracle, Trimenia barbu de Sicile, Xerés, Belotourka, Médéah, Pologne, Epeautre blanc, Amidonnier blanc, Amidonnier noir, Engrain commun, Engrain double

Eugène Graff, from Phillipe de Vilmorin, *Supplement aux Les Meilleurs Blés: Description et Culture des Principales Variétés de Froments* (Paris, 1909), 27 black-and-white plates, 12 ¼ x 8 ¾ inches

Bordier hybride, Stand-up, Blanc á paille raide, Richelle blanche hative, white stiff straw, Gironde, Bon Bermier hybride, Trésor hybride, Japhet, Gros bleu, Champlan hybride, Hâtif inversible hybride, Briquet jaune hybride, Massy hybride, Grosse tête hybride, Dattel hybride, Rouge d'Altkirch, Pithiviers, Lamed hybride, Teverson, Riéte barbu, Barbu à gros grain, Perle du Nuisement barbu, Rouge prolifique barbu, Mars de Swède rouge barbu, Poulard à six rangs, Giant du Milanais, Rouge d'Egypte

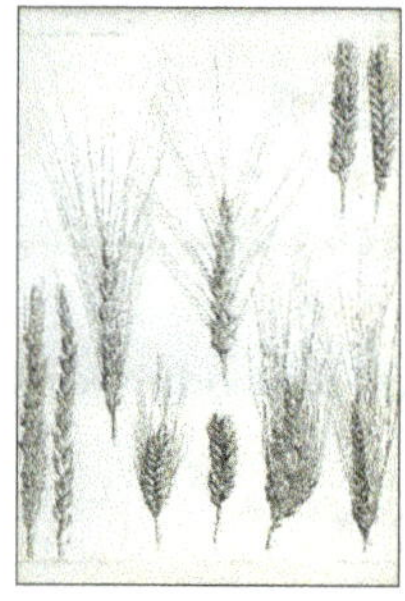
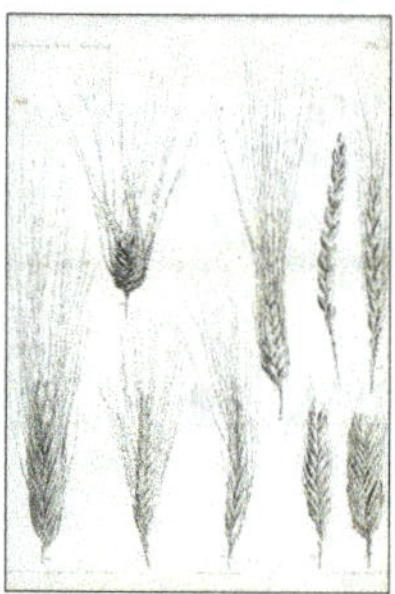

F. L. Meyer (copperplate engravings, Berlin: A. Bohl), in Friedrich Körnicke, *Arten und Varietäten des Getreides* (Berlin: Paul Parey, 1885), 9 ¼ x 11 ⅞ inches

Getreidearten [Cereal Grains] (Leipzig: F. A. Brockhaus, 1895), 10 ¾ x 8 ½ inches

Deborah Passmore, *Drought-Resistant Wheats—Hard Winter Varieties*, and Proctor, *Drought-Resistant Wheats—Macaroni Varieties*, in Mark Carlton, *Early Hybrids with their Parent Varieties* (Washington, D. C.: Government Printing Office, 1900), 5 ½ x 9 inches (page)

Triticum and *Hordeum*, in O. Thomé, *Flora von Deutschland, Österreich und der Schweiz* (Germany, 1904), 6 x 8 ⅝ inches (page)

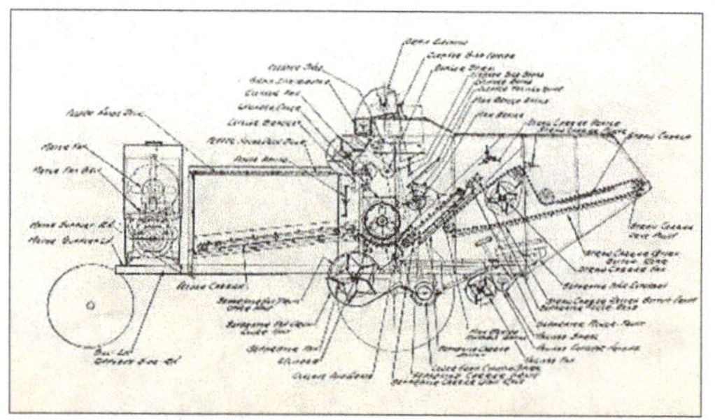
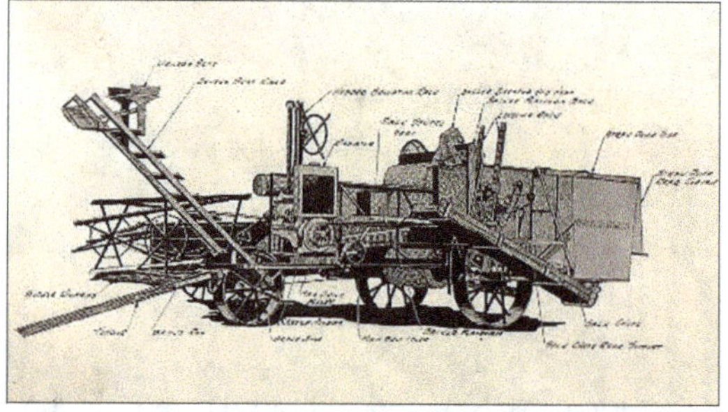

Advance-Rumley Combine Harvester Schematics, *Repair Parts List for the Advance-Rumley No. 1 Combine Harvester* (c. 1915)

10 plates with 48 cereals; first of each series: (I) *Triticum vulgare*, (II) *Triticum durum*, (III) *Triticum dicoccum*, (IV) *Hordeum distichum*, (V) *Avena sativa*, (VI) *Avena strigose*, (VII) *Panicum miliaceum*, (VIII) *Panicum miliaceum*, (IX) *Andropogon sorghum*, (X) *Panicum sanguinale*

D. Books and Gallery Publications

Gustave Doré, *The Ark Returned to Beth-shemesh; The Disciples Plucking Grain on the Sabbath; The English Bible* (1866), 10 ¼ x 14 ¼ inches;

Edwin Buckman, *Autumn—The Lusty Sheaves*, Birkit Foster, ed., *Poetry of the Year*, or *Pastorals from our Poets* (London, 1867)

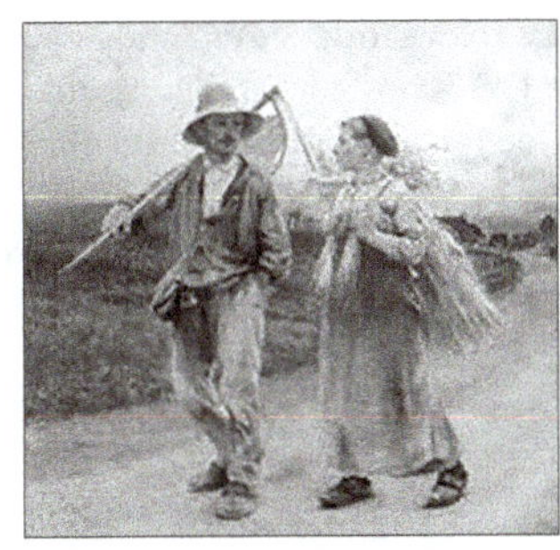

A.

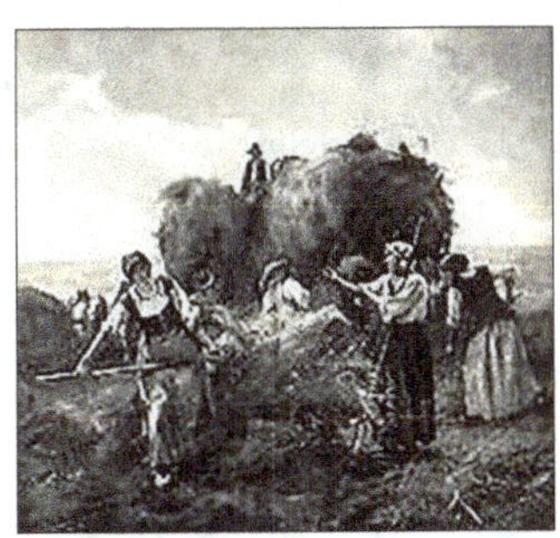

B.

John C. Ridpath, ed., *Art and Artists of All Nations: Photographic Reproductions of Great Paintings . . . Exhibited at the World's Columbian Exposition* (New York: Knight & Brown, 1894), 8 black & white reproductions: August Hagborg, *Evening* (A); Julien Dupre, *The Harvest* (B) and *The Balloon*; Theodore Mayan, *End of Day in Provence*; E. C. Penfold, *Gleaners*; Hans Dahl, *Toll Paid First*; Frederick Morgan, *The Tired Gleaners* and *Watching and Waiting*, 11 ½ x 7 inches and various

John Constable, J. and G. Nicholls (engravers), *The Corn-Field* (1816/1901), 7 ½ x 8 ¾ inches

Jean-François Millet, *Woodcuts: Reaping, Flax-Pulling, Mowing, Raking, Trussing*, 2 ⅞ x 5 ¼ inches; *Crayon Studies; Threshing, Winnowing*, 5 ¾ x 8 inches, Charles Holme, ed., *Corot and Millet* (London: The Studio, 1902)

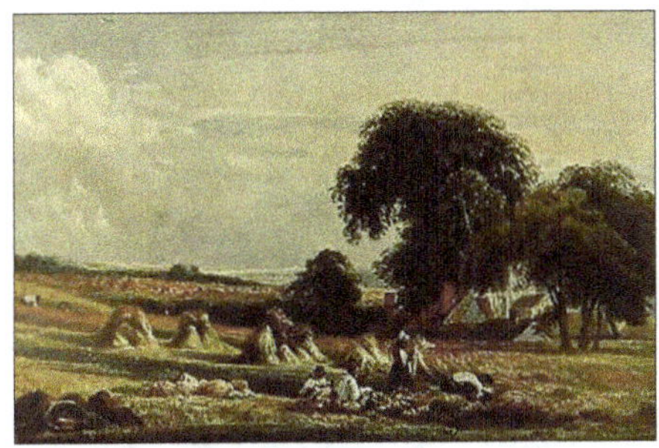

Peter de Wint, *Harvesting* (c. 1835), Charles Holme, *Masters of English Landscape Painting* (London, 1903)

James Orrock, *Holy Island Castle: Harvest-Time*, W. S. Crockett, *In the Border Country* (1906), 6 x 3 ⅞ inches

Jules Breton, *The Return from the Harvest*, c. 1860, *Bibby's Annual* (1910), 9 ¼ x 5 ¼ inches

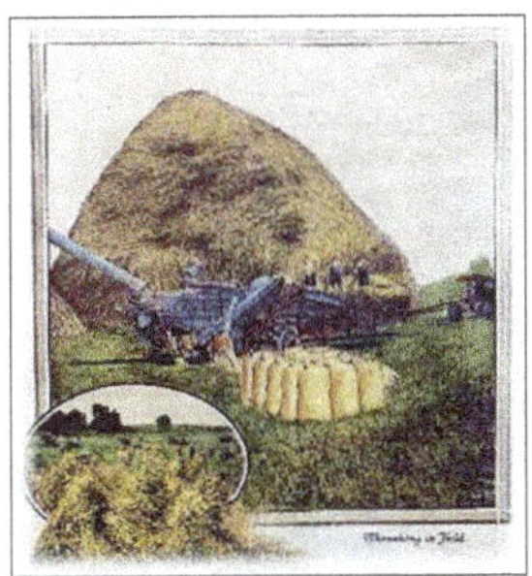

Threshing in the Field, *Filling the Silos*, and *Threshing Grain*, "Hearts Delight Farm, Chazy, New York," photogravures (1915), various sizes

Léon Lhermitte, *The Harvesters' Pay Day*, 1882, *Bibby's Annual* (1922), 8 ¼ x 7 ¾ inches

6 black-and-white plates from Thomas Craven, *A Treasury of American Prints* (New York: Simon and Schuster, 1939), approximately 8 x 10 inches: Arnold Blanch, *Along the Hudson*; (A) John Steuart Curry, *The Line Storm*; Adolf Dehn, *Threshing*; John S. De Martelly, *Give Us This Day*; (B) Joe Jones, *Missouri Wheat Farmers*; (C) Grant Wood, *Seedtime and Harvest*

Clare Leighton, *Lovers in the Wheatfield*, from *The Time of Man* (New York: The Viking Press, 1944), 5 ¾ x 8 ¾ inches

Raoul Dufy, *Threshing* (1943/1954, tipped-in plate), 9 ¾ x 8 inches

Maurice de Vlaminck, *The Harvest* (1958), 9 ⅜ x 7 ⅞ inches

Robert Blake, *Country Churchyard Reaper* watercolors (France, c. 1798/1971), 8 ¾ x 11 ½ inches

Ben Shahn, *Wheat Field* (c. 1958), from *Ecclesiastes or, The Preacher* (New York, 1971), 8 ⅞ x 12 inches

Camille Pissarro, *Harvest at Montfoucault, Brittany* (1896/1963), 8 ⅞ x 6 ¼ inches

Joachin Vayreda, *Harvest* (1880/c. 1975), 18 x 21 inches

Alexei Venetsianov, *The Threshing Barn* (1820), 8 ⅛ x 6 ⅜ inches; B. *Harvest Time, Summer* (1825), 8 x 10 ¼ inches; *Reapers* (c. 1828), 8 x 10 ¼ inches; from *Alexei Venetsianov* (Moscow: Aurora Publishers, c. 1985) [tipped-in color plates]

Charles Burchfield, *Study for "July"* (1941), *July Wheat and Fence* (1943), from *The Drawings of Charles Burchfield* (New York: Frederick A. Praeger, 1968), 10 ¼ x 6 ½ inches

Mary Azarian, *Tale of John Barleycorn*, 6 woodcut color reproductions (1982), 7 ¾ x 8 ¾ inches

C. W. Jefferys, *Western Harvesters* (1898), *A Field of Oats in Manitoba* (1906), *Afternoon in the Wheatfields* (1906), *Untitled [Harvested] Field, Manitoba* (1906), *Conquering Western Canada series* (3, 1908), *The Homesteader No. 4, The First Harvest* (c. 1907), *Wheat Stacks on the Prairie* c. 1906), *Wheat Stacks on the Prairie* (1907), from Robert Stacy, *Western Sunlight: C. W. Jefferys on the Canadian Prairies* (1986), various sizes

Lavern Kammerude, *Belgian Power*, 6 x 9 ½ inches; *Threshing*, 9 ⅝ x 5 ½ inches; *Steam Power*, 9 ⅝ x 6 ⅛ inches; from Chester Garthwaite, *Threshing Days: The Farm Paintings of Lavern Kammerude* (Mount Horeb, Wisconsin, 1990)

IV. PAINTINGS, LIMITED EDITION PRINTS, AND OTHER ORIGINAL FINE ART

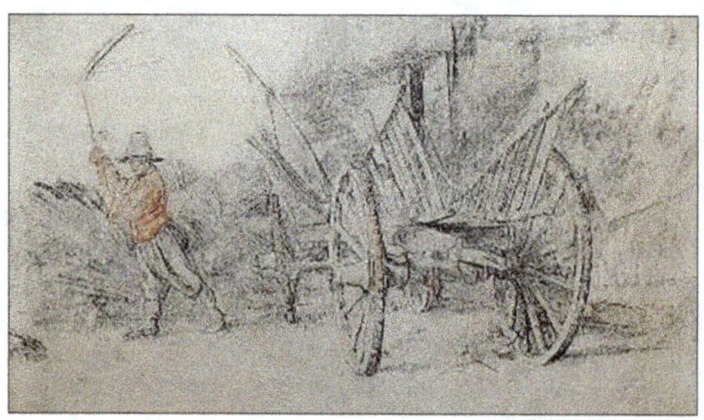

Peter Paul Rubens, (c. 1625/1908), Oxford: Vasari Society, lithograph on paper, 11 ¾ x 7 ½ inches

Marc-Henri Meunier, *Untitled Sheaves* (8/30, c. 1910), signed aquatint etching on paper, 16 ½ x 11 ¾ inches

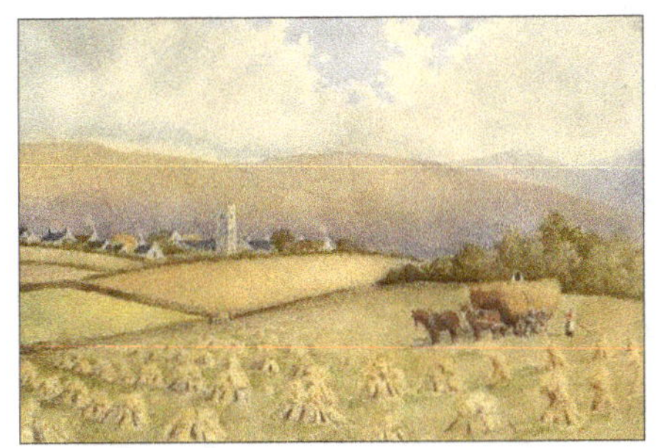

A.

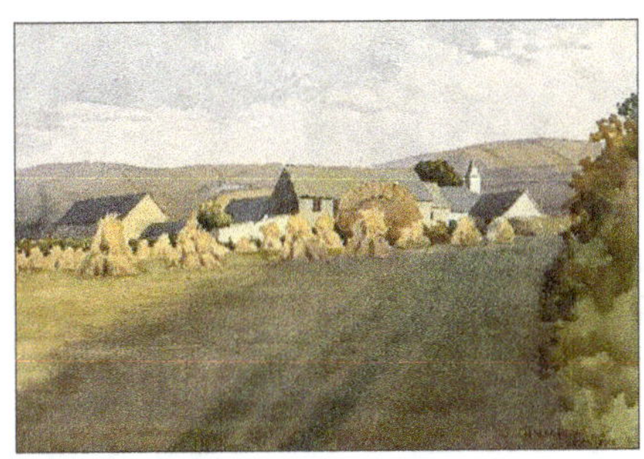

B.

A. Raymond Turner Barker, *Harvesting at Trevalga* (1914) [with Constable Art Club comment card]

B. Raymond Turner Barker, *Halgabron Farm* (1913), watercolor on paper, 10 x 14 inches

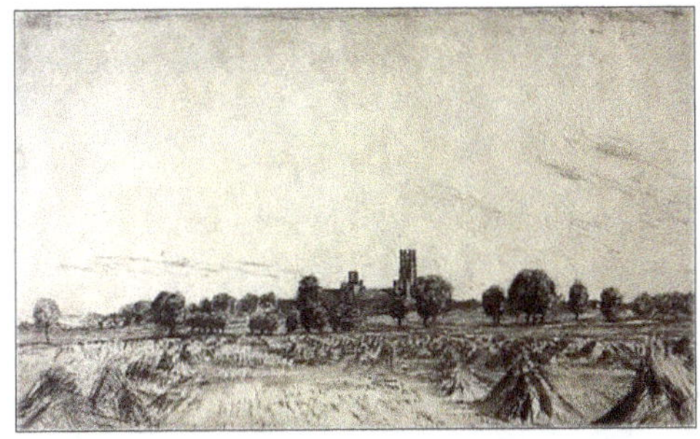

Job Nixon, *Harvest Field, Ely: Evening* (1/6, 1926), signed proof on wove paper, 10 ⅞ x 5 ½ inches

J. Elwood McDougan, Untitled Harvest Scene (1947), oil on canvas, 19 ½ x 25 ½ inches

Nils Munke Plum, Untitled Farm Scene (probably near Fredericksgav/Hagensov, Denmark, 1953), pencil and watercolor on paper, 10 x 14 inches

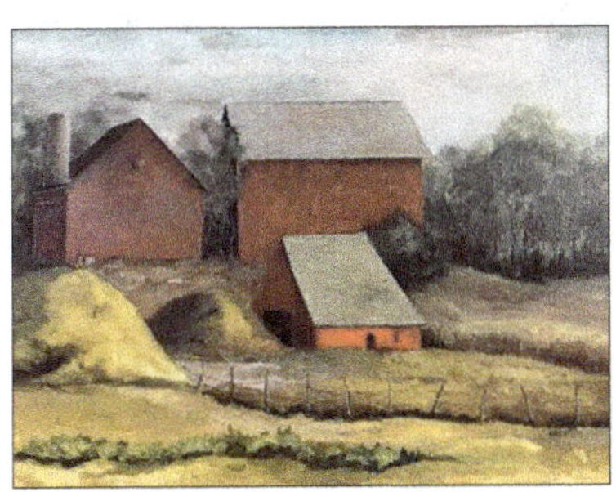

A.

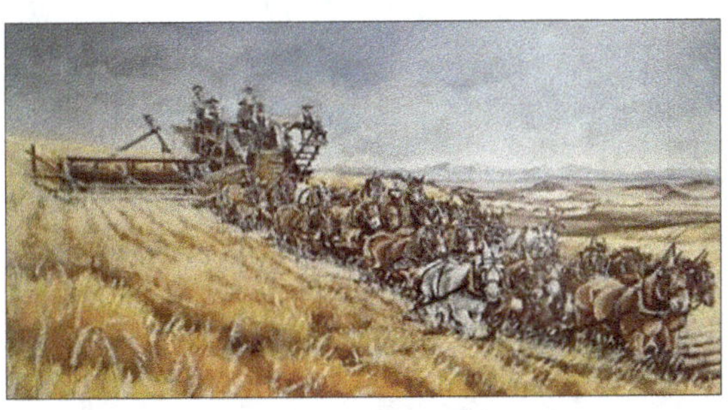

B.

A. James Valentin, *Barns and Threshings* (c. 1955), oil on board, 18 x 25 inches

B. R. J. Butler, *Harvest Home, Shropshire* (c. 1965), oil and watercolor on paperboard, 4 ¼ x 6 ¼ inches

Sherryl Evans, *26-Mule Hitch with Holt Combine* (1978), 10 x 20 inches [300/750]

John Wheat, *September Harvest* New York Graphic Society (1953/1957), 14 x 36 inches

Raymond Hosford, *The Amber Waves of Grain* The Franklin Mint (1973), 23 ½ x 30 inches

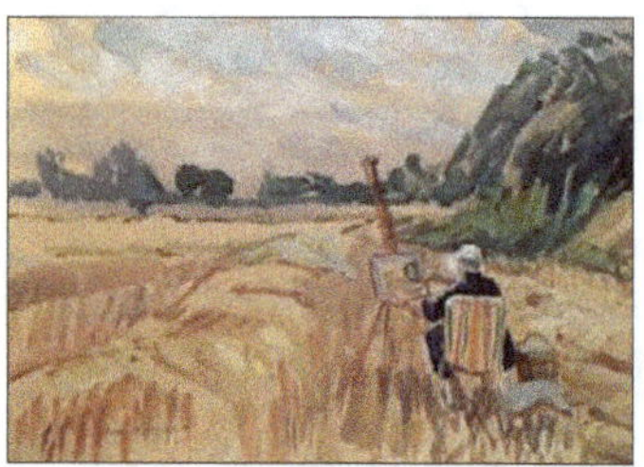

Jean Dryden Alexander, 5 watercolors (c. 1980s-90s), various sizes, including *Painting at Thelveton* (1991), gouache on paper, 10 ½ x 7 inches

Ruth Freeman (1903-1994), *Golden Wheat* (c. 1980), monoprint on paper, 6 ¼ x 4 ½ inches

Mary Azarian, (A) *Harvest the Wheat, Flail the Grain* and (B) *As Ye Sow Shall Ye Reape* (1983), woodcuts on paper, 11 ¼ x 18 ¼ inches

William Nelson, (A) *The Harvesters* (34/500, c. 1985), and (B) *No More Harvest* (361/500, c. 1985), serigraphs on paper, 18 x 25 inches

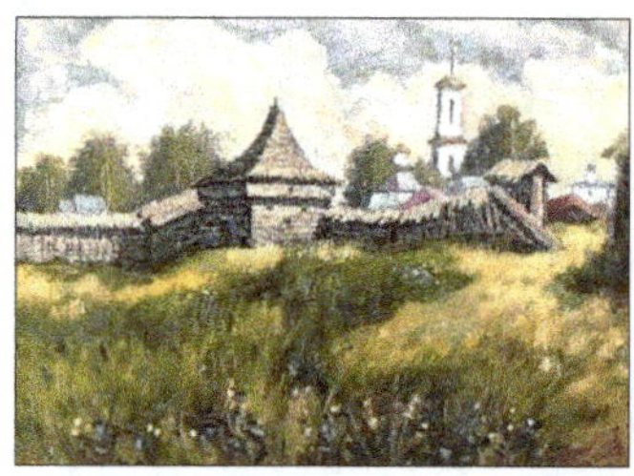

Yuri Peganov, Untitled Village Scene (1988), oil on canvas, 11 ½ x 12 ¾ inches

Dmitri L. Slobodin, Untitled Donbas Harvest Scene (1989), oil on paper, 17 ½ x 22 inches

Alan Fearnley, *Harvest in the Cotswolds* (1989, 385/500) and *The Threshing Team* (1989, 41/500), lithographs on paper, 17 x 23 ½ inches

David Brandt Thompson, *Harvest* (c. 1990, 18/200) and *McCoy Creek [Oregon] Summer* (c. 1990, 19/200, lithographs on paper, 6 x 7 ½ inches and 6 x 4 ¾ inches

J[ohn] Enkeboll, Untitled Farm Scene (1991), oil on hardboard, 16 ½ x 19 ¾ inches

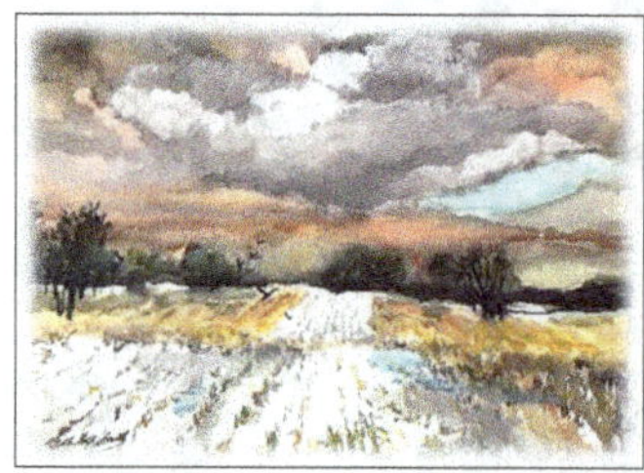

Mildred Gift Smith, *After the Harvest* (c. 1995), watercolor on paper, 22 ½ x 30 inches

John Rankin, *Home Place Harvest* (1995), silkscreen serigraph on paper, 20 x 32 inches

Jack Dorsey, *Harvester at Rest* (2002), watercolor on paper, 16 x 22 inches

Robert Rosenberg, Wheat Field (c. 2000, 62/250), serigraph on paper, 13 ¾ x 18 ¾ inches

J. K. Clausen, *The Gleaners, after Millet* (2004), pastel on paper, 8 ½ x 11 inches

Craig Whitcomb, *Amber Waves* (Colfax, Washington: Colfax Arts Council, 2008), 6 ½ x 12 inches

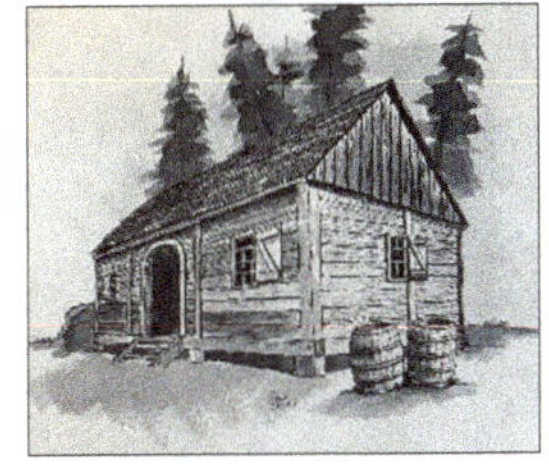 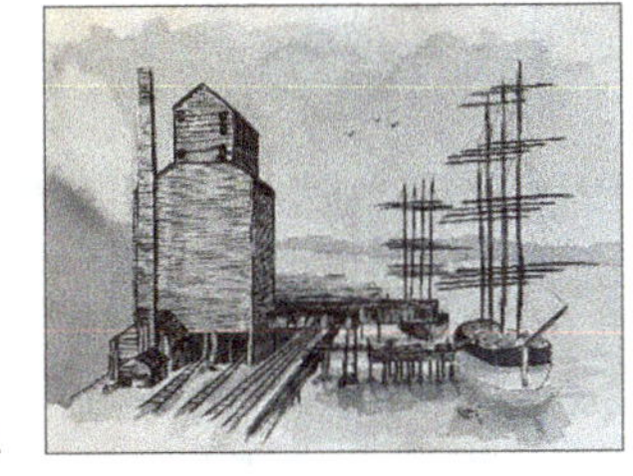 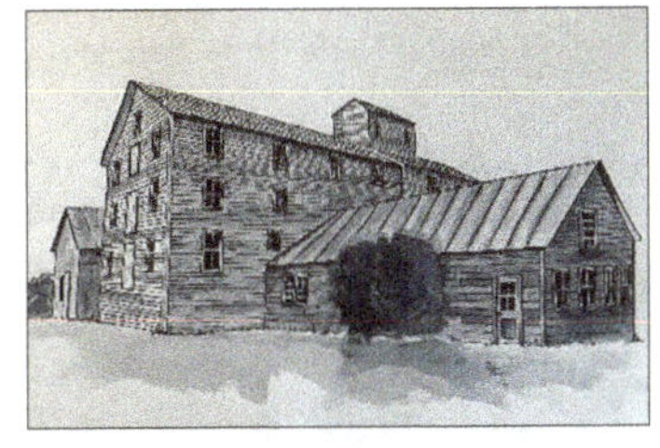

A. B. C.

A. Robert Smith, *Ft. Nisqually Granary* (c. 1840)

B. Robert Smith, *Elevator A, Tacoma Waterfront* (c 1900)

C. Robert Smith, *Barron Flour Mill, Oakesdale, c. 1960* (2012), pen and ink, 10 x 16 inches

Robert Smith, *Morrill Hall, Washington State University, Pullman* (c. 1915)

Charles Fritz, *Harvest Under the Big Sky (When Toil Brings Bounty)* (c. 2010), dye sublimate on aluminum panel,
8 x 10 inches

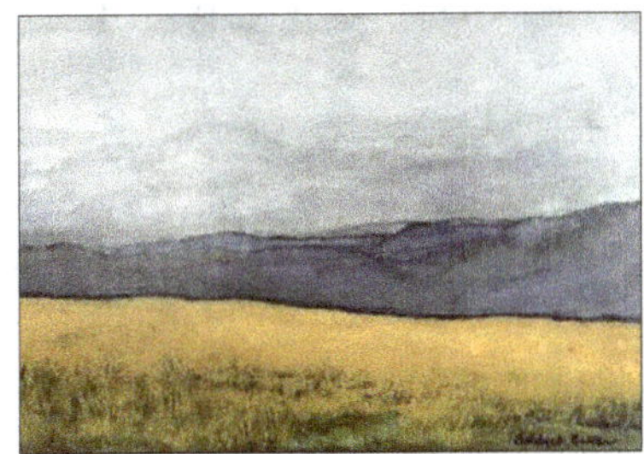 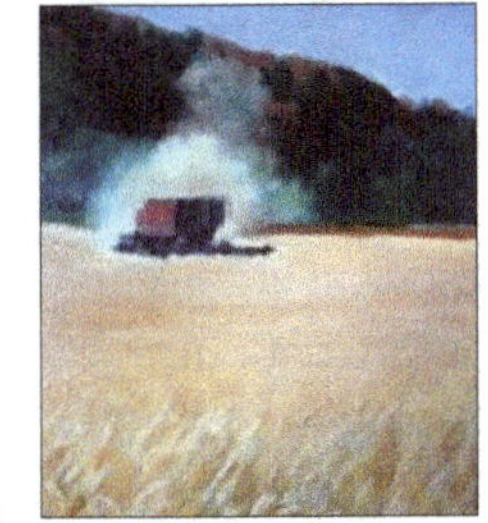

A. B.

A. Bridget Baker, *Olympic Gold* (2015), oil on canvas, 23 ½ x 17 ½ inches

B. Jim Gerlitz, *Palouse Colony Harvest* (2017), acrylic on canvas, 17 ¾ x 24 ¾ inches

Jacqueline Daisley, *Palouse Hills—End of Harvest* (2018), oil on canvas, 42 ½ x 32 inches

Steve Henderson, *Harvest Time, Touchet Valley* (2018), oil on board, 4 x 8 inches

Maja Shaw, *Palouse Harvest* (2017), watercolor print on paper, 13 x 16 inches

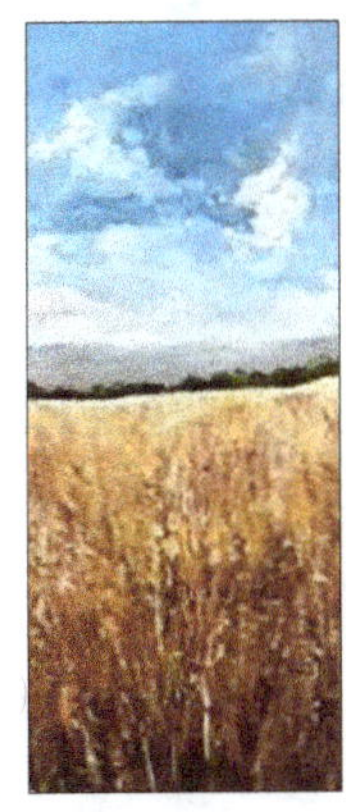 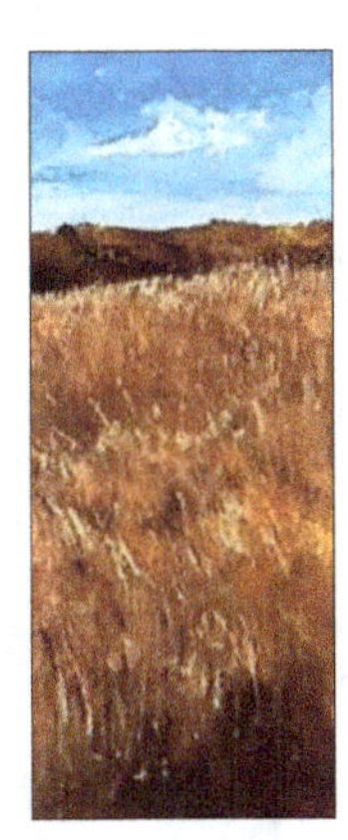 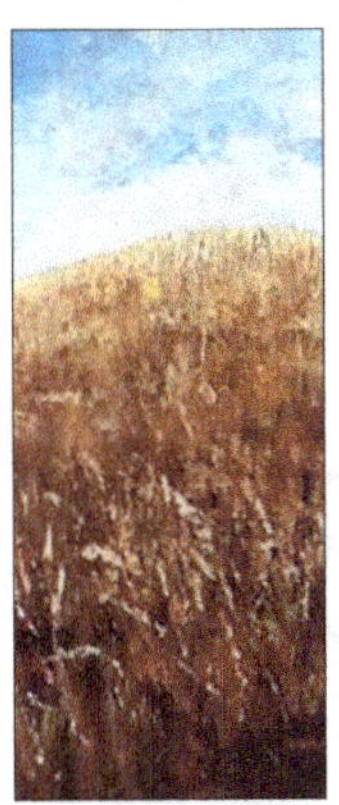

A. B. C.

Louise Smith, Grainfield Series (detail), *Flaxen Field, Tawny Meadow, Palouse Rising* (2018)

Katherine Nelson, 2 Grain Head Silhouettes (2018), pencil on paper, 9 x 11 ⅞ inches

Sue Nash, *Wheatfield III* (2019), pastel on paper, 10 x 14 inches

Andy Sewell, *Doubletime Before the Storm* (2021) [99/400] watercolor on paper, 13 x 18 inches

Washington Association of Wheat Growers Harvest Lithograph Series

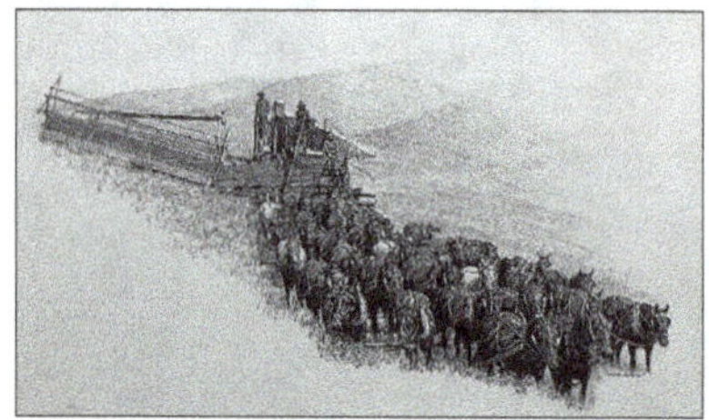

Lisa Specht-Urbat, Untitled Sidehill Horse-drawn Combine (1991), 14 ½ x 23 inches

Lisa Specht-Urbat, Untitled Header Tender and Mule Skinner (1994) [153/630] 15 ½ x 17 ¾ inches

Lisa Specht-Urbat, Untitled Steam Engine and Stationary Thresher with Grain Wagons (1993) [132/700] 14 ½ x 23 inches

Lisa Specht-Urbat, Untitled Reaper and Four-Horse Team (1995) [167/602] 14 x 20 inches

Lisa Specht-Urbat, Untitled Reaper with Two-Horse Team and Binder (1994) [445/515] 14 x 23 inches

Karen Reffett, Untitled Barn, Roadster, and Grain Wagons [n. d., 148/800] 18 x 24 inches

Karen Reffett, Untitled Pull-Combine with Tractor [n. d., 241/800] 18 x 24 inches

Karen Reffett, Untitled 32-Horse-Hitch Combine [n. d., 188/800] 18 x 24 inches

Karen Reffett, Untitled Stationary Thresher with Steam Engine and Water Wagon [n. d., 245/500] 18 x 24 inches

Karen Reffett, Untitled Ritzville Flour Milling Town Scene [n. d., 218/500] 18 x 24 inches

V. FOLK ART

Anna "Grandma Moses" Robertson, *In Harvest Time* (1948), lithograph, 14 x 20 inches

Signed "Bimson," 3 Midwest Harvesting Scenes plus "The Golden Text" Psalm 139:9-10 title card (c. 1955) [inscribed: The night cometh when no man can work. —John 9:4], watercolor on paper, 4 ½ x 12 inches

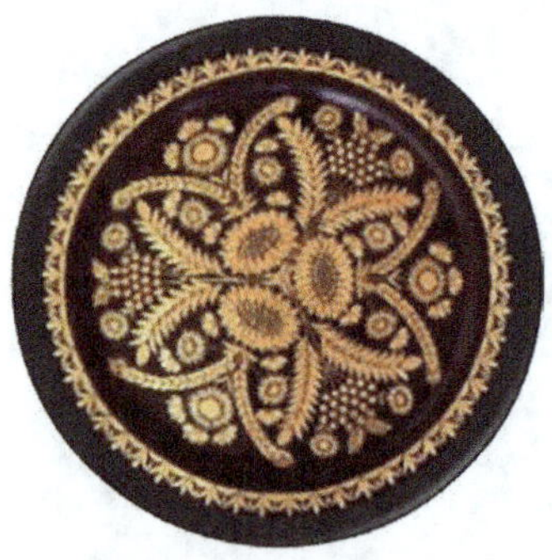

Cut straw overlays (*salomki*) on wood plate, Zagorsk, Russia, 17 ⅜ inches

Boxes, Kirov, Russia, 4 ⅞ x 7 ½ and 4 ⅜ x 4 ⅜

VI: POSTCARDS AND NOTECARDS

A. Postcards

(Dates indicate year of year of issue if known, year of postmark, or other date indicated on card.)

1. International Harvester Company of America, Chicago, "Harvest Scenes Around the World" Series (1909-1910)

Algiers—In this far off country; Chili—American harvest machines; France—The modern self-binder; Finland—A mower with reaping attachment; Finland—The women folks assist; India—Primitive methods are still in use (2 views); Mexico—Modern American binders near the Pyramids of Cholula; Roumania—An American harvesting machine; Russia—camels are used as draft animals; Scotland—A modern self-binder; Siberia—A harvest scene near Tobolsk; Siberia—A harvest scene on the tundra; Spain—In this land of romance and song; United States—In the land of sunshine and plenty; United States—The header is used quite generally

IHC "Prosperity" (1909)

2. Pacific Northwest

"Combined Harvester Scene," Holt Combines near Walla Walla, Washington (1906)

Harvest black and white photograph scenes (1 view, c. 1900-1920)

Harvest color photograph scenes (10 views, c. 1910-1960)

"Combined Harvester Scene," Holt Combines near Walla Walla, Washington (2 views, 1906)

3. Other United States and Canada Postcards

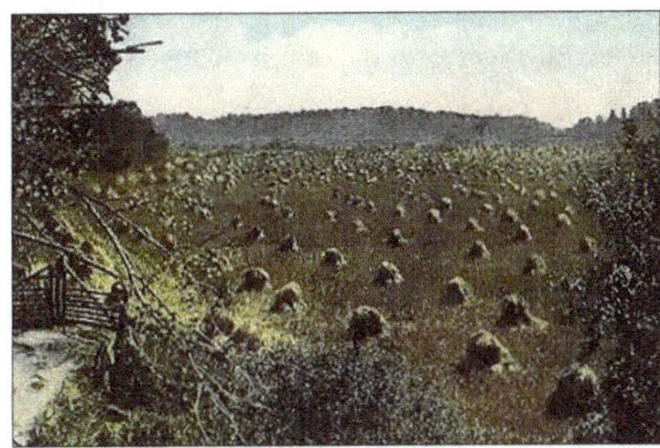

Harvest black and white photograph scenes (6 views, c. 1900-1920)

Harvest color photograph scenes (15 views, c. 1900-1920)

"Canadian Harvest Scenes" color series (5 views, c. 1920)

"Scenes Along Country Road" color series (2 views, c. 1915)

Harvest Festival—Lutheran Church (1906), *Our Farmers' Golden Harvest Fields* (Wisconsin, c. 1910)

Pope John Paul at Living History Farms (Iowa, 1979)

Our Farmers' Golden Harvest Fields (Wisconsin, c. 1910)

4. Foreign Postcards

Australia: *Wheat Threshing, Killarney* (c. 1920)

Great Britain: Untitled Harvest Painting (1908),

Germany: A. Schwartz, Untitled Mountain Scyther (Austria?), E. Henseler, *Last Sunbeams* (1920)

Eastern Europe: J. R. Wehle, *And They Followed Him* (Russia, 1900/1920), S. Bertz, *The Farmer* (Latvia, 1924), Z. Stryjenska, *Harvest* (Poland, 1940), E. Markachev, *Harvest* (1925/1982)

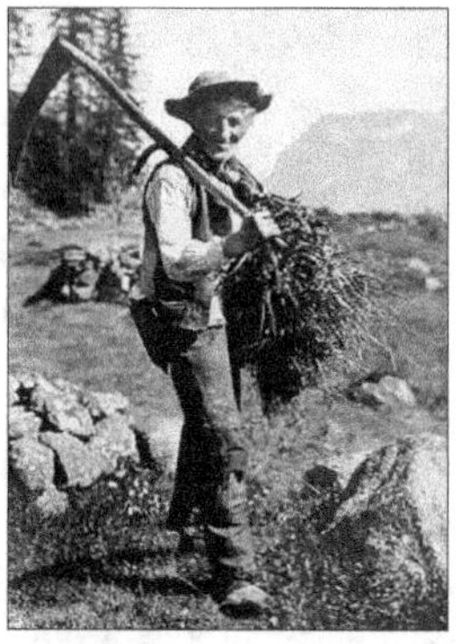

Arthur Schwartz, Untitled Mountain Scyther (c. 1910), Neue Photographische Gesellschaft, Berlin

Italy: Alberto Bianchi, *Plowing, Sowing, Reaping, Threshing, Milling* (c. 1910)

5. Thanksgiving Postcards with Harvest Motifs (3)

B. Notecards

"Art of the Palouse" series: Karla Matzke, *After the Storm*; Jack Dorsey, *Rhody's House*; Cheri O'Brien, Untitled; Katherine Nelson, *Spring Light on the Wheatlands* (by the artist, 2017)

Sybil Andrews (Campbell River, British Columbia), *Haysel* (1936); *The Mowers* (1937) [Glenbow Museum, Calgary, c. 2000 reprints]

Rockwell Kent, *To Thee, America!* (1946), *This Is My Own* (1940/Dover Publications, c. 2000 reprint)

Don Tiller (Port Townshend, Washington), *Field and Clouds, Open Gate, Spring Creek, Straw Hat, Suppertime* (by the artist, c. 2005-2015)

K. K. Schultz, *Imperial Farm Cottage* (c. 1835/Tsarskoe Selo near St. Petersburg, c. 2015 reprint)

VII. PHOTOGRAPHS AND PHOTOGRAPHIC REPRODUCTIONS

A. Photographic Reproductions

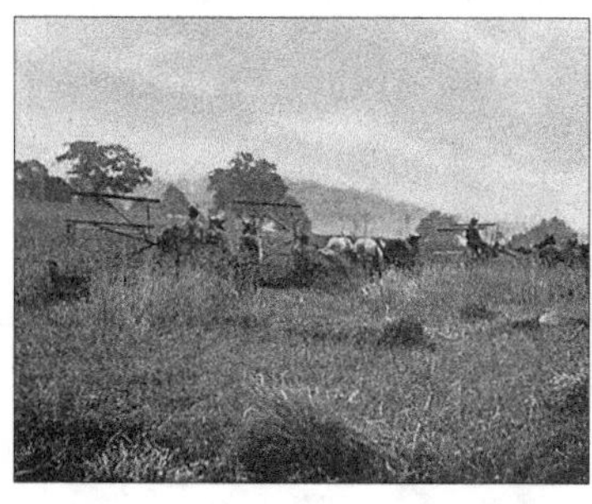
A.

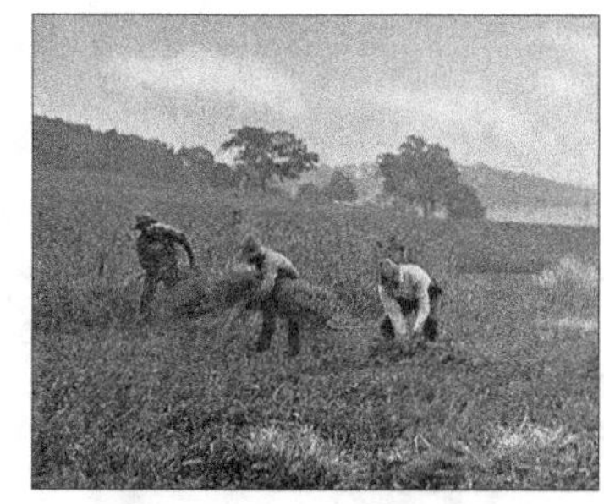
B.

Henry Troth, "Harvesting"—Reaping and Binding, *Country Life in America* (December 1901)

W. H. Hall, *A Bounteous Yield* (half-tone), *Penrose's Pictorial Annual, 1906-1907* (New York: Tennant & Ward, 1907), 9 ½ x 6 ¾ inches

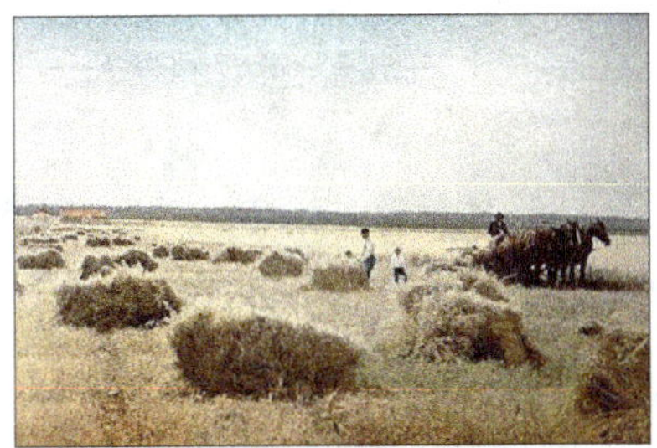

Paysans dalécarliens revenant de la Moisson (Suéde) and *La Moisson au Manitoba*, color plates from [?] Gravure fin XIXème (Paris, c. 1910), 8 ¼ x 5 ¾ inches

John M. Whitehead, *White Unto Harvest, The Amateur Photographer* and Photography (November 6, 1918)

Decoration of Font for Harvest Thanksgiving (c. 1920), 5 ⅜ x 4 inches

Edward S. Curtis, *Washing Wheat—San Juan, The Apache Reaper* (c. 1906), duotone sepia plates from *Portraits from North American Indian Life* (Toronto: New Press, in association with the American Museum of National History, 1972), 14 ½ x 19 ¼ inches

Horse-drawn Wheat Harvest (William Davies Threshing Outfit, c. 1920) and Sack Sewing, centerfold from *Small Farmer's Journal* 4:1 (Winter 1979), 8 ¼ x 21 inches

Larry Kanfer, *Under the Rainbow* and *The Harvest*, from *Prairiescapes* (Champaign: University of Illinois Press, 1987), 9 ¼ x 7 ½ inches

Lynn R. Miller, "The Anatomy of a Threshing" (illustrated McIntosh Threshing Bee article), *Small Farmer's Journal* 45:2 (Fall 2021); "Setting Up a Binder," *Small Farmer's Journal* 45:3 (Winter 2022)

Khoke and Ida Livingston, "Rebuilding a 7-Sweep Horsepower Unit," *Small Farmer's Journal* 15:4 (Spring 2022)

B. Keystone and Other Stereoviews (c. 1910)

Farms Along the Upper Rhine, Germany

A Telemarken Harvest Scene near Saude, Norway

Harvesting Wheat near Corinth, Greece

Harvesting Grain Among the Olive Trees, Spain

Glimpses of Swiss Life in the Village

Harvesting Wheat in Old England

Evolution of the Sickle and Flail—33-Horse Team Combined Harvester, Walla Walla, Washington

Harvesting in the Great West—Combined Reaper and Thresher, Washington

C. R. R. Hutchison Harvest Photographs

Harvest and grain storage views (9), Endicott, Washington (c. 1910-1920), various sizes

D. John Clement Calendars and Publications

The McGregor Company Calendars (2011-2024): *Harvest Patterns, Ripe Wheat Palouse (North from Steptoe Butte), North of Dixie, Horse Heaven Hills, North of Potlatch, North of Walla Walla, Golden Light, Morning on the Palouse, Cutting the Hillside, Palouse, Ready to Cut, The Palouse, Wheat Harvest*

Schweitzer Engineering Laboratory Calendars (1990-2001): *Waiting for Harvest, Palouse Harvest, Harvest Sunset, Palouse Summer, Gathering Storm, Untitled Wheat Stalks*

E. Richard Scheuerman Digital Images

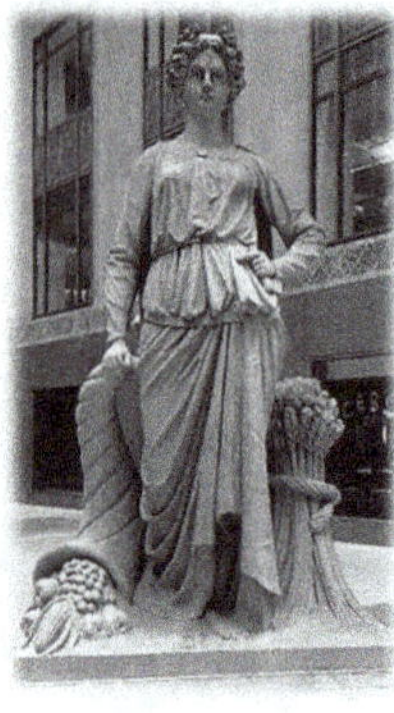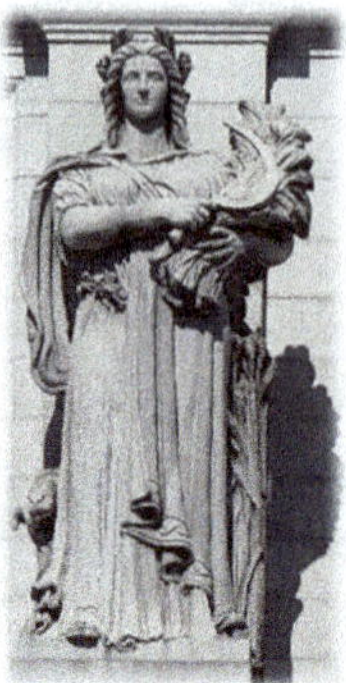

Left: *Agriculture—Ceres* (1885), Chicago Board of Trade Building; Right: Louis St. Gaudens, *Ceres* (c. 1914), Union Station, Washington, D. C.

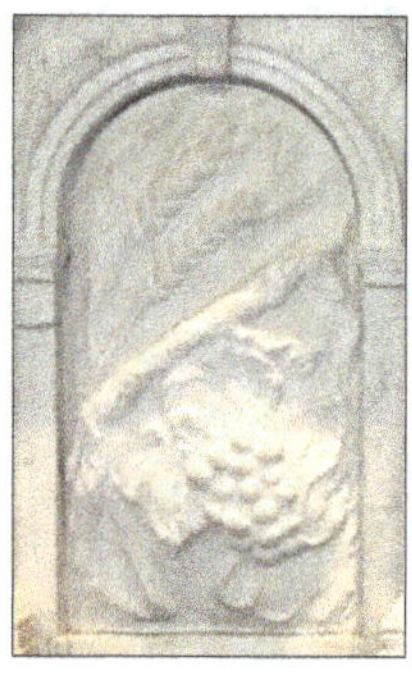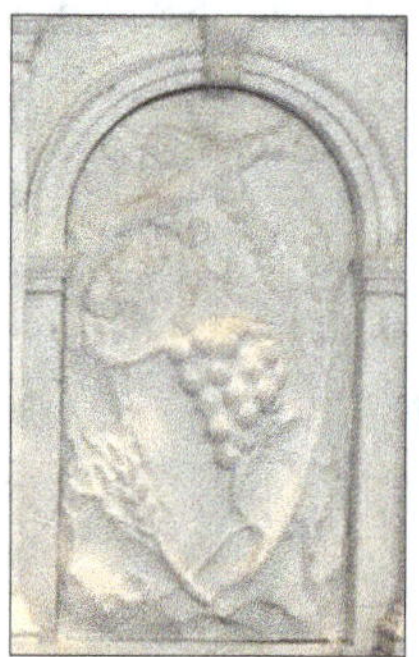

Mary Jo Anderson, Wheat and Grape Marble Altar Panels, St. James Cathedral; Seattle, Washington

Pat Boyer, Connell City Hall Murals, Connell, Washington

Benny Fountain, untitled Palouse Hills summer view (2008), oil on canvas mural, 9 feet, 6 inches x 5 feet, 10 inches, The 1912 Center; Moscow, Idaho

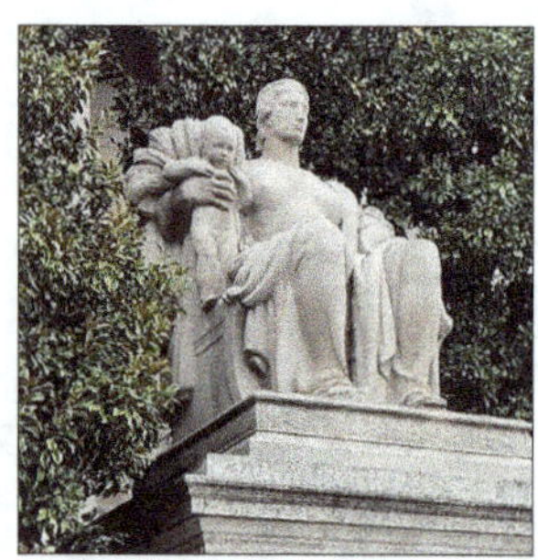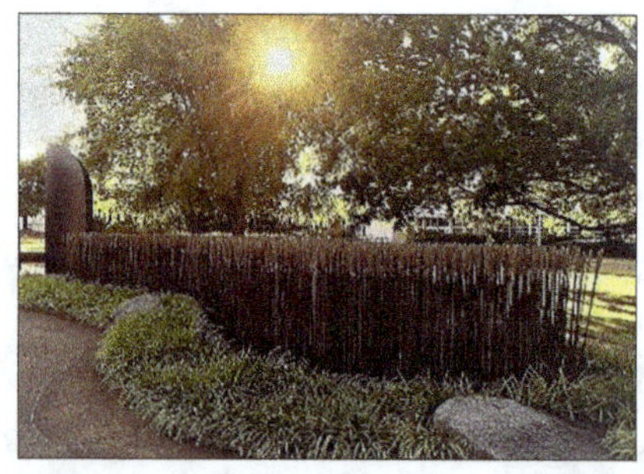

A. James E. Fraser (designer), Edward H. Ratti (sculptor), *Heritage*—Allegorical Matriarch with Child and Grain Sheaf (1935), National Archives Entrance, Washington, D. C. [inscription: "The heritage of the past is the seed that brings forth the harvest of the future"]

B. Simon Kogan, *Washington World War II Memorial* (1999), bronze, steel, granite, and concrete installation

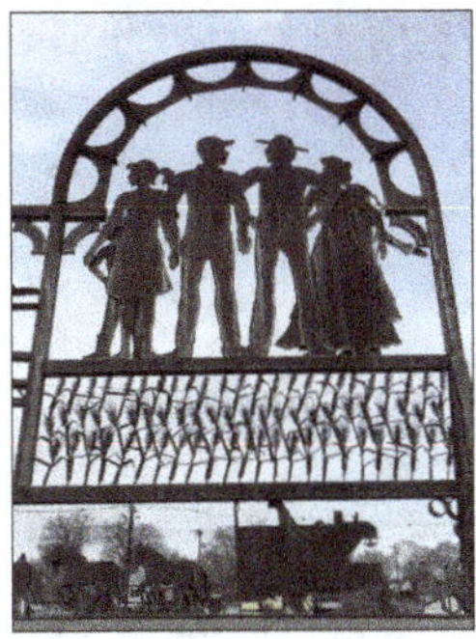

Grain Sheaf and *Men at Work* Stone Bas-reliefs (1941), Adams County Courthouse; Ritzville, Washington

David Govedare, *Dawn of a New Century* (1988), steel sculpture, Othello City Park; Othello, Washington

Jeffrey Hill, *Harvest Memories* bronze sculpture (2018); Land Title Plaza, Walla Walla, Washington; *Pioneer Harvest* mural (detail, 2011), Marcus Whitman Hotel, Walla Walla, Washington

Edwin Molander, Grain Sheaf Window Base Panel (1949), Trinity Lutheran Church, Endicott, Washington

Mara Smith, *Palouse Round Barn and Grainlands* (1992), Brick Bas-Relief, Neill Public Library; Pullman, Washington

Mission San Antonio Threshing Floor and Grist Mill, near Jolon, California

Our Mother of Perpetual Help Icon with Brass and Milky Glass Grain Stalk and Grape Candelabras (c. 1850), Mission San Carlos Borromeo, Carmel, California

Colonial Era Log *Ambar* Granary, Yágodnaya Polyána, Russia

Sheaf and Sickle Stained Glass Window (1914), Emmanuel Lutheran Church; Ritzville, Washington

Sheaves and Crown of Thorns, Ruth the Gleaner, Christ the Bread and Wine, Stained Glass Ensemble (c. 1910), Christ the Redeemer Church; Spokane, Washington

Esteban Camacho Steffensen and Jessily Brinkerhoff, *Working Forward, Weaving Anew* (2017), oil on masonry, 38 ½ x 72 feet, Prairie Line Trail Historic Interpretation Project, Tacoma, Washington

Thomas Gilbert White, *Happy Are the Men of the Fields* (detail, 1934), oil on canvas, 40 x 13 feet, USDA Whitten Building, Washington, D. C.

VIII. BOOKS WITH ART PLATES

Barringer, T. J. *Men at Work: Art and Labour in Victorian Britain*. New Haven: Yale University Press, 2005.

Barry, Roxana. *Land of Plenty: Nineteenth Century American Picnic and Harvest Art*. Princeton, New Jersey: Princeton University Press, 1982.

Bartels, Adolf. Der Bauer in *der deutschen Vergangenheit*. Leipzig: Eugen Diederichs, 1900. Woodcut reproductions: *Alte Darstellung des bäuerlichen Lebens, Schnitter mit zwei Frauen in der Ernte, Ernteleben, Schnittner mit Frau.*

Blackburn, Henry and Randolf Caldecott. *Breton Folk: An Artistic Tour in Brittany*. Boston: James R. Osgood and Company, 1881.

Carleton, Mark Alfred. *The Basis for the Improvement of American Wheats*. Washington, D. C.: Government Printing Office, 1900 (with "Ex Libris Philippe de Vilmorin" bookplate).

Cohen, Benjamin R. *Notes from the Underground: Science, Soil & Society in the American Countryside*. New Haven: Yale University Press, 2009.

Cox, David. *A Treatise on Landscape Painting in Water Colours*. London: The Studio, 1922. David Cox drawings and tipped-in watercolor plates: *Studies, Mid-Day, Effect—Mid-Day (Harvest).*

Curtis, Shelley. *The Bountiful Place: Art About Agriculture—The Permanent Collection*. Portland: Oregon Historical Society, 2006.

Hagenstein, Edwin C., Sara M. Gregg, and Brian Donahue, eds., *American Georgics: Writings on Farming, Culture, and the Land*. New Haven: Yale University Press, 2011.

Jacobs, Michael. *The Good and Simple Life: Artist Colonies in Europe and America*. Oxford: Phaidon Press, 1985.

Körnicke, Friedrich. *Die Arten und Varietäten des Getreides*. Berlin: Paul Perey, 1885.

Kosinski, D. *Van Gogh's Sheaves of Wheat*. New Haven: Yale University Press—Dallas Museum of Art, 2006.

Logsdon, Gene. *The Mother of All Arts: Agrarianism and the Creative Impulse*. Lexington: University of Kentucky Press, 2007.

Lord, Russell, Kate Lord, illustrator. *The Care of the Earth: A History of Husbandry*. New York: Thomas Nelson & Sons, 1962. (Wheeler McMillan's personal copy.)

Marx, Leo. *The Machine in the Garden: Technology and the Pastoral Idea in America*. Oxford: Oxford University Press, [1964] 1999.

McMillen, Wheeler. *Ohio Farm*. Columbus: Ohio State University Press, 1974.

Nikolayeva, Galina. *Harvest*. Moscow: Foreign Languages Publishing House, 1952.

Oppé, A. E. *The Water-Colours of Turner, Cox & De Wint*. London: Halton & Truscott, Smith, 1925. Tipped-in Peter de Wint watercolor plates: *A Cornfield, Harvesting, A Harvest Field with a Distant View of Lincoln*.

Pavord, Anna. *Landskipping: Painters, Ploughmen and Places*. New York: Bloomsbury, 2016.

Payne, Christiana. *Toil and Plenty: Images of the Agrarian Landscape in England, 1780-1890*. New Haven: Yale University Press, 1983.

Ritter, William. *La Moisson de Max Švabinský*. Prague: Jan Štenc, 1929.

Scott, James C., and Nina Bhatt. *Agrarian Studies: Synthetic Work at the Cutting Edge*. New Haven: Yale University Press, 2001.

Stites, Richard. *Serfdom, Society, and the Arts in Imperial Russia: The Pleasure and the Power*. New Haven: Yale University Press, 2005.

Theuriet, André. *La Vie Rustique*. Paris: Librairie Artistique—H. Launette, 1888. Léon Lhermitte illustrations, engraved by Clément Bellenger: *La Fenaison, Le Blé, La Moisson, Le Pain*.

Theuriet, André. *Rustic Life in France*. Boston: Thomas Y. Crowell and Company, 1896. Léon Lhermitte engravings: *Ploughing, Sowing the Seed, Reaping, Bread-Making*.

Viardi, Liana. "Imagining the Harvest in Early Modern Europe." *The American Historical Review* 101:5 (December 1996).

Weisberg, Gabriel P. *Jules Breton and the French Rural Tradition*. Omaha, Nebraska: Joslyn Art Museum, 1982.

Wolff, Barbara, and Roger Wieck. *The Book of Ruth: Medieval to Modern*. New York: The Morgan Library & Museum, 2020.

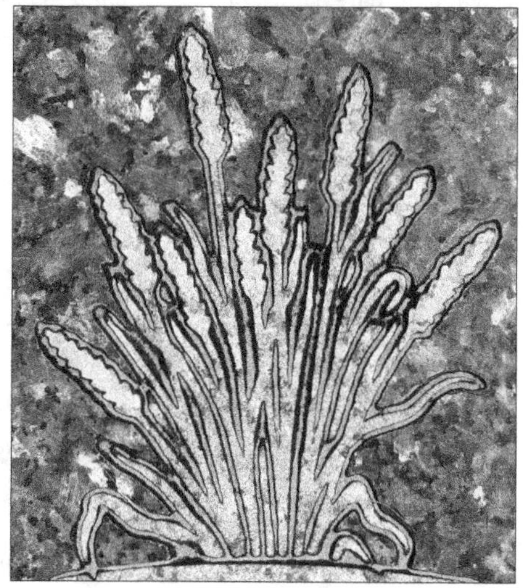

Grain Sheaf Headstone Carving (2023)
Delight Cemetery; Sandhills, Washington
Columbia Heritage Collection Photograph

APPENDIX B

Representative Works of Agrarian Art
with Museum, Gallery, and Library Locations

NORTH AMERICA

UNITED STATES

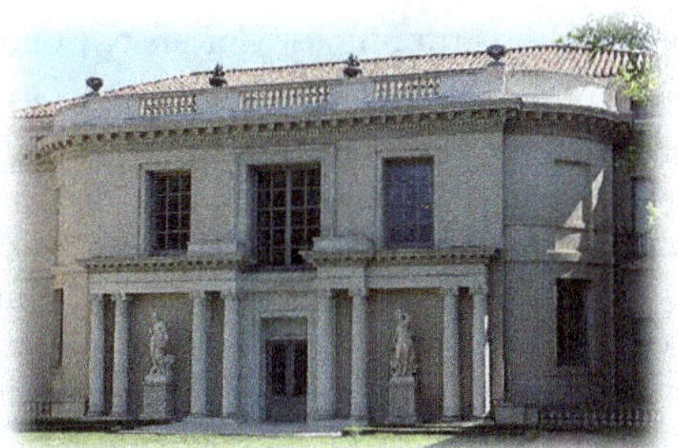

Huntington Gallery, San Marino

California San Francisco: De Young Museum (W. Halm, *Harvest Time*; A. Glendening, *Harvest Time*); San Francisco Stock Exchange (R. Stackpole, *Earth's Fruitfulness* sculpture); Museum of Modern Art (J. Wilson, *Wheat*); Sacramento: Crocker Art Museum (P. Tchelitchew, *The Harvest*); San Marino: Huntington Gallery (J. Breton, *The Last Gleaning, Love Token*; J. Dupre, *The Harvester*); Los Angeles: Getty Museum (C. Monet, *Grainstacks, Snow Effect*; Netherlandish Harvest tapestries), Los Angeles Museum of Art (F. Morgan, *The Gleaners*), Norton Simon Museum (E. Degas, *Wheat Field and Green Hill*), David Hockney Foundation (D. Hockney, East Yorkshire Series); Ventura: Santa Paula Art Museum; San Simeon: Hearst Castle (Flemish *Labors of the Months Tapestry—August: Harvest*); Long Beach: American Museum of Straw Art; Santa Rosa: BJ's Restaurant (M. Webb and M. McIntosh, *Harvest* mural); Exeter (C. Fletcher, *Golden Harvest* mural)

Colorado Denver: Denver Art Museum (K. Adams, *Reapers*; O. Berninghaus, *Indians Threshing Wheat—Taos*; J. Breton, *The Gleaner*; V. van Gogh, *Edge of a Wheatfield with Poppies*, D. Coen, *Family Harvest*; B. West, *Harvesting at Windsor*); Loveland: Benson Sculpture Park (S. Vandable, *Harvest*; D. Lamphere, *Windswept*); Fort Morgan: Morgan Community College (S. Furini, *Harvest* sculpture)

Connecticut New Haven: Yale Center for British Art (J. Breton, *The End of Harvest*; S. Palmer, J. Linnell, G. Stubbs); Farmington: Hill-Stead Museum (C. Monet, *Grainstacks*); Stamford: Ukrainian Museum and Library (L. Wolynetz, Harvest sketches)

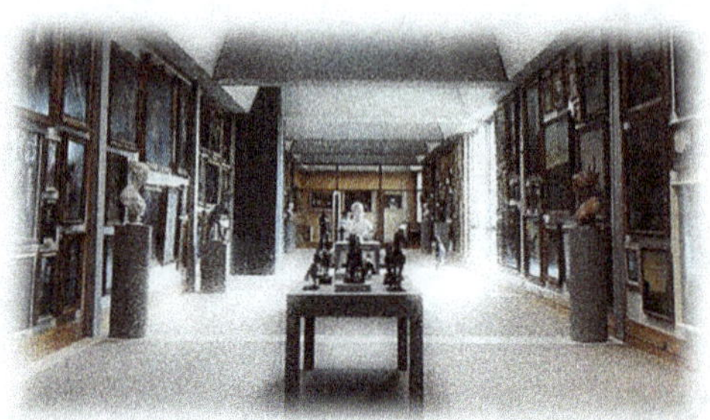

Yale Center for British Art, New Haven

Florida Ocala: Appleton Museum of Art (J. Dupré, *The Harvesters*)

Georgia Macon: National Society of Colonial Dames Headquarters (R. Rogers, *Ruth* sculpture)

Hawaii Honolulu: Honolulu Museum of Art (V. van Gogh, *Wheat Field with Sheaves*)

Idaho Moscow: University of Idaho (WPA Farm Murals, University Commons Building); Aberdeen: USDA National Small Grains Germplasm Research Center (C. A. Erickson, *East Idaho Harvest* stained-glass)

Illinois Chicago: The Art Institute of Chicago (C. Monet, Grainstacks Series [6]; J. Breton, *Song of the Lark*; P. Sérusier, *The Harvest of Buckwheat*; J. Storrs, *Ceres* sculpture), Chicago History Museum (Gordon Coster Photograph Collection), Board of Trade Plaza (Grain Sheaf Fountain, Ceres statue), Terra Foundation (J. Breck, *Studies from an Autumn Day*); Moline: John Deere Pavillion (John Deere Art Collection; Robin Moline, *Harvest Panorama*; "Reflections of an Era" exhibit); Naperville: Century Walk (K. Farrell and K. Scarboro, *Heartland Harvest* mosaic mural); Peoria: Arts for Relief in Missions Collection (R. Adcock, *Ruth and the Barley Harvest*); Bishop Hill: Olaf Krans Museum; Elmhurst (G. Smith, *There Was a Vision*)

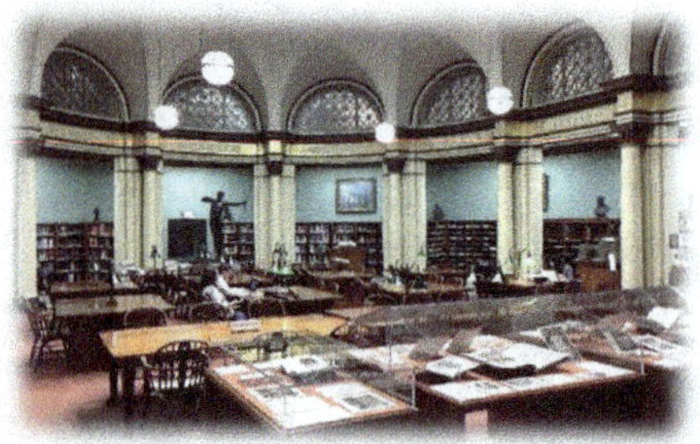

Ryerson Art Library, The Art Institute of Chicago

Indiana Indianapolis: Indianapolis Museum of Art (V. van Gogh, *Wheat Field with Peasant, Landscape at Saint-Remy*); Muncie: Owsley Museum of Art (J. Adams, *Gleaners Rest*, Victor Coleman Anderson, *Wheat*); Terre Haute: Sheldon Swope Art Museum (G. Wood, T. Benton, *Threshing Wheat*; J. Cox, *White Cloud*); Waveland: Waveland-Brown Township Public Library (T. Steele, *In Harvest Time*)

Iowa Ames: Iowa State University Library (G. Wood, Other Arts Follow murals); Davenport: Figge Art Museum; Iowa City: University of Iowa Museum of Art (F. Marc, *Sheaves of Grain*, G. Münter, W. Kandinsky)

Kansas Topeka: Capitol Building (J. Curry, *Kansas Pastoral* mural); Kansas Art Commission (Prairie Printmakers' Collection), Kansas Museum of History (Wheat People: Celebrating Kansas Harvest exhibit), Kansas State Capitol (L. M. Winter, *Threshing*; J. Haskell, *The Harvest* mural); Kansas City: Nelson-Atkins Museum of Art (P. Brueghel, *Summer Harvest*; L. Valtat, *Wheat Field with Poppies*), Board of Trade (J. Marquardt, *Heartland Harvest* sculpture), Wyandotte-Nichols Road Plaza (P. Romanelli, *Ruth* sculpture); Manhattan: Marianna Kistler Beach Museum of Art (Associated American Artists Print Collection);

North Newton: Bethany College Administration Building (Threshing Stone); Lawrence: Spencer Museum of Art (T. Benton, *Kansas Wheat Scene*); Russell: Deines Cultural Center (E. Hubert Deines Collection); Bonner Springs: National Agricultural Museum Center for Rural Art (G. Stone, *The Harvest Queen*, Lewis Watkins, National Farmers Memorial sculpture, Frederick James, *Farmland Cooperative* mural, R. L. Krouse, *Harvest* triptych); Goessel: Mennonite Heritage and Agricultural Museum; Lindsborg: Birger Sandzén Memorial Gallery (B. Sandzén, *Red Farm and Wheat Shocks*); Kingsville; Kingsville: Powell Gardens (J. Bown, *Sheaves of Wheat*); Wichita: Wichita Art Museum (G. Fuller, *The Harvesters*; W. Ufer, Hurd House Harvest mural)

Maine Rockland: Farnsworth Art Museum (N. C. Wyeth and Andrew Wyeth)

Maryland Baltimore: Museum of Art (G. Hartigan, *August Harvest*; A. Calder, *La Récolte*; C. Beauverie, *Les Blés*; A. Mouillon, *Wheat Fields*); Walters Art Center (J. Breton, *Returning from the Fields*; J. Veyrassat, *Harvest Scene*; W. Collins, *A Harvest Shower*); Beltsville: National Agricultural Library (USDA Botanical Arts Collection)

Massachusetts Boston: Museum of Fine Arts (J. Millet, *Harvesters Resting—Ruth and Boaz*; V. van Gogh, *Enclosed Wheat Field with Ploughman*, C. Monet, *Grainstack at Sunset*; W. M. Hunt, *Man in Wheat Field*; E. Hédouin, *Gleaners at Chambaudoin*); Springfield: Museum of Fine Arts (C. Monet, *Grainstack in a Field*); Williamstown: Clark Art Institute (J. Constable, *The Wheat Field*)

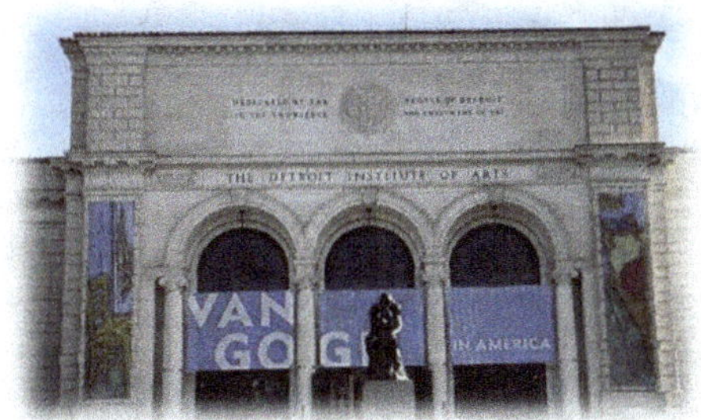

Detroit Institute of Arts

Michigan Ann Arbor: University of Michigan Museum of Art (W. Wendt, *Harvest by the Sea*; F. Cassara, *Grain*); Detroit: Detroit Institute of Arts (L. Lhermitte, *The Harvest*; J. Veyrassat, *Last Load of Wheat*; D. Rivera, *Woman Holding Grain*)

Minnesota Minneapolis: Ramsey County Courthouse (L. Lawrie, *Voice of the People* sculpture); Institute of Art (J. Patinir, *The Miracle of the Wheatfield*; C. Monet, *Grainstacks, Sun in the Mist*); The Museum of Russian Art (T. Khitrova, *Harvest*; V. Kondratyuk, *Collective Farm Harvest*); The Westin Hotel, formerly Farmers and Mechanics Saving Bank (W. Mossman, *Farmer* sculpture); Mill City Museum (K. Lawler, *Wheat Emporium*); St. Paul: State Capital Senate Chamber (E. Blashfield, et. al., *Minnesota, Granary of the World*); Elysian: Le Sueur County Museum (A. Dehn Collection)

Missouri St. Louis: Albrecht-Kemper Art Museum (L. Lhermitte, *The Harvest*); Jefferson National Expansion Memorial Museum (J. Jones, Grain Series *Riverport at St. Louis*); Kansas City: Thomas Hart Benton State Historic Site, Nelson-Atkins Museum of Art (T. H. Benton, *Cradling Wheat*); Branson: Harold Bell Wright Museum

Nebraska Omaha: Joslyn Art Museum (J. Breton, *Departure for the Fields*), Durham Museum (Bostwick-Frohardt Photograph Collection); Douglas County Courthouse (W. Rau, *Prairie Harvest* mural); Lincoln: State Capitol Building (L. Lawrie, *Maturity*; H. Meiere, *Harvest* mosaic), Pinnacle Arena (E. Carpenter, *Harvest* sculpture), Nebraska State Historical Society (S. D. Butcher Nebraska Homestead Experience Photograph Collection); David City: Bone Creek Museum of Agrarian Art (Dale Nichols, *Valley Farm*; Adolf Hofner, *At the Harvest*); Beatrice: Homestead National Monument (V. Armstrong Collection)

New Jersey New Brunswick: The Salgo Trust (I. Szőnyi, *Harvest*); Princeton: Princeton University Art Museum (S. Vranx, *Summer Landscape with Figures*)

New Mexico Santa Fe: Museum of Spanish Colonial Art (Grain Straw Applique Collection), Albuquerque: La Poblanos Historic Inn (H. G. Miller, *Reaping*), Chavez Park (J. Navarrete, *Wheat* sculpture)

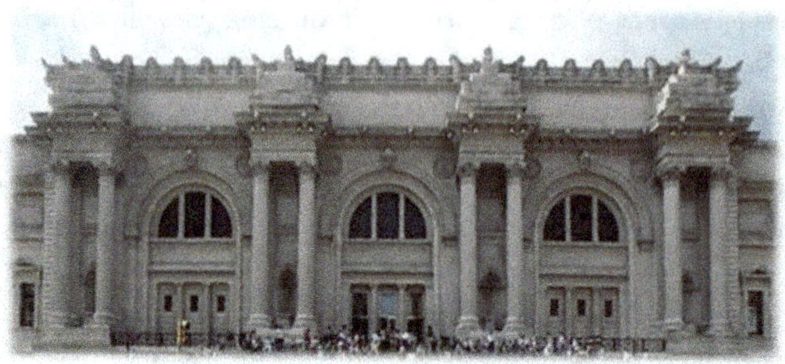

Metropolitan Museum of Art, New York

New York New York: Metropolitan Museum of Art (Amarna Barley Relief; Great Eleusinian Relief; Notre Dame-du-Bourg Chapel; G. David, *Nativity with Donors*; P. Bruegel, *The Harvesters*; V. van Gogh, *Wheat Fields with Cypresses*); Monet, *Grainstack in the Sun*; Rockwell Kent engravings), Morgan Library & Museum (B, Wolf, *The Book of Ruth*; Crusader Bible illuminations), New York Public Library (A. Anderson Proof Book Collection), Whitney Museum of American Art (M. Cohn, Wheat Series); Brooklyn: Dedalus Foundation (R. Motherwell, Summer Light "Harvest" series); Buffalo: Burchfield Penney Museum (C. Burchfield, *Wheatfields*); Syracuse: Syracuse University Library (Margaret Bourke-White Collection); Canajoharie: Arkell Museum (E. Gay, *Mother Earth*); Cooperstown: The Farmers' Museum (Plowline Photography Collection)

North Carolina Winston-Salem: Reynolda Museum of American Art (*Grant Wood and the American Farm*)

North Dakota Grand Forks: North Dakota Museum of Art (E. Lund, *Threshing Time, Threshers' Meal*); Williston: D. Njos, *Monument to Agriculture (Wheat)*

Oklahoma Oklahoma City: Oklahoma State Capitol (J. Dodd, *We Belong to the Land*); Tulsa: Philbrook Museum of Art (A. Delobbe, *Rest During the Harvest*); Alva: Co-op Building (R. Cooke, *Threshing Time*)

Ohio Cincinnati: City School District (H. Mosler, *Harvest Festival*); Cleveland: Museum of Art (J. Breton, *The Tired Gleaner*; G. Inness, *The Harvest Field*; J. Cox, *Wheat Shocks*; Gobelin Harvest tapestry); Dayton: Dayton Art Institute (D. Nichols, *While the Sun Shines*; L. Lhermitte, Sundown); Toledo: Toledo Museum of Art (V. van Gogh, *Wheat Field with Reapers*; L. Lhermitte, *Rest During Harvest*; J. Bassano, *The Flight into Egypt*); Canton: Canton Museum of Art (A. Biehle, *Wheatfield in Zoar*; K. Cox, *The Harvest*; F. English, *Wheat Harvest*; E. Gay, *Summer Landscape*)

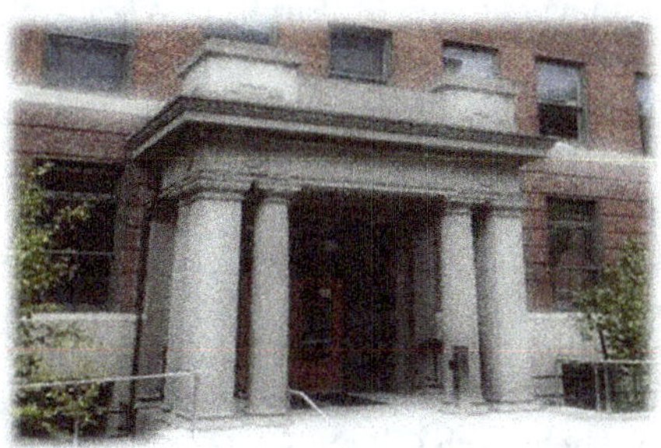

Art About Agriculture Gallery, Oregon State University, Corvallis

Oregon: Corvallis: Gallery 440, Oregon State University (Art About Agriculture Collection); Hillsboro: Oregon Civic Center (D. Huebsch, *Harvest*); Salem: Capitol Rotunda Staircase (B. Faulkner, *Wheat*), Revenue Building (R. Weller, *The Fields of August*); Irrigon: Skinny Bull Farm Museum (R. Blue, *Oregon Holt Harvest*); The Dalles (R. Thomas, *Where Wheat Was King* mural)

Pennsylvania Philadelphia: Philadelphia Museum of Art (V. van Gogh, *Enclosed Wheat Field in the Rain*; M. Chagall, *Wheatfield on a Summer's Afternoon*; Wierux Family, *The Harvest*, *Threshing Wheat*, *The Decision to Grow Wheat*; J. van Almoloveen, *Landscape with Harvesters*; J. de Boissieu, *Wheat Field*), Barnes Foundation (P. Picasso, *The Peasants*; R. Lotiron, *The Harvesters*); Pittsburg: Carnegie Museum of Art (V. van Gogh, *Wheat Fields After the Rain*); Brandywine: Brandywine River Museum of Art (*Rural Modern: American Art Beyond the City*); University Park (J. T. Biggers, *Day of the Harvest* mural)

South Dakota Brookings: South Dakota Art Museum (Harvey Dunn Collection)

Tennessee Memphis: Brooks Museum of Art (C. Gutherz, *The Gleaners*); Nashville: Tennessee State Museum (M. Avent, *Wheat Field off Hillsboro Pike*)

Texas Dallas: Dallas Museum of Art (V. van Gogh, *Sheaves of Wheat*), Centennial Fair Park Food and Fiber Building (C. Ciampaglia and H. Serbaroli, *The Wheat Harvester*); Fort Worth: Amon Carter Museum (P. Moran, *Threshing Wheat, San Juan*; *Indians Winnowing Wheat*); Houston: Museum of Fine Arts (J. Breton, *Recall of the Gleaners*); San Antonio: San Antonio Museum of Art (J. Cropsey, *Harvest Scene*; C. Williamson, *The Harvest*)

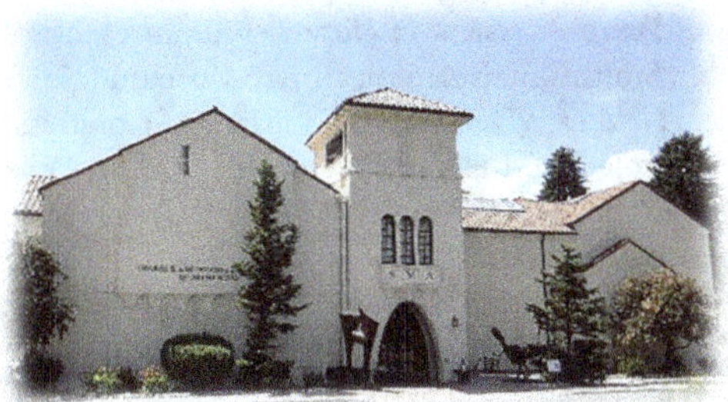

Springville Art Museum, Springville, Utah

Utah Provo: Brigham Young University Museum of Art (E. Evans, *Grain Fields*; J. Wier, *Connecticut Grainfield*); Springville: Springville Museum of Art (O. Campbell, *Timpanogos Harvest*; E. Evans, *Stacking Grain in Salt Lake Valley*; J. B. Fairbanks, *Harvest in Utah Valley*; J. L. Fairbanks, *Utah Valley Harvest*; S. Jepperson, *Harvest Time, Utah County*; L. Pratt, *Harvest Time in Cache Valley*, *Harvest Scene in France*; M. Sawyer, *French Farm Scene near Giverny*; D. Smith, *The Binder*; G. Smith, *Farmer with Grain Sack*; L. Stewart, *Threshing Wheat in Porterville*; M. Young, *Agriculture: The Farm Worker*; V. Beklemishev, *Young Rural Love*; V. Fedorov, *The Rye is Almost Ready*; A. Korolev, *Plenty*; V. Nechitailo, *The Threshing Floor*)

Vermont Burlington: Robert H. Fleming Museum (A. Farnham, *Wheat Field*); Shelburne: Shelburne Museum (C. Monet, *Grainstacks, Snow Effect*)

Virginia Richmond: Virginia Museum of Fine Arts (V. van Gogh, *Wheat Field Behind Saint-Paul*); Norfolk: The Chrysler Museum (D. R. Knight, *Harvest Scene*); Williamsburg: Colonial Williamsburg Museum; Troutdale: Ripshin Farm National Historic Site

Washington Seattle: Frye Art Museum (A. Lhermitte, *In the Valley*; W. Homer, *The Wheat Gatherer*; S. Bongart), Seattle Art Museum (P. Guigou, *Wheat Field*); Pullman: Washington State University Library (Columbia Heritage Collection, R. R. Hutchison Studio Photographs Collection); Tacoma: Pacific Avenue Museum District (E. C. Steffensen & J. Brinkerhoff, *Working Forward, Weaving Anew*); Richland: Hanford History Project (Dillman Grain Elevator Photograph Collection); Olympia: Capital Campus (S. Kogan, World War II Memorial); Walla Walla: Land Title Plaza (J. Hill, *Harvest Memories* sculpture); Fort Walla Museum Agricultural Hall (C. Poppenga, Harvest Mural)

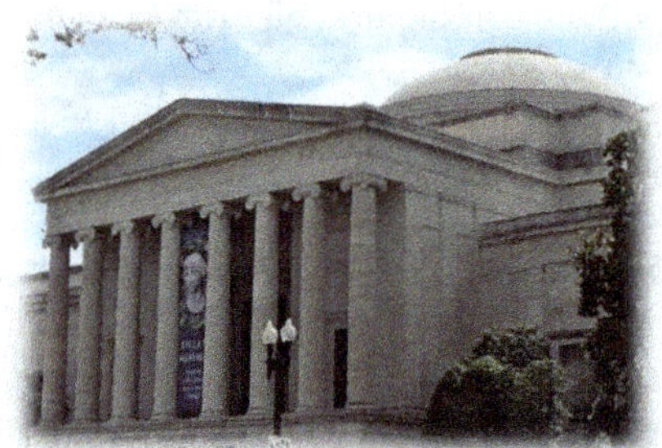

National Gallery of Art, Washington, D. C.

Washington, D.C. National Gallery of Art (S. Palmer, *Harvesters by Firelight*; V. van Gogh, *Farmhouse in Provence*; *Green Wheat Fields, Auvers*; A. Watteau, *Ceres*; G. Inness, *Harvest in the Delaware Valley*; L. Toeput, *Summer Harvest*; E. Field, The Ark of the Covenant; J. Miró, The Farm); Corcoran Gallery of Art (J. Breton, The Rapeseed Harvest); National Archives South Entrance (J. Fraser and E. Ratti, *Heritage* sculpture); Department of Agriculture Whitten Building (A. Weinman, *Cereals* pediment sculpture); Clinton Building Mellon Auditorium (E. Walters, *Columbia—Agriculture* pediment sculpture) George Washington Memorial Parkway (*J. Fraser, The Arts of Peace—Music and Harvest* sculpture); The Phillips Collection (B-Z. Weinmann, *Paths in a Wheat Field*); Smithsonian American Art Museum (New Deal Post Office Mural Studies; T. Benton, *Wheat*; W. Holmes, *Field of Wheat Shocks*); Library of Congress (Farm Security Administration Photographs, Federal Writers' Project and Historical Records, Kansas Wheat Harvest [F. Bauman] Photographs, Archives of American Art, F. Palmer Currier & Ives Lithographs, S. Proudkin-Gorskii Photographs, E. Matson Photographs); General Services Administration New Deal Arts Collection (J. Roman, *Harvest Time*; Paul Weller, *Harvest Hands*; Max Kahn, *Harvest*; Francis Stephen, *Harvesting Wheat*); The Phillips Collection (V. van Gogh, *Wheat Field at Auvers*); Union Station (Ceres sculpture)

Wisconsin Madison: Wisconsin Historical Society (R. Hodgell, *From Cradle to Combine* murals; E Nordfeldt, Harvest murals; McCormick-International Harvester Collection), University of Wisconsin College of Agriculture (J. Curry, *The Good Earth*); Milwaukee: Marquette University Haggerty Museum of Art (J. Jones, Grain Series panels), Grohmann Museum (*Midwest Murals: Joe Jones and J. B. Turnbull*; J. Dupré, *Stacking Wheat Sheaves*), Froedtert Malting Company (G. Bohland, *The Reaper* sculpture)

CANADA Ottawa: National Gallery (C. Schaefer, *The Wheat Field*; J. E. H. MacDonald, *In a Wheat Field, Evening Shadows*; H. Watson); Montreal: Museum of Fine Arts (W. Eaton, *Harvesters at Rest, The Harvest Field*), Toronto: University of Toronto Art Centre (C. Jefferys, *Wheat Stacks on the Prairie*); Calgary: Glenbow Museum (F. Shook, *Four Year's Effort*, A. F. Kenderdine, Sybil Andrews); Regina: Saskatchewan Arts Board (J. Didur, *Custom Combining*, J. Geddes, *Wheat Field*, W. McCargar, *Old Threshers*), Regina Plains Museum (J.

Berting, *The Wheatfield* glass sculpture); Assiniboia Gallery (R. Pawson, *Cutting Oats and Wheat, Beginning to Cut, Harvest Time*); Whitby: Brooklin Landmark Square (Wheat Sheaf sculpture); Winnipeg: Winnipeg Art Gallery (L. L. FitzGerald, *Harvest*; H. V. Fanshaw, *Wheat Harvesting Near Winnipeg*)

MEXICO Guanajuato: Diego Rivera House Museum (*The Threshing Floor*); Morelia: Government Palace of Michoacán (A. Zalce, *People and Landscape of Michoacán* mural)

EUROPE

ALBANIA Tirana: National Gallery (I. Sulovari, *Bringing in the Harvest*)

AUSTRIA Vienna: Kunsthistorisches Museum (P. Bruegel, Seasons Collection; J. D. de Heem, *Communion Cup with Fruit and Wheat*; D. Teniers, *Village Fête*; F. Bassano, *Summer*); National Library (*Calendar of Salzburg*); Belvedere Gallery (V. van Gogh, *Wheat Field Near Auvers*; J. Dobrowsky, *Field at Harvest*; Stefan Simony, Reapers; F. Grabmyr, *Grain Stooks*); Zwettle: Cisterian Abbey (Jörg Breu the Elder, Altar Panel Harvest Scene with St. Bernard)

BELARUS Minsk: Museum of National Art (Alla Zamai, *Source of Life*); Raubichi: Museum of Belorussian Folk Art (Orthodox Holy Gates straw weavings)

BELGIUM Brussels: Meunier Museum (C. Meunier Monument of Labor Series—*Harvest*), Cinquantenaire Park (C. Meunier, *The Reaper*); Deinze: Museum of Deinze and the Leie (A. Servaes, C. Permeke); Ghent: Museum of Fine Arts (A. Servaes, *The Harvest*; V. Servvancks, *A Sea of Grain*)

BULGARIA Sofia: Museum of Fine Arts (I. Angelov, *Harvesting in Chepino*), National Gallery: V. Dimitrov, *Singing Girls Returning*); Kyustendil: Vladimir Dimitrov Gallery

CROATIA Zagreb: Croatian Museum of Naïve Art (I. Generalic, *Harvesters*)

CYPRUS Nicosia: Zampelas Art Museum (M. Kassialos paintings)

CZECH REPUBLIC Prague: Municipal Building (J. Obrovský, *Harvest in Bohemia* mosaic), V. J. Rott House (M. Aleš, Harvester murals); Kroměříž: Kroměříž Town Hall (M. Švabinský, *Harvest*); Budweis: St. Mary's Church (*Maria im Ährenkleid*)

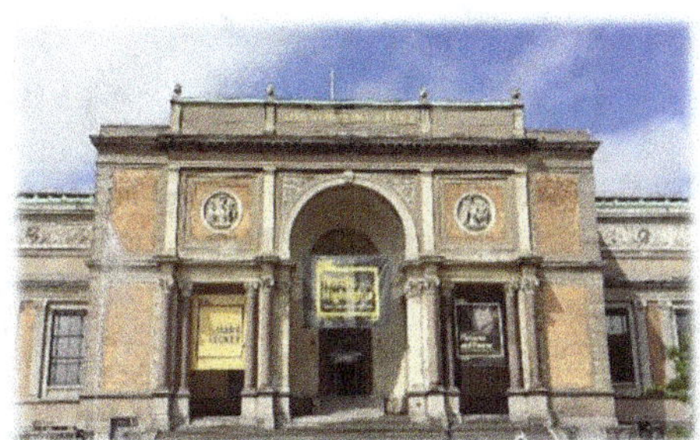

National Gallery of Denmark

DENMARK Copenhagen: National Gallery (L. A. Ring, *The Harvester*), Rosenborg Castle (Netherlandish landscapes); Faaborg: Faaborg Museum ((P. Hansen, *Threshing with Oxen, Winnowing Wheat, Harvest*); Skagens: Skagens Museum (A. Anchor, *Harvest Time*)

FRANCE Paris: Louvre (L. Robert, *Return of the Harvesters*; V. van Gogh, N. Poussin, *Four Seasons—Summer* [*Ruth and Boaz*]; Maktar Harvester Stele), Musée d'Orsay (J. Millet, *The Gleaners*; J. Breton, *The Blessing of the Wheat*; *Recalling the Gleaners*; C. Monet, *Grainstacks, End of Summer*; C. Daubigny, *Harvest*; P. Gaugin, *The Yellow Harvest*; E. Bernard, *Harvest by the Sea*; B. Morisot, *Wheatfield at Gennevilliers*), Musée des Beaux-Arts (P. Baudoüin, *History of Wheat* series); Musée de Cluny (Puy d'Abbeville, *Virgin with Wheat*); Musée Rodin (V. van Gogh, *Reapers*), Musée Marmottan-Monet; Place Rhin-et-Danube (L. Deschamps, *The Harvester* sculpture); Musée National d'Art Moderne—Centre Pompidou (A. Yves, *Master of the Harvest*; A.

Servaes, *Harvest*; Raoul Dufy, Jean Roubier photographs), Four Seasons Fountain (E. Bouchardin, *Summer Harvest* bas-relief); Versailles Palace (L. Lagrenée, *The Harvest*; T. Regnaudin, Ceres Fountain; J-P Poultier, Ceres statue): Barbizon: Musée L'Ecole de Barbizon, J. Millet Workshop-Museum; Rouen: Musée des Beaux-Arts (A. Rigolot, *The Threshing Machine*); Nantes: Museé de Beaux-Arts (G. Courbet, *The Wheat Sifters*); Nice: Musée Marc Chagall; Chantilly: Musée Condé (*Les Très Riches Heures*); Nozeroy: Collegiate Church of St. Antoine (*The Good Shepherd* and *Jesus and the Samaritan* altarpieces); Arles: Museon Arlaten; Reims: Mars Gate (Gallic-Roman Reaper Decorative Panel)

GERMANY Berlin: National Gallery (A. Lier, *Grainfield at Harvest Time*); Old National Gallery (Casa Bartholdy Murals; K. F. Schinkel, *Harvest Festival Procession*; L. Schwanthaler, *Ceres and Prosperpina*); Bonn: State Art Museum (H. Nauen, *Harvest*); Füssen: Schranneplatz Sheaf Sculpture; Hanover: Sprengel Museum (P. Klee, *The Ripe Harvest*); Munich: Bavarian National Museum (*Maria im Ährenkleid*); Kassel: Neue Galerie (C. Bantzer, Harvest Worker from Hesse); Trier: Archaeological Museum (Gallo-Roman Reaper Bas-Relief); Schotten: Vogelsberger Heimatmuseum (F. Leiber, *Jesus the Bread and the Wine*); Munich: Bavarian State Art Museum (J. Becker, Peasants *Alarmed by Lightning*); New Pinakothek (C. Lohse, *Ripe Grain*); Weinigerode: Harzmuseum (W. Pramme, *Summer Evening*); Willingshausen: Willingshausen Art Colony Museum

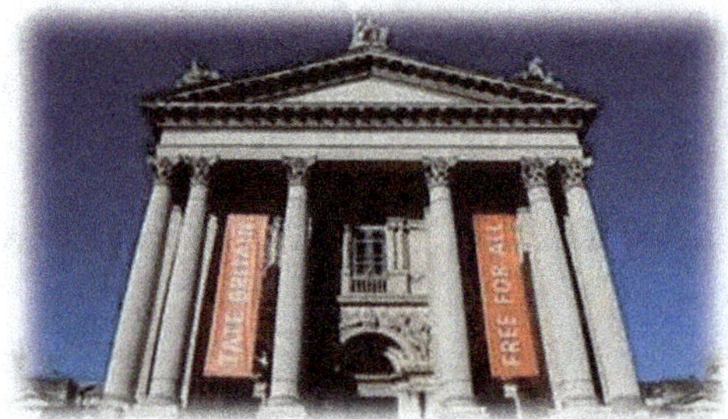

Tate Britain (formerly the National Gallery of British Art), London

GREAT BRITAIN London: Tate (National Gallery) Britain (J. Turner, G. Lewis, G. Stubbs, G. Mason, J. Linnell, G. Clausen, J. Nash), Victoria and Albert Museum (Medieval Stained Glass Harvest Labors of the Months, P. De Wint, *The Cornfield*), British Library (Luttrel Psalter), Imperial War Museum (Ministry of Information War Photograph Collections), Institute of Chartered Accountants (H. Thornycroft, *Agriculture*), Foreign and Commonwealth Office, Whitehall (John Birnie Phillips, *Agriculture*), Victoria Memorial, Buckingham Palace (T. Brock, Ceres/Agriculture statue), Royal Society of Arts (J. Barry, *A Grecian Harvest Home*); Museum of Childhood, Bethnal Green (*Wheat Harvest* and *Mowing* exterior mosaic panels); Cirencester: Royal Agricultural University Art Collection (J. Voelcker, *Landscape with Harvesters*; J. Linnell, *Landscape with Harvesters*); St. Paul's Cathedral Quire Aire (W. Richmond, *Christ with Sheaves of Wheat* mosaic); Farnham: Museum of Farnham (W. Allen, *Harvest Field with Elms*); Birmingham: Barber Institute of Fine Arts (T. Gainsborough, *The Harvest Wagon*); Oxford: Ashmoleon Museum (S. Palmer, *The Valley Thick with Corn, A Pastoral Scene*), Pitt Rivers Museum (corn dollies); Reading: University of Reading Museum of English Rural Life; East Riding: Beverley Art Gallery (W. H. Williams, *Harvesting*); East Riding: Beverley Art Museum (W. H. Williams, *Harvesting*); Leeds: Granary and Brewery Wharf (G. Wilson, *Cornucopia*, C. Knight, *The Sheaf Tree*); Liverpool: Walker Art Gallery (H. Thornycroft, *The Mower*); Siddington: All Saints Church (R. Rush Corn Dolly Collection); Truro: Royal Cornwall Museum (L. Kemp, *Harvesting*); Manchester: Whitworth Art Gallery, University of Manchester (J. Linnell, *Harvest Scene*); Perth: Museum and Art Gallery (D. Cameron, *Harvest Time in Lorne*); Edinburgh: National Gallery of Scotland (W. McTaggart); Glasgow: Scottish National Gallery of Modern Art (J. Eardley, *Harvest*), Hunterian Gallery (W. Gillies, *Harvest Field*; M. Greiffenhagen, *Harvest* mural), National Museum of Rural Life; Musselburgh: Brunton Hall (J. Bellany, *The Harvester*); Devises: Wiltshire Museum (Robin Tanner Print Collection), Winchester: Hampshire Cultural Trust (W. Allen, *Harvest Field with Stooks, After the Harvest*); Wrexham: Glyndwr University Creative Industries Building: M. Eldrige, *The Dance of*

Life mural Harvest panel)

GREECE Athens: National Museum (Great Eleusinian Relief); Museum of the Ara Pacis (Mother of Earth "Tellus" Panel)

HUNGARY Budapest: Hungarian National Gallery (Nagybánya Colony paintings); Zebegény: Szőnyi István Museum; Kaposvár: Rippl-Rónai Museum (M. Goszthony, *Harvest, Wheat Sheaves*; L. Knuffy, *Wheat Field, Pitching Sheaves*); Debrecen: Déri Museum (Z. Pálnagy)

IRELAND Dublin: The National Gallery (N. Hone, *A Harvest Scene, North Dublin*; *Grain Stooks Under Trees*; G. Barret, *The Harvest Waggon*; J. Breton, *The Gleaners*; J. Veyrassat, *Loading the Grain*)

ITALY Rome: Trajan's Column (Harvest Scene); Vatican City: Sistine Chapel Ceiling (Michelangelo, *Ruth, Obed, and Boaz*); Genoa: Academy of Fine Art (P. Nomellini, *Harvest*); Naples: National Museum (F. Rossano, *Field of Wheat*), Lucca: State Library (Hildegard of Bingen, Codex Latinus *Liber Divinorum Operum*); Florence: Palantine Gallery (P. Rubens, *Return from the Harvest*), Uffizi Gallery (L. da Vinci, *Landscape with River [Arno]*), Crusca Academy Library (grain related furnishings), Cesiomaggiore, Veneto (M. Farina, *Barley Harvest* collage mural); Pescara (Basilio Cascella Civic Museum); Sienna: Palazzo Pubblico, Sala dei Novo (A. Lorenzetti, *Effects of Good Government in the Country*); Caserta: Royal Palace (J. P. Hackert, *Harvesting at San Lieucio*); Trento: Castello del Buonconsiglio (Cycle of the Months fresco—*August*)

LUXEMBOURG National Museum (J. Schaak, *Sheaves of Wheat*)

THE NETHERLANDS Amsterdam: Rikjsmusum (U. Leyniers, *Harvest* tapestry, Northern Renaissance harvest engravings), Van Gogh Museum (*Wheat Field with a Reaper*, Drawings), Beurs van Berlage/Grain Exchange (J. Toorop, *Sower and Reaper*); Kröller-Müller Museum (V. van Gogh Drawings); Rotterdam: Museum Boijmans Van Beuningen (J. Wierix, Cultivation of Grain series)

NORWAY Oslo: National Gallery (J. Dahl, *Hjelle in Valdres*; T. Kittleson, *Grain Stooks in Moonlight*; J. de Momper, *Summer*); City Hall (A. Revold, *Farming and Fishing* Mural)

POLAND Warsaw: National Museum (W. Tetmajer, Harvest); Kraków: National Museum (W. Tetmajer, *Fields of Grain*; K. Alchimowicz, *Harvest*)

PORTUGAL Lisbon: Museum of Contemporary Art (J. Pomar, *The Reaper*)

ROMANIA Bucharest: Romanian National Museum of Art (R. Schweitzer-Cumpăna Collection); Baia Mare: Baia Mare (Nagybánya Colony) Art Museum; Pitesti: Rudolf Schweitzer-Cumpăna Art Museum

RUSSIA Moscow: State Tretyakov Museum (V. Perov, *Reapers Returning from a Field Near Ryazan*; A. Venetsianov, *Harvest Time, Summer*; I. Shishkin, *A Rye Field*; K. Rozhdestvensky, *Landscape with Sheaves*; A. Plastov, *Harvesting*; T. Yablonskaya, *Grain*), Pushkin Museum (C. Friedrich, *View of Schmiedeberger Ridge*; L. Lhermitte, *The Reapers*); St. Petersburg: State Hermitage Museum (D. Teniers II, *The Reaping*, S. Vrancx, *Harvest*), State Russian Museum (G. Myasoyedov, *Time of Toil—The Reapers*; I. Repin, *The Barge Haulers*; I. Levitan, *Grain Stacks and Village*; A. Plastov, *Harvest Festival*); Kostroma: Kostroma State Museum (N. Dubovsky, *Harvested Field*); Ryazan: Ivanov Art Museum (V. Ivanov, *Harvesting Near Ryazan*); Samara: State Art Museum (V. Rozhdestvensky, *The Harvesters*); Omsk: Vrubel Museum of Fine Arts Museum (N. Goncharova, *Harvest*)

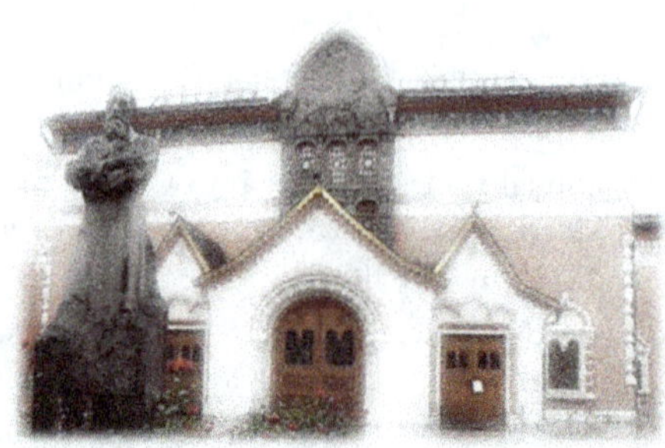

State Tretyakov Museum, Moscow

SERBIA Novi Sad: Pavle Beljanki Collection (M. Konjović, *Harvest* [1938]); Ruma: Ruma Municipal Museum (F. Soretić paintings); Sombor: Konjovic Gallery (M. Konjovic, *Harvest* [1937]); Vršac: Vršac City Museum (P. Jovanović, *Vršac Triptych—Sowing and Harvesting and Market*)

SLOVAKIA Bratislavia: Slovak National Gallery (J. Collinásy, *Harvest*; F. Studený, *Harvest*; M. Benka, Harvest paintings; M. Martinček Harvest photographs); Martin: Martin Benka House Museum

SPAIN Madrid: Museo del Prado (H. Bosch, *The Path of Life*; J. Patinir, *The Miracle of the Wheatfield*, F. Goya, *The Threshing Floor*); Thyssen-Bornemisaa Museum (P. Picasso, *The Harvesters*; P. Renoir, *The Wheatfield*; A. Lhermitte, *The Harvest*); Barcelona: Picasso Museum (P. Picasso, *House in a Field of Wheat*); Malaga: Museo Carmen Thyssen (G. Martinez, *Farmers Resting*); Girona: Girona Art Museum (J. Vayreda, *The Harvest*); Leon: World Agricultural Museum

SWEDEN Stockholm: National Museum (A Zorn, *Our Daily Bread*; H. Salmson, *The Little Gleaner*); Modern Museum (H. af Klint, Primordial Chaos Series); Sundborn: Spardavet Farm (Carl Larsson Collection)

SWITZERLAND Basel: Fondation Beyeler (V. van Gogh, *Field with Wheat Stacks*, *Wheat Field with Cornflowers*), Kunstmuseum (C. Pissarro, *The Gleaners*); Geneva: Museum of Art and History (V. van Gogh, *Wheat Fields with Auvers in the Background*); Zürich: Kunsthaus (C. Monet, *Grainstacks*)

UKRAINE Kyiv: National Museum of Art (V. Orlovsky, *Harvest*; A. Plastov, *Kolkhoz Threshing*; M. Pymonenko, *The Reaper*); St. Volodymyr's Cathedral (M. Vrubel, *Wheat and Flowers* fresco)

ISRAEL

Jerusalem: The Jerusalem Museum (V. van Gogh, *Harvest in Provence*, N. Gutman, *Sheaving the Harvest*); Haifa Archaeological Museum of Grain (M. Gumpel, *That You May Gather Your Grain* mosaic)

AUSTRALIA

Brisbane: Queensland Art Gallery; Canberra National Library of Australia (E. Spowers, *Harvest*; S. T. Gill Collection); Melbourne Melbourne Museum (H. V. McKay and W. Boyd Photograph Collections); Sydney Art Gallery of New South Wales (A. Streeton, *Sussex Harvest*)

NEW ZEALAND

Auckland: Auckland Art Gallery (P. Brueghel the Younger, *A Village Fair*; E. Spowers, *Harvest*); Wellington: Museum of New Zealand (W. Crane, *Harvest Field, Normandy*; C. Savage, *Mountain Harvest*)

ENDNOTES

Introduction

[1] M. Inge, 1969:3-6; S. Bellow, "They Shall Overcome" (1965), cited in Z. Leader, 2018:31; R. Hoftadter, 1956:42-53. Sources on Agrarian egalitarianism include James A. Montmarquet, *The Idea of Agrarianism: From Hunter-Gatherer to Agrarian Radical in Western Culture* (1989); and Henry Nash Smith, *Virgin Land: The American West as Symbol and Myth* (1950). On Neoclassical Arcadianism see Leo Marx, *The Machine in the Garden: Technology and the Pastoral Ideal in America* (1964); and Peter J. Schmidt, *Back to Nature: The Arcadian Myth in Urban America* (1969). Neoclassical Arcadianism was inspired by the ideas of nineteenth British writer John Ruskin and designer William Morris. They sought affirmation of country life virtues in literature and decoration by reviving the artistic sensibilities of the ancient and medieval eras. In the United States this Arts and Crafts style influenced the design of furniture, pottery, and other objects that incorporated floral embellishments, stylized grain sheaf symmetry, and other agrarian motifs. These masterful creations challenge modern distinctions between fine art and craft, creativity and design, painting and decoration. Country Life Progressivism is discussed in William Bowers, *The Country Life Movement in America, 1900-1920* (1974); and David Danbom. *The Resisted Revolution: Urban America and the Industrialization of Agriculture, 1900-1930* (1979).

[2] A. Pratt, 2019; J. Louk, 2022. For a critical view of the Midwest Regionalists that casts them as generally complicit in depictions of a racialized heartland, see Britt E. Halvorson, *Imagining the Heartland: White Supremacy and the American Midwest* (2022). On political repression throughout America during the period 1917-1921, see Adam Hochschild, *American Midnight: The Great War, a Violent Peace, and Democracy's Forgotten Crisis* (2022).

[3] Country Life Progressivism is discussed in William Bowers, *The Country Life Movement in America, 1900-1920* (1974); and David Danbom. *The Resisted Revolution: Urban America and the Industrialization of Agriculture, 1900-1930* (1979). Following a December 1908 government-sponsored Country Life Commission regional hearing in Spokane, the city's Chamber of Commerce provided the first widely published exposition of rural Progressive principles in its *Report of the Country Life Commission and Special Message of the President of the United States* (1909).

[4] J. Dadak-Kozicka, "Long-sounding Notes and Ornamentation as Characteristic Qualities in Musical Expression in Slavic Harvest Songs," in P. Dahling, 2009:97. Dadak-Kozicka's insightful research discusses the distortion of festive and ritual harvest songs by Eastern European socialist regimes after the Second World War. Her work is based in part on research described in Eugenia Jagiełło-Łysiowa's authoritative *Elementy Styló Życia Ludności Wiejskiej* (Elements of the Lifestyles of the Rural Population, 1978). On the grain trade of the ancient world see Peter Garnsey, *Famine and Food in the Graeco-Roman World: Responses to Risk and Crisis*, 1989. Concerns regarding modern-day food security due to global warming and geopolitics are explored in Thane Gustafson, *Klimat: Russia in the Age of Climate Change* (2022), and Karl A. Scheuerman, "Weaponizing Wheat: How Strategic Competition with Russia Could Threaten American Food Security," *Joint Forces Quarterly* 111 (October 2023).

[5] W. Cather, 1922:157.

[6] M. Jones, 1957:3. Lifetime Palouse Country resident Lucille G. Howard provided a first-person account of harvest cookhouse experience in "30 Days in a Cookhouse," *Bunchgrass Historian* 45:2 (2019): 5-7.

[7] J. Barzun, 1989:120-129.

[8] A. Barton, 2020. Degas' remark about art is in a December 1872 letter to his businessman-artist friend Henri Rouart. On Baudelaire and literature see Richard Sieburth's introduction to *Late Fragments* (Yale University Press, 2022), a collection of the French poet-critic's late fragmentary writings.

[9] V. van Gogh to É. Bernard, June 19, 1888; retrieved at vangoghletters.org/vg/letters/let628/letter. In 1814 Scottish artist Patrick Syme (1774-1845) published *Werner's Nomenclature of Colours*. Artists and scientists used this landmark taxonomic guide to the colors of the natural world based on classifications first devised by German mineralogist Abraham Gottlob Werner (1849-1817). The Werner-Syme system identified fourteen hues of yellow including gamboge, saffron, sulphur, sienna, and ochre.

[10] J. Korbelik, 2009; D. Vergauwen and I. De Smet, 2020:717-719. Kooser's poem, "So This Is Nebraska," was published a 1980 collection of his work, Sure Signs. Curiale's recordings feature the Royal Philharmonic and London Symphony.

Chapter I: *The Bread and the Wine*—Grain Scythes and Altar Cloths

[1] T. Crossley, 2021:52-55; S. Barnett and L. Lippard, 2020:31-32. The Washington Association of Wheat Growers' "Wheat Queen" program continued until 1987. In recent years the organization and affiliated Washington Wheat Foundation have sponsored the Wheat Ambassador Program for high school seniors to receive college scholarships and serve as association advocates. WAWG has also sponsored a series of limited edition "Harvest Print" lithographs by Lisa Specht-Urbat and Karen Reffett showing panoramic scenes from the pioneer era of animal- and steam-powered threshing.

[2] R. Peschel, 2012; R. Scheuerman, 2020: 44-49.

[3] A. Weitz, 1972. Palouse Colony immigrant life and farming is chronicled in R. Scheuerman, *Hardship to Homeland: Pacific Northwest Volga Germans* (2018). Old English *hærfest*—"autumn," is eloquently described by the tenth-century *Menologium* as ". . . [W]ela byð geywed fægere on foldan*, or when "Plenty is revealed, beautiful upon the earth." The term *Kerb* is derived from Proto-Germanic *kharbitas*, which is also the source of Old English "harvest" as well as Saxon *hervist*, German *Herbst*, and Norse *haust*. The Germanic root *kerp-* means "to gather, pluck, harvest," and related to Greek *karpos*, "fruit," and Latin *carpere*, "to cut, divide, pluck."

[4] J. Ruskin, XVII, 1905:12. Ruskin rejected the prevailing economic model of John Stuart Mill that favored mass production of cheap, disposable products for unending consumer demand. Among other contributions Ruskin made to the Arts and Crafts Movement was his conception of true economic value as durable, handmade goods that can be both functional and beautiful. His "art of living" was a personal discipline of wise consumption for simple pleasures rather than accumulation of wealth. Ruskin rejected socialism, but favored a capitalist system mediated by regulation for necessary constraint. He wrote about the limits of natural resources which if unwisely exploited would imperil the wellbeing of his children's generation. The ideas of Ruskin and other nineteenth-century English Romantics that anticipated modern environmentalism are discussed in Vicky Albritton and Fredrik Albritton Jonsson, *Green Victorians; The Simple Life in John Ruskin's Lake District* (2016). Among the many Victorian artists deeply influenced by Ruskin was Alfred Hunt (1830-1896) whose drawing *Autumn* depicts a family elder showing a child how a small sheaf is made. A grain sheaf, known as a "garb" in Middle English, were commonly featured in heraldry shields to symbolize prosperity. New England half-round sheaf signs based on carvings by Samuel McIntire and W. Clarke Noble commonly marked locations of bakeries and inns. Noble's carving bears remarkable resemblance to the ancient marble sheaves that decorated the entablature of the Lesser Propylon (Entry) Gate at Eleusis near Athens. A rare American folk art example of a wooden sheaf

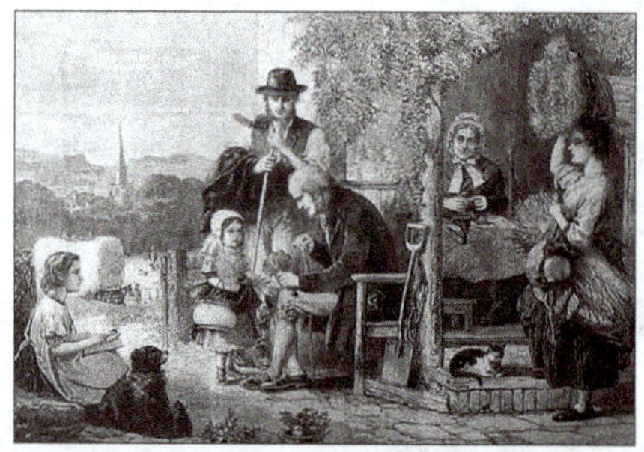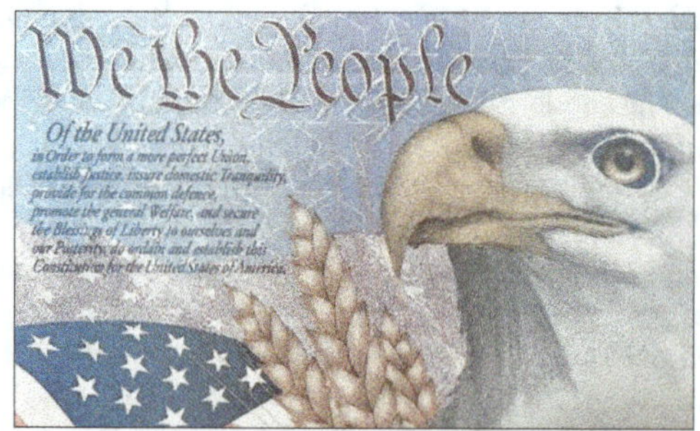

Alfred Hunt, *Autumn* **(Tying a Wheat Sheaf) United States Passport Signature Page (2020)**
The Illustrated London News **(October 3, 1863)**

carved in the round is the Metropolitan Museum of Art's Guennol Collection Sheaf. Present-day United States passports feature the 2007 "American Icon" intaglio design that includes a cluster of wheat between the flag and an eagle on the signature page. Although substantially self-taught, Ben-Zion Weinmann (1897-1987) mastered painting, printmaking, and sculpture, and also wrote poetry. His Phillips Collection *Path through the Wheat* (1950) is one of several mystical renderings showing a thick stand of leaning yellow-ochre grain that parts down the middle as if inviting the viewer into its mystery.

[5] S. Barnett and L. Lippard, 2020:35-36.

[6] S. Heaney, 1979:71-73. Heaney, who cited the influence of Virgil, Giovanni Pascoli, and T. S. Eliot, was awarded the 1995 Nobel Prize in Literature. His poem "The Wife's Tale" (1969) relates a farm woman's prodigious and welcomed mealtime service in the field to a group of harvesters.

[7] R. Peden, 1966:249. Ruskin extolled such luminous "colors of the Earth" as muddy roadside browns, gravel pit oches, and coal dust black, and preferred ground-up organic matter and mineral pigments to brighter sythentic chemical paints. Today soil scientist Karen Vaughan of Wyoming combines a range of colorful soil samples to mix with acacia gum, glycerin, and other natural materials create watercolor paints; see S. Goldstein, 2023:36-41. On the Palouse as a premier scenic farm region, see Matthew Kronsberg, "A Gateway to the 'Tuscany of America,'" *The Wall Street Journal*, July 27, 2018; and Barbara Austin, "A Paradise Called the Palouse," *National Geographic Magazine*, 161:6 (June 1982).

[8] P. Hamerton, 1885:375-379; S. Drinkard, 2006.

[9] J. DeWitt, 2017:2-3. Gerster described the Palouse as "the Louvre of farmlands."

[10] B. Lopez, 2006:83.

[11] G. Coates, 1919:14.

[12] Mary Sunwold Scrapbook, c. 1910-1930, author's collection.

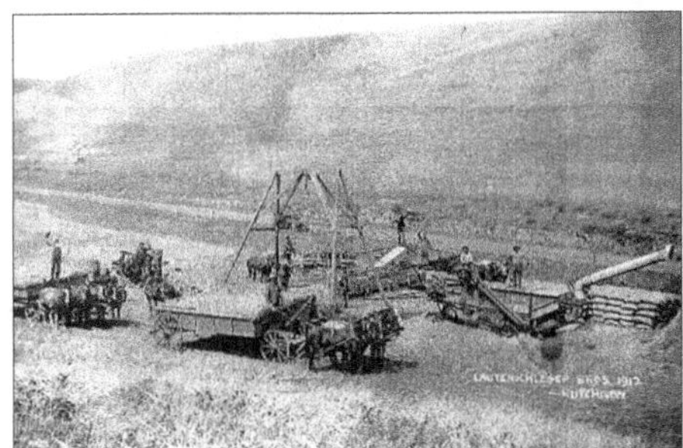

"Lautenschlager Bros. 1912" R. R. Hutchison Harvest Photograph
Henry B. Scheuerman Farm near Endicott, Washington
Dale Schneidmiller Collection

[13] Harvest wages across the Columbia Plateau in the first decade of the century reflected the value of specialized labor. Separator engineers earned about $5 per day, sack sewers and loaders $3, header tenders $2.75, grain loaders $2.25, and header-box drivers $1.75. In the course of this study, longtime friend Dale Schneidmiller of St. John, Washington, whose German ancestors had also immigrated to the Palouse from Yágognaya Polyána, told me of a "Lautenschleger Bros. 1912" Hutchison photograph passed down in his family that bore striking resemblance to my Grandfather Scheuerman's. His research on oral histories provided by family elders in the 1960s revealed that my Great Grandfather Scheuerman had leased his farm to neighbors Adam and Philip Lautenschlager since my Grandfather Karl was still in his teens and did not start farming full time until that year. The 1911 and 1912 photographs show the same Case separator and other threshing equipment with the latter image being the earliest known photograph of our family farm. The arduous tasks of sacking and transporting bagged grain gradually gave way from about 1920 to 1940 to use of combine bulk tanks that dropped grain by chutes or augers into wagons and trucks. Seattle author Albert Laughon's novel *Spirit of the Wild Rosebush* (1938) about 1880s Northwest farm life recounts a lively contest between separator pitchers on Rebel Flat in the central Palouse.

[14] Carl Penner in A. McGregor, 1982:165, 271; J. Payne, 2023. Caterpillar Company formed in 1925 with

the merger of Holt Harvester and Best, which primarily built steam-powered tractors. In 1939 the Caterpillar combine line was sold to John Deere.

[15] The region's first combine trial is mentioned without specifying the exact year in G. Kincaid and A. Harris, [1934] 1979:42; pioneering harvesting methods in the area are described in D. Meinig, 1968:392-397.

[16] Z. Grey, 1919:55-56; G. Topping, 1978:59-60; P. Lewis, 2021:46-50. Following the commercial success of several of his early novels, Zane Grey formed the motion picture company Zane Grey Productions that released a film version of *The Desert of Wheat* in 1920 that played in theatres as *Riders of the Dawn*. Grey eventually sold the company to Paramount Studios which also distributed the movie *Wild Harvest* (1947) starring Alan Ladd, Dorothy Lamour, and Robert Preston about a custom threshing crew traveling throughout the West. According to Gene Worth (2017), the harvest scenes were filmed in eastern Washington's Big Bend district near Connell and in California's San Joaquin Valley. The 1917 Grey family photo album recovered in 2019 by Zane Grey West historians Terry Bollinger and Ed Meyer is now at University of Oregon's Knight Library Special Collections in Eugene.

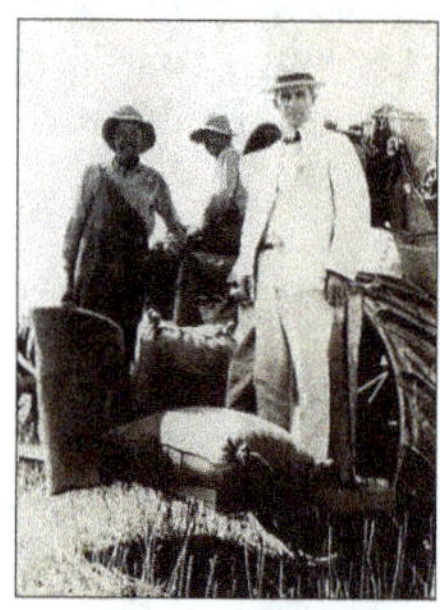

Zane Grey on Columbia Plateau Horse-drawn Thresher (July 1917)
Zane Grey West Society

[17] Paul Lewis's *Lost in the Hill* musical premiered in Salem, Oregon in August 2022.

[18] H. Davis poems and T. Ferril in H. Davis, O. Burmaster, ed., 1978.vi-v, 1, 7; H. Davis, 1935:356.

[19] E. Sloan, 1971:18. *In Legacy* (1979), one of Sloan's tributes to New England agrarian traditions, he wrote that the primitive reaping-hook's rare, delicate beauty represented American fine art worthy of museum exhibition.

[20] M. Lust, oral history, 2019; unpublished short stories, n. d., collection of the author, Republic, Washington.

[21] The Homer Laughlin China Company of Newell, West Virginia, was among several companies that produced the Golden Wheat pattern in the 1950s and '60s. Pieces featured five stalks of wheat with a 22k gold border and is still found in many Palouse Country households. The Kaysons Golden Rhapsody design was a cluster of gray oats with gold highlights.

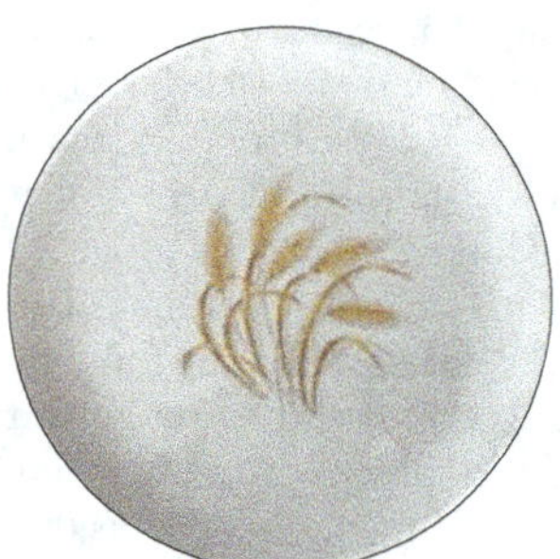

Homer Laughlin Golden Wheat Pattern Dinner Plate
Columbia Heritage Collection

[22] S. Turner, 2009:61-62.

[23] L. Morasch, 2019; D. Scheuerman, 2024.

[24] M. Nelson, 2016.

[25] V. Moberg, 1995:337.

[26] Francis Duggan verse retrieved from www.dreamagic.com/poetry.duggan.

[27] C. Luft, 1977; and C. Luft, "The Garden of My Youth," in R. Scheuerman, 1995:81-83.

[28] C. Blumenschein, 1980. Volga German speech was heavily seasoned with Russian loanwords, especially in areas like Yágodnaya Polyána that were located on the periphery of the colonial enclave and closer to ethnic Russian settlements. Our immigrant elders' word for granary, *ambar*, was from a Russian peasant term for barn, *ambary*, that is probably Persian and came to southern Russia through the region's Tatar tribes. Like inhabitants of many rural communities, the Volga Germans were very clannish and residents of Yágodnaya Polyána divided the village into the *Galmucka* and *Totten* sections. These names were derived from the native Buddhist Kalmyk and Muslim Tatar tribes.

[29] M. Morasch, 1971; who identified two plants used for processing into fabric—*Höneft* and *Fabel*, possibly localized Volga German terms for hemp and flax. Dominant Russian flax varieties of the era were *Slanets* (dew-retted) and *Motchenets* (water-retted). Volga German production of colorful *Sarpinka* gingham from cotton was a thriving business originally established by Moravian colonists from Sarepta near the Sarpa River. On Northwest American flax production origins, see A. W. Thornton, *European System of Flax Culture Americanized and Adapted to Local Conditions of U.S.A. (especially Puget Sound)*, c. 1917.

[30] R. Conquest, 1986; Younger Brother P's [Phillip Morasch] Children to Henry Morasch, Endicott, Washington, letter, c. 1932, Berry Meadow Archive Collection, Manuscripts, Archives, and Special Collections, Washington State University. Conquest's ground-breaking study has stood the test of time though it did not benefit from the opening of relevant Russian archives in the wake of glasnost. His estimate of 14.5 million casualties in Russia from 1926 to 1937 includes deaths from both the terror-famine and campaigns against the relatively prosperous property-owning *kulaks*. The expression "Memory Hole" is from Conquest's 1972 book *The Nation Killers* about Stalin's deportations of the Volga Germans, Crimean Tatars, and other minorities to Central Asia and Siberia during World War II.

[31] M. Poffenroth Geier, "My Life Story," 1963, author's collection.

[32] The 2021 McIntosh Threshing Bee held near Terrebonne, Oregon, featuring a stationary Case threshing machine powered, Oliver tractor belt drive, and John Deere header was the subject of an extensive photographic documentary by author-artist Lynn R. Miller, "Anatomy of a Threshing," *The Small Farmer's Journal* 45:2 (Winter 2021).

[33] H. Franke, 1933:32-33. On Franke's life and art, see Wolfgang Hartmann, "*Kunstmaler Hans Franke zum 70 Geburtstag*," *Badishe Heimat* 42 (1962).

[34] *Evangelical Lutheran Worship*, 2006:484, 689.

[35] *Service Book and Hymnal*, 1958:28, 114. The Festival of Harvest readings were Deuteronomy 11:8-21, Acts 14:11-18, and Luke 12:15-34.

Chapter II: *May Grain Abound*—Western Harvest Expressions

[1] W. Whitman, 1871:70. Studio Craft master Wharton Esherick (1887-1970) designed nine woodblocks including the harvest scenes *Reaper* and *Harvesting* for the Franklin Printing Company's 1928 limited edition of *As I Watched the Ploughman Ploughing*. Whitman's poem also influenced Virgil Thomson (1896-1989) to compose the ballet "The Harvest According" which was choreographed by Agnes de Mille (1905-1993) and premiered in 1952 at New York's Metropolitan Opera House.

[2] *Manual of Subordinate Granges*, 1906:37-39; *The Patron*, 1925.

[3] H. Helm, 1960:28.

[4] W. Kittredge, in M. Mayo, 1994:12. Northwest art historians Ryan Hardesty and Sid and Pat White provided valued insights on the art of Robert Helm.

[5] R. Sund, 1969. Sund was named a Washington Governor's Author in 1984 for *Ish Country*.

[6] I am indebted to Alex McGregor (Pullman, 2022), Dennis Solbrack (Colfax, 2020, 2022), Lee McGuire (Cashup, 2022), Bruce LePage (Pasco, 2021), brothers Jim and Mike Kroll (Colfax, 2022), Greg Druffel (Colton, 2022), and David Ruark (Pomeroy, 2023) for the chronology and descriptions of Palouse Country harvesting

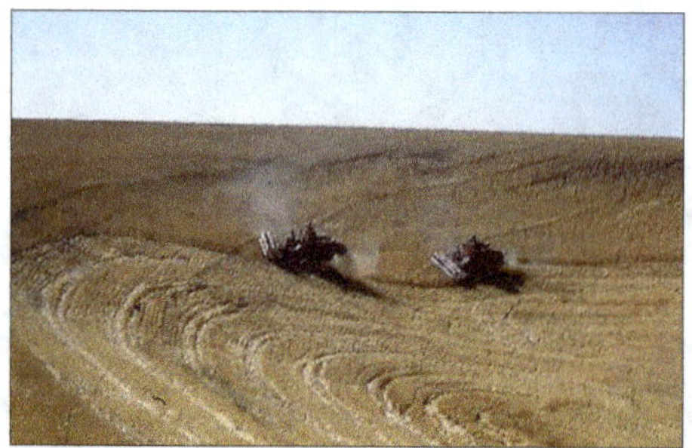

Prototype IH 125 and IH 141 Self-Leveling and Self-Propelled Combines
Herb Druffel Farm near Colton, Washington (c. 1955)
Greg Druffel Collection

equipment innovations; also see W. Overman, 1954:12-16 and C. Kroll, 2017:4. Various dates have been reported with these and other inventions since the development process takes place over years with formation of a concept, protype development, patent application and subsequent registration, and eventual manufacture. An early field trial of International Harvester's 141 hillside combine prototype on the William Boyd farm near Pullman, Washington, was reported in the July 2, 1954 issue of Idaho's *Genesee News*. Gleaner-Baldwin introduced its first self-propelled combine, the Model A, in 1951 and was acquired four years later by Milwaukee-based Allis-Chalmers which had produced its first self-propelled combine, the All-Crop SP 100, in 1953. Gleaner had long specialized in harvesting equipment and its line soon outsold AH models which were then discontinued. In 1959 AH introduced the self-leveling Gleaner AH combine. (Since 1990 Allis-Chalmers has been known as AGCO; in 1985 International Harvester and J. I. Case merged to form Case IH.) Garfield inventor Archie Neal worked with John Love's company to introduce an automatic header elevation control for John Deere combines in 1965 that prevented accidental "bulldozing" when harvesting crops low to the ground like peas and lentils. John Deere discontinued production of hillside combines in 1984 which led to a joint effort between the company and the R. A. Hanson Company of Spokane to manufacture the 7722 model in 1986 which leveled up to 45% and also featured improved horsepower to weight ratio. In the early 1990s Hillco Technologies of Nez Perce, Idaho, developed the first aftermarket mechanisms to convert new level-land combines for hillside operation. Palouse Country farmer-inventors also developed a range of innovative tillage and seeding equipment including

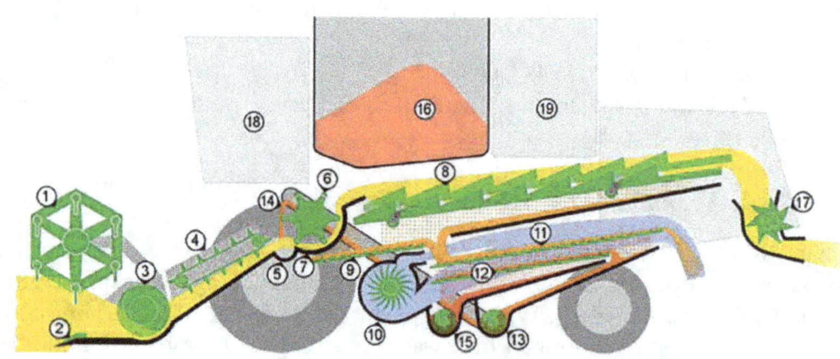

Conventional Self-Propelled Combine Components (c. 1960)
Wikimedia Commons

1. Header reel	7. Concaves	13. Secondary clean grain auger
2. Sickle bar	8. Straw walkers	14. Tailings elevator return
3. Platform auger	9. Forward sieves	15. Primary clean grain auger
4. Feeder housing	10. Fan	16. Grain bulk tank
5. Spout pivot	11. Rear top sieve	17. Straw chopper
6. Cylinder	12. Rear lower sieve	18. Cab

brothers Kyle and Cleve Wolfe's rotary rod weeder (c. 1910), the rotary subsoiler by Claude Caulkins and Lawrence Hunt (c. 1940), steel shank fertilizer applicator by Sherman McGregor, Cliff Rollins, and Chester Field (c. 1950), Arlie Hill's fold-up harrow cart (c. 1955), the deep-furrow split-packer drill by Robert Zimmerman (c. 1962), and Morton Swanson's no-till drill (c. 1974).

[7] G. Heffner, 2015:36-47; B. LePage, 2021. Massey-Harris changed its name to Massey-Ferguson in 1958. Ben Grant also built an 850-bushel capacity self-powered grain cart using a MF 760 combine frame that he modified to run backwards so the load was largely carried over the main wheels.

[8] D. Meinig, 1981:6-7.

[9] R. Papendick, 2023. The Washington, D.C.-based Institute for Alternative Agriculture published *The American Journal of Alternative Agriculture* (now *Renewable Agriculture and Food Systems* from Cambridge University Press) and featured an article in its inaugural issue (Winter 1986) coauthored by Robert Papendick with Lloyd F. Elliot and Robert B. Dahlgren, "Environmental Consequences of Modern Production Agriculture." For an overview of the sustainable agriculture movement in the United States, see John P. Reganold, Robert I. Papendick, and James F. Parr, "Sustainable Agriculture," *Scientific American* 262:6 (June 1990).

[10] D. Paarlberg, 1976:306.

[11] A. Eastland, in C. Ernst, I, 1937, 95-96.

[12] D. Day [1940], 2012:138-139.

[13] G. Wood, 1935:4. In contrast to the eastward trek of prominent Northwest Modernists like Still and Motherwell, sculptor-printmaker Jim Dine, a native of Ohio, participated in the 1962 inaugural Pop art exhibition in New York City. Dine later established a residence in Walla Walla where he continues to explore new aesthetic possibilities that include agrarian-related works. His print series *A History of Communism* (1914) combines Soviet symbolism—grain clusters and the hammer and sickle, with cast-off student work from the German Democratic Republic to explore interpretations of manual labor. His substantial sculpture *Wheatfields* (2018) was an inaugural piece for the Parrish Art Museum's Field of Dreams outdoor sculpture and performance initiative in Water Mill, New York. It resembles an abandoned farming implement with shovels and various hand tools, animal figurines, and other found objects mounted along the frame. The assembly is crowned in the middle by an immense human skull—one of the artist's favored expressions of mystery.

[14] Still letter to Weldon Schimke, August 13, 1927, retrieved at clyffordstillmuseum.org/blog/clyfford-stills-influences-the-west/. The story of Griffin, Still, and the Nespelem Art Colony is chronicled in J. J. Creighton, *Indian Summers: Washington State College and the Nespelem Art Colony, 1937-41* (2000). In the course of our time conducting oral histories on the Colville Reservation in the 1970s and '80s, Clifford Trafzer, Michael Finley, and I spoke with tribal elders who had participated in the work of the colony including Agnes Andrews Davis, Elijah Williams, and Frank Andrews; see R. Scheuerman and M. Finley, *Finding Chief Kamiakin: The Life and Legacy of a Northwest Patriot* (2008). In the 1950s Still became increasingly disenchanted with the art world and withdrew at the height of Abstract Expressionism's popularity in 1961 to a small farm near Westminster, Maryland. He established his studio in a barn and continued to paint until the end of his life. Still's will stipulated that his entire personal collection of 2,400 works was to be retained by an institution devoted solely to its exhibition and study, and in 2004 Still's widow, Patricia, announced that Denver, Colorado, had been selected to establish the Clyfford Still Museum. Sotheby's of New York auctioned one of Still's works in 2011 for a record price of $61.7 million along with three other paintings for a total sale of $114 million to support museum operations. Art treasures from the WPA era continue to emerge with one of the most remarkable found in a Skagit County farm shop in 2014. The owner of the property near Edison uncovered an enormous roll of linen sailcloth measuring 28´ x 7´ on which artist Northwest William "Bill" Cumming had painted a grand tapestry in earth tones of egg tempura featuring agrarian scenes with grain stacks, dairy cattle, and row crops. The grand tapestry had been installed in 1941 in nearby Burlington High School but removed and stored away in obscurity in the 1960s. Following restoration, the mural was appraised for $500,000 and installed at the Museum of Northwest Art in nearby LaConner.

[15] R. Motherwell, oral history, 1972. Motherwell named his innovative collage print series for the colorful German *Ernte* (Harvest) brand cigarette labels given to him by a neighbor. He described the creation of his hybrid media collages as arising spontaneously from a random assembly of banal objects and personal mementoes. The arrangements are then guided by poetic allusions and color and stylistic preferences. Mark Rothko emigrated in 1903 with his family from Dvinsk, Russia (present-day Daugavpils, Latvia) and settled in Portland before studying at Yale and the Art Students League in New York in the 1920s. His early work included

rural landscapes but the later compositions for which he became famous featured multi-layered oil painting of vertically aligned rectangular forms. Many of his works feature combinations of summertime colors—bright yellows, oranges, and reds, but Rothko associated these nonrepresentational forms with disturbance, tragedy, his personal struggles, and the human condition.

[16] Gene Kloss quote in P. Kloss, 2000:58; A. Adams and M. Austin, 1930. Adams and Mary Austin's seminal book, *Taos Pueblo*, was based on their visit to New Mexico the previous year. Both naturalist and mystic, Austin participated in Carmel Art Colony life from about 1907 to 1914. *Taos Pueblo* was published in a small edition of 108 copies with the harvest photographs among just twelve silver prints that Adams hand-printed on rag paper. Anthropologist Merton L. Miller (1898:22-23) described the traditional methods used for the Taos Pueblo grain harvest:

> *The harvesting of wheat is a most laborious task. It is done with a small sickle. . . . The threshing is an interesting sight. A circle of tall poles Is set up. Then the ground within the circle and for a space outside is wet and packed hard by a flock of sheep or goats. . . . A fence is built and the harvested grain heaped up within the enclosure. One or two men stand on the top of the pile to pitch the grain down into the circle just outside, while others drive the sheep or goats around and around till the grain is threshed. . . . It only remains then to remove the straw and sweep up ethe wheat and the chaff from the hard floor.*
>
> *. . . [Women] take care of the wheat and maize after it has once been harvested. . . . She washes [the wheat] in the creek to get rid of the chaff which remains after winnowing. This is done by partially filling a coarse basket made of yucca blades with the wheat. . . . The wheat is then spread out in the sun and allowed to dry, after which it is carefully picked over by hand to find the little pebbles and sticks which may still be left. When all this is done, it is ready to be put in sacks and loaded on a burro to be taken to the mill.*

Chapter III: *In the Valley*—Europe, Frye, and Rural Aesthetics

[1] A. Burks, 1934:68-69, 88.

[2] A. Dillard, "Total Eclipse," in *Teaching a Stone to Talk: Expeditions and Encounters* (1983):9-28.

[3] Rev. Carl Mau, whose father pastored Trinity Lutheran Church in Endicott in the 1930s, led the global Lutheran fellowship from 1974 to 1985. Growing up in a rural German-American community prepared him for postwar service in Hanover where he assisted in the rehabilitation of German Protestant churches. In 1984 he organized LWF's first World Assembly in Eastern Europe which convened in Budapest.

[4] S. Lloyd, 2014:219-221. In modern times Nashe's play inspired the brilliant British composer and conductor Constant Lambert (1905-1951) to write a vivacious 1935 orchestral cantata of the same name, premiering at Queen's Hall in London the next year. Lambert was enthralled by the combined power of music, dance, and design, and the score, which he characterized as "a curiously English affair," is considered his finest work.

[5] Maddy Prior lyrics from *Sails of Silver* (1980) and *Year* (1993) sleeve notes. A half-century after Steeleye Span appeared on the English folk rock scene, Prior remains the only original member and also administers the Stone Barns Art Centre in Bewcastle, Cumbria. The group's harmonic compositions continue to include such songs based on rural themes as "Harvest of the Moon," from their album *The Journey* (1999), and "Harvest," which begins with John Dryden's "Harvest Home" and is featured on Steeleye's 2019 recording *Est'd 1969*.

[6] E. Nesbit, 1908:60-61.

[7] K. McConkey, 1978; H. R. Haggard, 1899:307-308. From 1903 to 1911 La Thangue spent considerable time in the Italian region of Liguria where he painted numerous scenes of harvest and other country views.

[8] J. Colling, 1865:1-5, 132.

[9] E. Thomas, 1910:5-6.

[10] A. Chekhov, 1915:9.

[11] N. Gogol, 1915:199-200.

[12] L. Tolstoy quotation in A. Maude, II, 1910:653, L. Tolstoy, 1896:262-263.

[13] L. Tolstoy, 1886, VI:297-298.

[14] A. Tolstoy, 1896:262-263.

[15] I. Turgenev, 1906:133-134.

[16] S. Aksakov, [1858]1916:39, 179.

[17] I. Kallmeyer, 1967:12-22; 75-83.

[18] J. Dorsey, 2016; J. and L. Ochs, 2018, M. Balcomb, 2002; J. Dorsey, 2023:20-26. Aspiring California musician Gregory Frank Harris traveled to Spokane for a concert in 1977 and became enamored with the vast grainfields of the Columbia Plateau. The experience prompted him to read about French Impressionism and the Barbizon masters and studied art and theatre at the University of California-Long Beach and the Fechin Institute in Taos. Harris formulated an oeuvre of plein air naturalism using Old World techniques evident in paintings like *A Golden Harvest, Going Home from the Fields*, and *A Bountiful Harvest*. The Art of the Palouse Group, who also included Karla Matzke, Jerry McCollum, Cheri O'Brien, and Pat Tolle, exhibited agrarian works in oil, pastel, watercolor and digital photography inspired by the heartland of the Inland Pacific Northwest. Jason Dorsey's book *Vintage Watercolorists of Washington* (2023) profiles the careers of twenty prominent artists including Deanne Lemley and Jerry Stitt who also studied with Sergei Bongart.

[19] V. Swanson and others, 2013:357-411; J. Mansfield, 2023. Dr. Swanson received Russia's International Plastov Award in 2013 for outstanding contributions to global fine arts appreciation and education.

[20] V. Belikov quotation on art as autobiography retrieved at www.konst.se/belikov. For wheat as a Soviet national symbol and architectural theme of Moscow's All-Russian Exposition Center's *Golden Sheaf Fountain*, see Diana Kurkovsky, "Monumentalizing Wheat: Soviet Dreams of Abundance," *Gastronomica* 7:1 (Winter 2007), 15-17.

[21] J. Jovanov, 2007.

[22] A. Morasch, 1973.

[23] T. Riha, 1964:267-269.

[24] A description of this memorable trip appears in *Return to Berry Meadow and Other Stories of Our People* (Lincoln, Nebraska: Augstums Press, 1995).

[25] A. Arreguín, 2009. Sid White and S. E. Solberg were presented the 1991 Governor's Writers Award for coediting the Ethnic Heritage Committee's commemorative volume, *The Peoples of Washington: Perspectives on Cultural Diversity* (1989). Arreguín's art is the subject of Lauro Flores's beautiful book, *Afredo Arreguín: Patterns of Dreams and Nature* (2002).

[26] D. Joseph, 2018. Z. Z. Wei's paintings were featured in the Frye Art Museum's 1993 exhibit, "Three Chinese Painters."

[27] I am indebted to local historians Ron Colley, Kim Neff, and Larry Bafus for informing me about the filming of *Wild Harvest, The Desert of Wheat* (*Riders of the Dawn*), *The Basket*, and other aspects of Franklin County agrarian history.

[28] N. Schrager, 2007; W. Ruhlmann quote retrieved from allmusic.com/album/days-of-heaven-mw; T. Malick from "Making 'Days of Heaven'" at cinephilabeyond.org/terrence-malicks-days-of-heaven.

[29] Historian Benjamin S. Child discusses the significance of mechanization on Midwestern agriculture as seen in Malick's film in "Fields of Progress: The Mechanization of Agriculture in 'Days of Heaven,'" *American Studies* 57:1/2 (2018):79-102.

[30] B. Woolston, 1982, "Introduction;" B. Woolston, 2023.

[31] G. Evans, "History—*Art About Agriculture*," and L. Allan, "Art About Agriculture," in S. Curtis, 2006:1-3, 30. In a notable university interdisciplinary collaboration, Tom Allen was later designated the first artist-in-residence for OSU's College of Agriculture. In addition to the OSU collection, *Modern Farmer* contributing editor Andrew Amelinckx names four other notable American collections of agrarian art: Bone Creek Museum of Agrarian Art in David City, Nebraska; Joslyn Art Museum in Omaha, Nebraska; Santa Paula Art Museum, Ventura, California; and Shelburne Museum, Shelburne, Vermont. See "These Five Museums Put the 'Culture' in 'Agriculture,'" *Modern Farmer Magazine* (August 16, 2016).

[32] M. Escoubas, 2019:18-19; S. Henderson, 2019.

[33] K. Hooks, oral history, July 31, 2008.

[34] S. Nash, oral history, July 2, 2019; M. Shaw, oral history, June 20, 2021.

[35] J. Danzker, 2016:6-9. The Frye Museum is located in Seattle's Capitol Hill district adjacent to St. James Cathedral, home church of the Catholic Archdiocese. Architecture and art of the splendid structure, dedicated in 1907, include many agrarian themes with religious significance including wood carvings, marble sculpture and stained glass. The sanctuary altar is framed with marble panels interpreting the bread and wine as the Body and Blood of Christ in the form of grain sheaves and grape clusters carved by contemporary artists Mary Jo Anderson, Larry Ahvakana, Randal Rosenthal, and Harold Vogel.

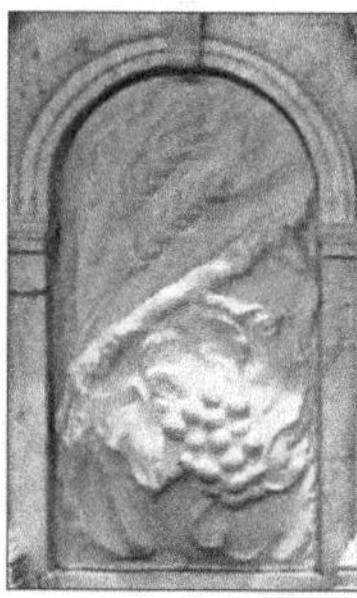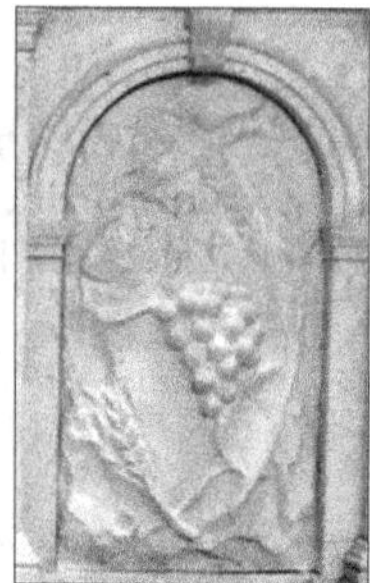

Mary Jo Anderson, Grain and Grape Marble Altar Panels
St. James Cathedral; Seattle, Washington
Columbia Heritage Collection Photographs

[36] N. Hengen, 1985:44-46.

[37] S. Turner, 2009:122-123.

[38] R. Beer, 2003:6; G. Morris, 1994:15.

[39] J. Rankin, oral history, 2017; K. Wheaton, 2012:3-5.

[40] H. Scott, 2018:56.

[41] Arthur Rothstein, oral history, May 25, 1964, Archives of American Art, Smithsonian Institution, Washington, D.C. Living history ("open-air") museums throughout the world have undertaken recent heritage grain demonstration projects and reconstruction of harvest equipment. The "Year on the Field" Project organized by the International Association of Agricultural Museums (AIMA) was launched in 2021 to facilitate a global exchange of ideas and experience on historical methods of grain cultivation and harvest and to promote sustainable agriculture. The project includes living history farms and museums where heritage crops have been raised and is coordinated through the Lauresham Open-Air Laboratory near Worms, Germany, which also operates Lorsch Abbey Manor Farm. Participating locations in North America include the Canada Agriculture and Farm Museum (Ottawa), Carter Historic Farm (Bowling Green, Ohio), Greenfield Village Firestone Farm (Dearborn, Michigan), Genesee Country Village (Rochester, New York), Howell Living History Farm (Titusville, New Jersey), Pleasant Hill Shaker Village (Harrodsburg, Kentucky), and Sterling College-Berry Center Farm (Port Royal, Kentucky). AIMA was founded in Prague in 1966 to facilitate East-West scholarly exchange and communication among agricultural museum staff worldwide. From 1968 to 1995 the organization published *Tools & Tillage: A Journal on the History of the Implements of Cultivation and Other Agricultural Processes*, proceedings of periodic congresses (*Acta Museorum Agriculturae*), *Newsletters* and *Special Reports*. Headquartered in Rochdale, Massachusetts, AIMA's North American affiliate, the Association for Living History, Farm & Agricultural Museums (ALHFAM) was founded in 1970 and publishes a quarterly *Bulletin* and annual *Conference Proceedings*.

[42] J. Mullan, 1863:27-30, 40-41; R. Scheuerman and A. McGregor, 2013:47-48; Alan G. Marshall, "Unusual Gardens: The Nez Perce and Wild Horticulture on the Eastern Plateau," in D. Goble and P. Hurt, 1999:173-187. The native peoples of the Mid-Columbia and Lower Snake Rivers had intimate knowledge of regional plant resources and gathered and processed numerous nutritious species. Among traditional Snake River-Palouse root dietary staples have been *škúlkul* and *lukš* (varieties of *Lomatium canby*), generally gathered in rocky soils in March, followed by *xáwš* (biscuitroot, *Lomatium cous*) and *piyaxí* (bitterroot, *Lewisia rediviva*) in similar locations from April to May, and the blue-flowered *xmáš* (camas, *Camassia quamash*) harvested on fertile bottomlands and upland meadows during early summer. Other important foods included springtime Lomatium "Indian celeries" and various mountain berries and seeds gathered in fall. Nez Perce-Palouse elder Allen Pinkham identifies an area near the hamlet of Dixie along present Dry Creek northwest of Walla Walla as the location of one of area Indians' first cultivated fields of grain. See Richard D. Scheuerman and Clifford E. Trafzer, *River Song: Naxiyamtáma (Snake River-Palouse) Oral Traditions from Mary Jim, Andrew George, Gordon Fisher, and Emily Peone* (Pullman: Washington State University Press, 2015).

[43] J. Deans, "Rustic Rhyms by a Rural Rhymster: Poetry and Recollections of Life at Craigflower [1853]," Royal British Columbia Archives, Provincial Museum, Victoria.

[44] O. Rackham, 1986:25-26. The Clinesmiths' remarkable heritage work has also included restoration of

the Lund family's enormous barn, recently designated a Washington Heritage Barn. The couple also safeguards a substantial seed collection of heirloom grains that have been raised in the vicinity over the past seventy years.

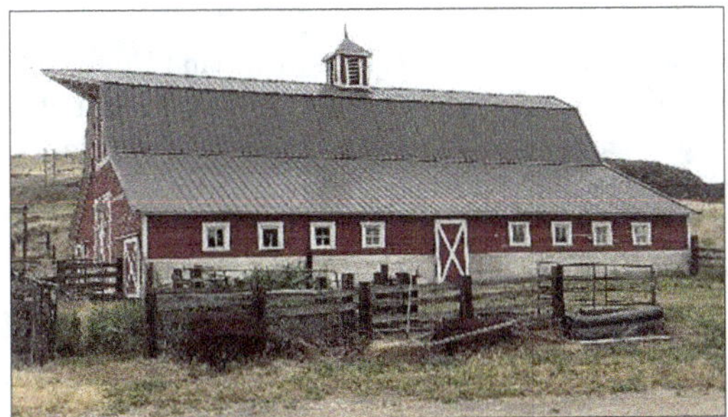

Clinesmith-Lund State Heritage Barn near Benge, Washington
Columbia Heritage Collection Photograph

Chapter IV: *Mirabilia*—New Agrarians and Renewed Community

[1] J. Wilson, 2002:1. Varda's artistic inspiration for the film came after she viewed Pierre Edmond Hédouin's *The Gleaners of Chamboudouin* (1857) at the Musee Paul Dini in Villefranche-sur-Saône. The 2009 film *The Book of Ruth: Journey of Faith* directed by Stephen Patrick Walker offers a modern depiction of the traditional biblical account.

[2] B. Bailie, oral history, 2014.

[3] R. Houston, 2008:25-26.

[4] J. LemMon, 1979:30-33, 60-61. Known since 2013 as the Catholic Rural Life, the National Catholic Rural Life Conference was founded in 1923 in St. Louis, Missouri, to better meet the needs of underserved rural parishes and schools, and to promote appreciation for agrarian endeavors. In the 1960s the NCRL also promoted land reform in the developing world by encouraging service in the Peace Corps and Papal Volunteers for Latin America. In recent years the organization has partnered with a wide range of groups for environmental justice. On CRL/NCRLC history, see Michael Woods, *Cultivating Soil and Soul: Twentieth-Century Catholic Agrarians Embrace the Liturgical Movement* (2009). On Thomas Merton's spirituality, environmentalism, and poetics see Monica Weiss, *The Environmental Vision of Thomas Merton* (2011).

[5] *Seattle Post-Intelligencer*, September 29, 1994; Governor Mike Lowry, oral history, 2007. The "Moscow: Treasures and Traditions" exhibition of notable art and artifacts from Soviet collections was held in conjunction with Seattle's 1990 Goodwill Games and organized by the USSR Ministry of Culture, Smithsonian Institution, and Seattle Art Museum. Works included many based on agrarian themes including *The Reaper* (c. 1928) by Kazimir Malevich—a milestone of Modernist expression, and sculptor Vera Mukhina's model for *The Industrial Worker and Collective Farmer* (1936) later executed in the 80-foot steel statue that served as a symbol for the Soviet state. The accompanying eponymous large-format book published by University of Washington Press (1990) included twelve essays by Soviet and American scholars exploring five hundred years of Russian history, art, and culture.

[6] Oren Long in G. Logsdon, 2007:208-209.

[7] Quotations from J. Crace, 2013.

[8] W. Berry, 1978; 1981:170-75; R. White, "Are You an Environmentalist or Do You Work for a Living?" in W. Cronin, ed., 1996:181.

[9] A. Sewell, 2022; R. Feldman, 2017. The English term "threshold" also connotes a place of entry into a new state or special place, and referred originally to the slightly raised area at the edge of an area where grain was threshed to protect kernels from spreading outside.

[10] N. Wirzba, 2011:17.

[11] T. Dearborn, 1996:77.

[12] G. Evans, 1969:156-157; 1961:236-237.

[13] K. Nelson, 2018.

[14] G. Evans, 1961:238-243.

[15] W. Cronin, 1991:120; W. Berry, 1978:33; 1969:204.

[16] P. Joyce, 2023:22-27; B. McKibben, 2018:4-8.

[17] M. Weaver, 2010; A. Blomfield, 2022.

[18] In 2022, two Just Stop Oil climate protesters at London's National Gallery glued their hands to the gilt frame of one of the country's most celebrated paintings, John Constable's *The Hay Wain* (1821). The pair also taped posters to the canvas showing a dystopian view of the rural landscape with asphalt highway, smokestacks, and jet contrails. They told reporters that the famed artist's "'green and pleasant land'" was at serious risk from continued environmental damage.

[19] I. Stevens [1855], XII:I, 1860:199-200. A single selection of English White "Hudson Bay" Lammas wheat collected in 1916 by a USDA "plant explorer" in Oregon's Willamette Valley kept vital this historic variety.

[20] R. E. McDole and S. Vira, 1980.

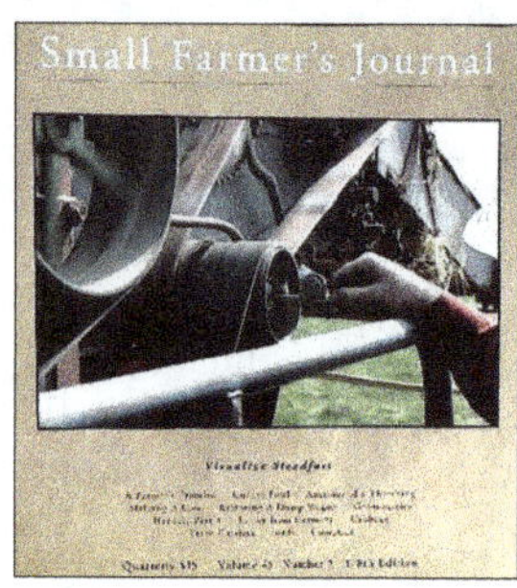

Small Farmer's Journal **45:2 (November 2021)**
Cover Photo: Measuring a Case thresher operating speed

[21] K. and I. Livingston, 2022: 48-58; L. Miller, 2021: 50-59; L. Miller, 2022: 3-10.

[22] S. Conn, 2023:162-169. The Agricultural Adjustment Act of 1938 championed by Franklin Roosevelt and Agriculture Secretary Henry Wallace introduced the concept of parity through government price supports for commodities, converted marginally productive farmland to grass, and established the "Ever-Normal Granary" concept to prevent regional food shortages. Wallace's Ever-Normal Granary idea was rooted in the experience of ancient Egypt and China where rulers stored surplus grain to curb famine. See Derek Bodde, "Henry A. Wallace and the Ever-Normal Granary," *The Far East Quarterly* 5:4 (August 1946), 411-426.

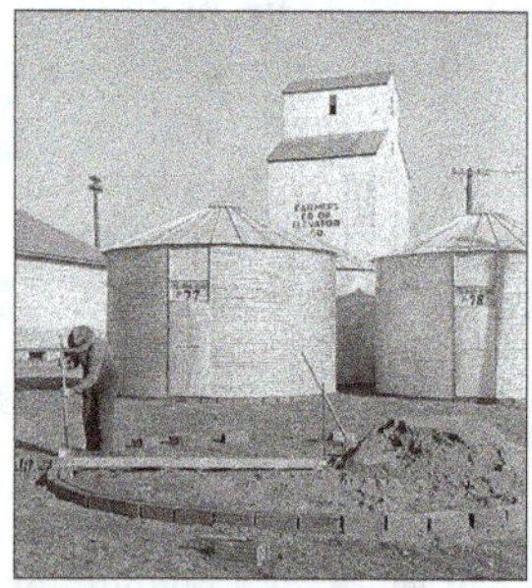

Arthur Rothstein, *Constructing an Ever-Normal Granary Bin; Grundy Center, Iowa* (1939)
Prints and Photographs Division, Library of Congress

[23] E. Schumacher quote in B. Wood, 1984:116-117. Although born in Bonn and educated there and in Berlin, Schumacher fled to England before World War II and experienced country life firsthand as a worker at Eydon Hall Farm in Northampshire for nearly two years. After considerable formal study of modern economic theory, he considered his time at Eydon (pronounced *eden*) to have been a turning point in his thinking. *Small is Beautiful* made a significant contribution to the rise of the environmental movement. The dire economic consequences of America's 1970s farm crisis are profiled in Osha Gray Davidson's classic study, *Broken Heartland:*

The Rise of America's Rural Ghetto (1996).

[24] C. Hedges, 2018:26-27; G. O'Brien, 2021:23-25.

[25] One aspect of small-town life in the Palouse as elsewhere across rural America is common knowledge of the fate of school graduates going back many decades. Community elders and younger residents readily provided information for the demographics cited here regarding classes at Endicott, Winona, St. John, and Pine City. These towns have been combined into a single public school cooperative since 1988.

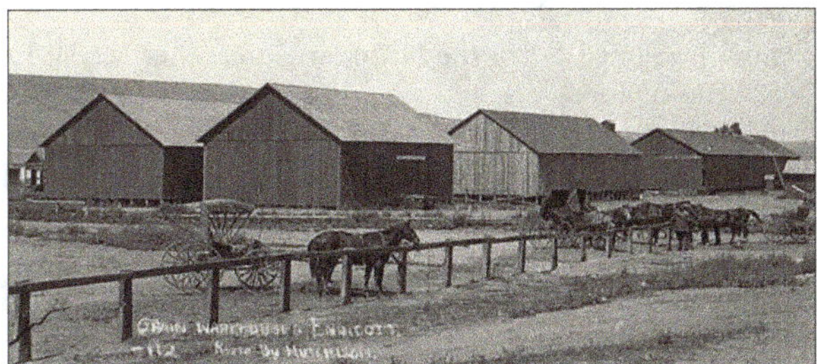

Columbia & Palouse Railroad Grain Storage Flathouses, Endicott, Washington (c. 1910)
Columbia Heritage Collection Photograph

[26] Among the most recent developments in farm mechanization is the application of autonomous driving capabilities to tractors and combines. Technology has also transformed bulk grain handling facilities at strategically located places like Endicott and elsewhere along railroads and river ports. Endicott was platted in 1882 by the Oregon Improvement Company, a subsidiary of the Northern Pacific Railroad, on its Columbia & Palouse branch line. This route tapped the fertile grain district along a route from the main NPRR transcontinental line at Palouse Junction (present Connell, Washington) eastward to Endicott, Colfax, and eventually Pullman and Moscow. An extensive network of feeder lines later fanned across the northern and southern Palouse to other farming communities.

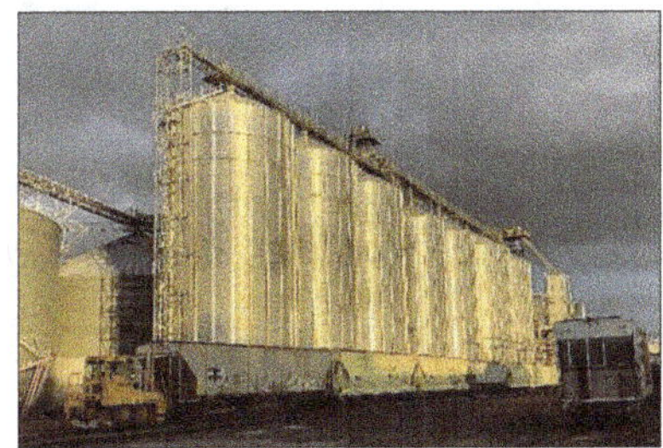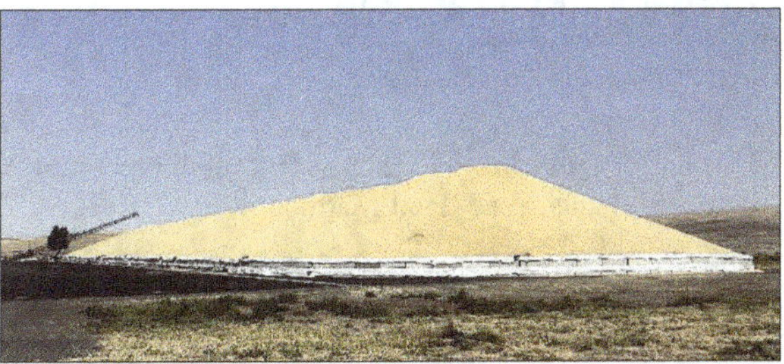

Left: Northwest Grain Growers High-Capacity Unit Train Loading Facility, Endicott
Right: 1.5 Million Bushel Wheat Pile near Endicott (August 2023)

With the merger of Whitgro cooperative with Walla Walla and other locations to form Northwest Grain Growers in 1998, the new entity constructed a storage and 110-car unit train loading facility in Endicott since the central Palouse rail line had been constructed with heavier weight capacity. The project called for construction of seven immense steel silos located to bring total capacity there to approximately 3,100,000 bushels. The new facility, which became operational in 2020, is designed for rapid one-day loading of trains which hold 100 tons of grain per car for a total capacity of 420,000 bushels. Grain is trucked from farms and other elevators for rail shipment and barging along the lower Snake and Columbia Rivers to Portland and Kalama for distribution worldwide. Other prominent Columbia Plateau grain storage, seed handling, and marketing cooperatives include Waterville, Washington-based HighLine Grain Growers, Ritzville Warehouse Company, Pendleton Grain Growers, Morrow County (Oregon) Grain Growers, and Genessee, Idaho-based Pacific Northwest Farmers Cooperative.

[27] B. Holbert, oral history, 2021; B. Holbert, "Ordinary Days," unpublished manuscript (2021). I am in-

debted to the author—my longtime friend, former teacher colleague, and fellow harvest truck driver, Bruce Holbert, for permission to reprint this selection.

[28] W. Johnson, in R. McFarland, 1990:95.

[29] H. Nemerov [1955], 1997:97.

[30] D. Montgomery and A. Bilké, 2016:237-238.

[31] W. Berry, 1969:44. On potential benefits and risks of gene editing, see Natalie de Souza, "Editing Humanity's Future," 2021.

[32] J. Crouse, 2010; W. Berry, 1987:54-56.

[33] W. Berry, 1986:107; and 2002. President Obama presented Berry the National Humanities Medal in 2010 for lifetime achievements in writing, farming, and conservation.

[34] W. Berry, 1993:119-20; C. James and J. Fitzgerald, 2008:x-xi.

[35] Van Gogh letters to E. Bernard (June 24, 1888) and Theo van Gogh (July 1889) in J. van Gogh-Bonger and C. de Dood, 1958.

[36] Tolstoy's fable "The Grain" (1886) relates the unusual experience of a king who obtains an enormous kernel of grain as a curiosity. He summons his advisors to determine its origin and eventually encounters three generations of farmers responsible for its propagation. The oldest members of the family appear more youthful than the younger, and the eldest explains, "In the old time, men lived according to God's law. They had what was their own, and coveted not what others had produced" (L. Tolstoy, IX, 1904:447-451).

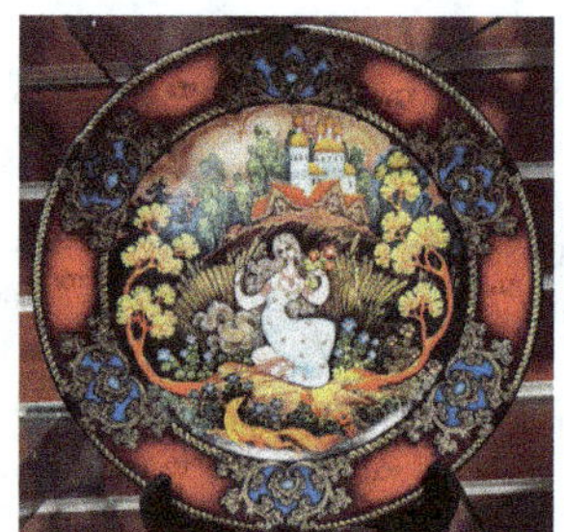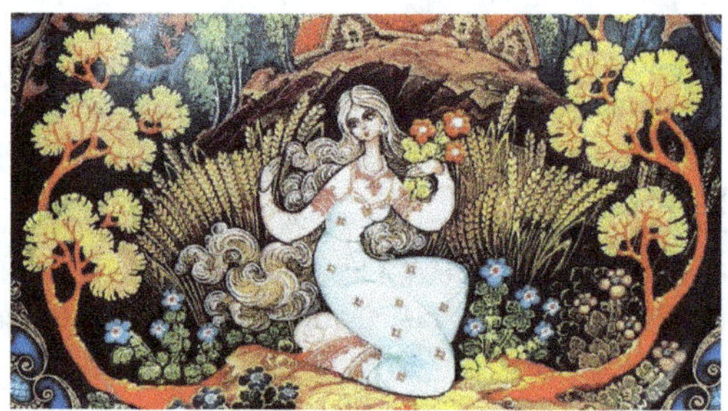

Sergey Devyatkin, *Gifts of Summer* (1993)
Byliny Porcelain; Kholui, Russia
Columbia Heritage Collection

[37] On the hard choices involving significant fossil fuel reduction and implications for global political and commodity market changes, see Geoff Mann and Joel Wainright, *Climate Leviathan: A Political Theory of our Planetary Future* (2018).

[38] L. Tolstoy, IV, 1886:257-258.

[39] L. Tolstoy, 1896:262-263.

[40] Brian Wren's entire hymn of five stanzas is included in many contemporary hymnals including Donald P. Hustad, ed., *The Worshiping Church: A Hymnal* (1990). One of the twentieth century's most popular English Methodist hymnodists, Fred Pratt Green (1903-2000), composed several songs for Harvest-Home and Thanksgiving including *Harvest Hymn* (*For the Fruit of All Creation* [1970]), *Come, Sing a Song of Harvest* (1976), and *Now That Harvest Crops are Gathered* (1989).

Epilogue

[1] T. Owens, "Wheatfield—An Inspiration." *Explore Big Sky*. April 19, 2024: Jenny Moore and Larry Bamburg, 2024

[2] L. Jacobs, 2023; N. Roach, 2019:26-28.

[3] A. Banks, and others, 2017:i-iii.

BIBLIOGRAPHY

Original Works

Adams, Ansel, and Mary Austin. *Taos Pueblo*. San Francisco: Grabhorn Press, 1930.

Aksakov, Sergey. *Years of Childhood*. London, [1858] 1916.

Berry, Wendell. "The Agrarian Standard." *Orion* 21:3 (Summer 2002).

Berry, Wendell. *The Gift of Good Land: Further Essays Cultural and Agricultural*. San Francisco, 1981.

Berry, Wendell. *A Timbered Choir: The Sabbath Poems, 1979-1997*. Washington, D.C., 1998.

Berry, Wendell. *The Unsettling of America: Culture & Agriculture*. San Francisco, 1978.

Bailey, Liberty Hyde. *The Holy Earth*. New York, 1915.

Beer, Ralph. *In These Hills*. Lincoln, Nebraska, 2003.

Beers, Lorna Doone. *Prairie Fires*. New York, 1925.

Bryant, William Cullen. *Thirty Poems*. New York, 1864.

Evangelical *Lutheran Worship*. Minneapolis, Minnesota, 2006.

Burks, Arthur J. *Here Are My People*. New York, 1934.

Cather, Willa. *One of Ours*. New York, 1922.

Coates, Grace Stone. *Portulacas in the Wheat*. Caldwell, Idaho, 1919.

Colling, James K. *Art Foliage for Sculpture and Decoration*. London: By the author, 1865.

Crace, Jim. *Harvest*. London, 2013.

Davis, H. L. *Honey in the Horn*. New York, 1935.

Davis, H. L., Orvis C. Burmaster, ed. *The Selected Poems of H. L. Davis*. Boise, Idaho: Ashanti Press, 1978.

Day, Dorothy. *All the Way to Heaven: Selected Letters of Dorothy Day*. New York, 2012.

Deans, James. "Rustic Rhyms by a Rural Rhymster: Poetry and Recollections of Life at Craigflower [1853]." Royal British Columbia Archives, Provincial Museum, Victoria.

DeWitt, John L. *World Food Unlimited: Producing Abundant, Safe, Food, Sustainably Using Modern Agricultural Technologies*. N. p., 2017.

Dillard, Annie. *Teaching a Stone to Talk: Expeditions and Encounters*. New York, 1983.

Egan, Timothy. *The Worst Hard Time: The Untold Story of Those Who Survived the Great American Dust Bowl*. New York. 2006.

Ernst, Charles, ed. *Told by the Pioneers: Tales of Frontier Life*. 3 Vols. 1937-1938.

Escoubas, Michael. *Steve Henderson in Poetry and Paint*. Normal, Illinois, 2019.

Evans, George Ewart. *Ask the Fellows who Cut the Hay*. London: Faber and Faber, 1961.

Evans, George Ewart. *The Farm and the Village*. London: Faber and Faber, 1969.

Fleener, Dora Otter. *Palouse Country Yesteryears*. Moscow, Idaho: By the author, 1978.

Franke, Hans. *Die Sieben Sphären*. Caritas, 1933.

Geier, May Poffenroth. "My Life Story." 1963. Palouse Regional Studies Collection, MASC, WSU.

Gogol, Nikolai. *Dead Souls*. London, 1854.

Goncharov, Ivan. *Oblomov*. London, 1915.

Grey, Zane. *The Desert of Wheat*. New York, 1919.

Grove, Noel. "North with the Wheat Cutters." *National Geographic Magazine* 142:2 (August 1971).

Haggard, H. Rider. *The Farmer's Year: Being His Commonplace Book for 1898*. London, 1899.

Hamerton, Philip. *Landscape*. London, 1885.

Heaney, Seamus. *Fieldwork*. New York, 1979.

Helm, Harry. *Love Singer in Paradise*. Spokane, Washington, 1962.

Hengen, Nona. *Plodding Princes of the Palouse*. Woolwich, Maine, 1984.

Hentzner, Paul, and Robert Naunton. *Travels in England and Fragmented Regalia*. London, 1797.

Holbert, Bruce. *The Hour of Lead*. Berkeley: Counterpoint Press, 2014.

Holbert, Bruce. "Ordinary Days." Unpublished manuscript, 2021.

The Holy Bible: English Standard Version. Wheaton, Illinois, 2001.

Hunter, Paul. *Come the Harvest*. Eugene, Oregon: Silverfish Review Press, 2008.

Jones, Mathilda Siegmund. "The Harvest Cycle on the Jacob Siegmund Farm." *Marion County History* 3 (1957).

Kirby, Don. *Wheatcountry*. Tucson: Nazraeli Press, 2001.

Lambert, Constant. *Summer's Last Will and Testament: A Masque*. London, 1935.

Laughon, Albert J. *Spirit of the Wild Rosebush: A Novel of the Pioneer Period of the Inland Empire*. Spokane: Shaw & Borden Company, 1938.

Lauk, Jon K. *The Good Country: A History of the American Midwest, 1800-1900*. Norman: University of Oklahoma Press, 2023.

Le Sueur, Meridel. *Harvest Song: Collected Essays and Stories*. New York: West End Press, 1990.

Le Sueur, Meridel. *North Star Country*. New York: Duell, Sloan & Pearce, 1945.

Lewis, Paul. "'Lost in the Hills'—A Musical." Unpublished typescript. Seattle, Washington. 2021.

Lopez, Barry. *Home Ground: Language for an American Landscape*. San Antonio, 2006.

Lust, Mike. Unpublished short stories, n. d.

Manual of Subordinate Granges of the Patrons of Husbandry. Philadelphia, 1906.

McCaw, Martin. *The Low Road*. Boston: Big Table Publishing Company, 1917.

McFarland, Ron, Franz Schneider, and Kornel Skovajsa. *Deep Down Things: Poems of the Inland Pacific Northwest*. Pullman, Washington, 1990.

Meyer, Alison. *Palouse Perspective: Landscape Photographs*. Flying Cedar Press, 2008

Miller, Lynn R. "Anatomy of a Threshing." *Small Farmer's Journal* 45:2 (Winter 2021), 50-59.

Miller, Lynn R. "Seedbeds & The Rooted Mend." *Small Farmer's Journal* 46:1 (Spring 2022), 2-10.

Moberg, Vilhelm. *The Settlers*. New York, 1996.

Motherwell, Robert, with Paul Cummings. Oral history. Smithsonian Archives of American Art, Washington, D.C. March 30, 1972.

Motherwell, Robert, and Kenneth E. Tyler. *Robert Motherwell: Summer Light Series*. Los Angeles: Gemini G. E. L, 1973.

Mullan, John. *Report on the Construction of a Wagon Road from Fort Walla-Walla to Ft. Benton*. Washington, D.C.: U.S. Government Printing Office, 1863.

Nemerov, Howard. *The Collected Poems of Howard Nemerov*. Chicago, 1997.

Nesbit, Edith. *Ballads and Lyrics of Socialism, 1883-1908*. London, 1908.

Osborn, John. *The Harvest: Five Generations of Wheat Farming in Eastern Oregon*. San Francisco: Blurb, 2020

Paarlberg, Don. "Agriculture Two Hundred Years from Now." *Agricultural History* 50:1 (January 1976).

Radishchev, Alexander N. *A Journey from Moscow to St. Petersburg [1790]*. Cambridge, 1958.

Report of the Country Life Commission and Special Message of the President of the United States. 60th Cong. 2nd Sess., SED 705. Spokane: Spokane Chamber of Commerce, 1909.

Robinson, Marilynne. "A Theology of the Present Moment," *The New York Review of Books* LXIX:20 (December 22, 2022).

Ruskin, John; E. T. Cook and Alexander Wedderburn, eds. *The Complete Works of John Ruskin*. 39 vols. London, 1905.

Schumacher, E. F. *Small is Beautiful*. New York, 1973.

Sloane, Eric. *I Remember America*. New York, 1977.

Sloane, Eric. *Legacy*. New York, 1979.

Stevens, Isaac I. *Narrative and Final Report of Explorations for a Route for a Pacific Railroad Near the Forty-*

Seventh and Forty-Ninth Parallels of North Latitude from St. Paul to Puget Sound. Washington, D.C.: Government Printing Office, 1860.

Sund, Robert. *Bunch Grass.* Seattle: University of Washington Press, 1969.

Thomas, Edward. *The Last Sheaf.* London, 1928.

Thomas, Edward. *Rose Acre Papers.* London, 1910.

Tisdale, Sallie. *Stepping Westward: The Long Search for Home in the Pacific Northwest.* New York: Henry Holt and Company, 1991.

Tolstoy, Leo. *Anna Karénina.* London, 1896.

Tolstoy, Leo. Leo Wiener, trans. *The Complete Works.* 8 vols. Boston, 1904.

Tolstoy, Leo. *War and Peace.* 9 vols. New York, 1886.

Turner, Steve. *Amber Waves and Undertow: Peril, Hope Sweat, and Downright Nonchalance in Dry Wheat Country.* Norman: University of Oklahoma Press, 2009.

Van Gogh, Vincent. J. van Gogh-Bonger and C. de Dood, trans. *The Complete Letters of Vincent van Gogh.* 3 vols. Greenwich, New York, 1958.

Whitman, Walt. *Leaves of Grass.* Washington, D.C.: By the author, 1871.

Wirzba, Norman. *Food and Faith: A Theology of Eating.* Cambridge: Cambridge University Press, 2011.

Wood, Grant. *Revolt Against the City.* Iowa City, Iowa: Clio Press, 1935.

Woolston, Bill. *Harvest: Wheat Ranching in the Palouse.* Genesee, Idaho: Thorn Creek Press, 1982.

Secondary Sources

Albritton, Vicky, and Fredrik Albritton Jonsson. *Green Victorians: The Simple Life in John Ruskin's Lake District.* Chicago: University of Chicago Press, 2016.

Ament, Deloris Tarzan. *Iridescent Light: The Emergence of Northwest Art.* Seattle: University of Washington Press, 2002.

Anderson, David. *Images of the Land: Washington Wheat Country.* Ritzville, Washington: Washington Association of Wheat Growers, 2005.

Balcomb, Mary N. *Sergei Bongart.* Seattle: Cody Publications, 2002.

Banks, Anne, and others. *The [Washington] Arts Learning Standards: Kindergarten-12th Grade.* Olympia: Office of Superintendent of Public Instruction, 2017.

Barnett, Susan Floyd and Lucy L. Lippard. *Tracy Linder—Open Range.* Billings, Montana: Yellowstone Art Museum, 2020.

Barton, Alexis. "Self-taught Painter and Sculptor Honors Ancestors." *The Washington Post,* November 30, 2020.

Barzun, Jacques. *The Culture We Deserve.* Hanover, New Hampshire: Wesleyan University Press, 1989.

Bowers, William. *The Country Life Movement in America, 1900-1920.* Port Washington, New York: Kennikat Press, 1974.

Broehl, Jr., Wayne G. *John Deere's Company: A History of Deere & Company and Its Times.* New York: Doubleday & Company, 1984.

Carlson, Brad. "Ag Mural Brightens Idaho Statehouse." *Capital Press.* April 1, 2019.

Conn, Steven. *The Lies of the Land: Seeing Rural America for What It Is—and Isn't.* Chicago: The University of Chicago Press, 2023.

Conquest, Robert. *The Harvest of Sorrow: Soviet Collectivization and the Terror-Famine.* New York: Oxford University Press, 1986.

Cronin, William, ed. *Uncommon Ground: Rethinking the Human Place in Nature.* New York: W. W. Norton & Company, 1996.

Crossley, Trista. "Holding Court—Once Upon a Time, Queens Ruled Over Washington's Wheat Industry." *Washington Wheat Life* 64:1 (January 2021).

Curtis, Shelley. *The Bountiful Place: Art About Agriculture—The Permanent Collection*. Portland: Oregon Historical Society, 2006.

Dahlig, Piotr, ed., and John Comber, trans. *Traditional Musical Cultures in Central-Eastern Europe: Ecclesiastical and Folk Transmission*. Warsaw: University of Warsaw Institute of Musicology, 2009.

David Danbom. *The Resisted Revolution: Urban America and the Industrialization of Agriculture, 1900-1930*. Ames: Iowa State University Press, 1979.

Danzker, Jo-Anne Birnie. *Beloved: Pictures at an Exhibition*. Seattle: Frye Art Museum, 2012.

Danzker, Jo-Anne Birnie. *Cris Bruch: Others Who Were Here*. Seattle: Frye Art Museum, 2016.

Dearborn, Tim. *Taste & See: Awakening Our Spiritual Senses*. Downer's Grove, Illinois: InterVarsity Press, 1996.

de Souza, Natalie. "Editing Humanity's Future." *The New York Review* LXVII:7 (April 29, 2021).

Dorsey, Jason. *Vintage Watercolorists of Washington*. Camano Island, Washington: Sunnyshore Studio, 2023.

Drinkard, Susan. "Marilynne Robinson—Pulitzer Prize-Winning Author." *Sandpoint Magazine* (Winter 2006).

Eastwood, Harland. Adams County, *"The Bread Basket of the World"—A Photographic History*. Moses Lake, Washington: Grain Belt Publishing Company, 2018.

Edwards, Robert W. *Jennie V. Cannon: The Untold Story of the Carmel and Berkeley Art Colonies*. Oakland: East Bay Heritage Project, 2012.

Flores, Lauro. *Alfred Arreguín: Patterns of Dreams and Nature*. Seattle: University of Washington Press, 2002.

Garnsey, Peter. *Famine and Food Supply in the Graeco-Roman World: Responses to Risk and Crisis*. Cambridge: Cambridge University Press, 1989.

Geoff, and Joel Wainright, *Climate Leviathan: A Political Theory of our Planetary Future*. New York: Verso Books, 2018.

Gerster, George, and Joyce Diamanti. *Amber Waves of Grain: America's Farmlands from Above*. New York: Harper, Weldon, Owen, 1990.

Ghosh, Amitav. *The Great Derangement: Climate Change and the Unthinkable*. New York: Penguin Books, 2016.

Gilbert, Cyrus L. *Old Trails and Other Verse*. Burien, Washington: By the author, 1948.

Goble, Dale, and Paul Hurt. *Northwest Lands, Northwest Peoples: Readings in Environmental History*. Seattle: University of Washington Press, 1999.

Goldstein, Shelly. "The Art and Heart of Soil." *American Lifestyle* 121 (Spring 2023).

Goodman, Ruth Ellen, Alex Langlands, and Peter Ginn. *Victorian Farm: Rediscovering Forgotten Skills*. London: Pavilion Books, 2009.

Gustafson, Thane. *Klimat: Russia in the Age of Climate Change*. Cambridge: Harvard University Press, 2022.

Halvorson, Britt E. *Imagining the Heartland: White Supremacy and the American Midwest*. Berkeley: University of California Press, 2022.

Hedges, Chris. *America: The Farewell Tour*. New York: Simon & Schuster, 2018.

Heffner, Gary. "Ben Grant: Combine Pioneer." *Massey Harris Ferguson Legacy Quarterly* 23 (July 2015).

Herbert, Linda, and others. *The 1919 Walla Walla Tractor Show*. Walla Walla: The Blue Mountain Land Trust, 2019.

Hochschild, Adam. *American Midnight: The Great War, a Violent Peace, and Democracy's Forgotten Crisis*. New York: HarperCollins, 2022.

Hofstadter, Richard. "The Myth of the Happy Yeoman." *American Heritage* 7:3 (April 1956).

Houston, Ron. 55th *Anniversary International Folk Dance Workshop*. Dayton, Ohio: Miami Valley Folk Dancers, 2008.

Hustad, Donald. P., ed. *The Worshiping Church: A Hymnal*. Carol Stream, Illinois: Hope Publishing. 1990.

Inge, M. Thomas. *Agrarianism in American Literature*. New York: Odyssey Press, 1969.

James, Christopher, and Joseph A. Fitzgerald, eds., *Of the Land and Spirit: The Essential Lord Northbourne on Ecology and Religion*. Bloomington, Indiana: World Wisdom, Inc., 2008.

Joseph, Dana. "Z. Z. Wei: Shadow Stories." *Cowboys & Indians* 26:4 (May/June 2018).

Jovanov, Jasna. Srdja Jancović, trans. *Feja Soretić: Five Decades Without Pause*. Novi Sad, Serbia: Museum of Vojvodina, 2007.

Joyce, Patrick. "Remembering Peasants." *Natural History* 131:8 (September 2023).

Kallmeyer, Ilse. "Adolf Heinrich Lier, 1826-1882: Ein Beitrag zur Münchner Landschafterschule des neun zehnten Jahrhunderts." Doctoral Dissertation. University of Innsbruck, 1967.

Kincaid, Garret D., and A. H. Harris. *Palouse in the Making*. Fairfield: Ye Galleon Press [1934], 1979.

Kloss, Phillip Wray. *Gene Kloss Etchings*, Santa Fe: Sunstone Press, 2000.

Kroll, Chris. "Combine's Convoluted Leveler History." *Washington Wheat Life* 60:4 (April 2017).

Lauck, John K. *The Good Country: A History of the American Midwest*. Norman: University of Oklahoma Press, 2022.

Leader, Zachary. *The Life of Saul Bellow: Love and Strife, 1965-2005*. New York: Alfred K. Knopf, 2018.

LemMon, Jean, ed. *John Paul II Visits Rural America*. Des Moines: Meredith Publishing Services, 1979.

Livingston, Khoke and Ida. "The Harvest of Grain." *Small Farmer's Journal* 46:1 (Spring 2022), 48-58.

Lloyd, Stephen. *Constant Lambert: Beyond the Rio Grande*. Woodbridge, Suffolk: The Boydell Press, 2014.

Logsdon, Gene. *The Mother of All Arts: Agrarianism and the Creative Impulse*. Lexington: University of Kentucky Press, 2007.

Magaret, Patricia and Donna Slusser. *Inspiration · Imagination · Interpretation: Quilting Artistry of Karen Schoeflin Hagen*. Lewiston, Idaho: Western Printing, 1999.

Marx, Leo. *The Machine in the Garden: Technology and the Pastoral Ideal in America*. New York: Oxford University Press, 1964.

Mayo, Marti. *Robert Helm, 1981-1993*. Houston: Blaffer Gallery, 1994.

McConkey, Kenneth. *A Painter's Harvest: Works by Henry Herbert La Thange, R. A., 1859-1929*. Oldham, England: Oldham Art Gallery, 1978.

R. E. McDole, R. E., and Shiraz Vira, *Five-Point Program: Soil Erosion Control Under Dryland Crop Production*. Information Series No. 483. Moscow: University of Idaho College of Agriculture Cooperative Extension Service, July 1980.

McGregor, Alexander C. *Counting Sheep—From Open Range to Agribusiness on the Columbia Plateau*. Seattle: University of Washington Press, 1982.

McKibben, Bill. *Falter: Has the Human Game Begun to Play Itself Out?* New York: Henry Holt, 2019.

McKibben Bill. "A Very Grim Forecast." *The New York Review of Books* LXV:18 (November 22, 2018), 4-8.

Meinig, Donald W. "The Geography of the Humanities, and an Example of Humanistic Geography." Remarks to the New York Council for the Humanities, Annandale-on-Hudson. February 1, 1981. Palouse Regional Studies Collection, Manuscripts, Archives, and Special Collections, Washington State University, Pullman.

Meinig, D. W. *The Great Columbia Plain: A Historical Geography, 1805-1910*. Seattle: University of Washington Press, 1968.

Menard, Louis. *The Free World: Art and Thought in the Cold War*. New York: Farrar, Straus and Giroux, 2021.

Meyer, Garth. "The Harvest Watcher." *Whitman County Gazette* 140:34 (August 24, 2017).

Miller, Merton L. *A Preliminary Study of the Pueblo of Taos, New Mexico*. Chicago: University of Taos Press, 1898.

Montgomery, David R., and Anne Bilké. *The Hidden Half of Nature: The Microbial Roots of Life and Health*. New York: W. W. Norton & Company, 2016.

Montmarquet, James A. *The Idea of Agrarianism: From Hunter-Gatherer to Agrarian Radical in Western Culture* Moscow: University of Idaho Press, 1989.

Moseman, Mark and Carol. *Agrarian Spirit in the Homestead Era: Artwork from the Moseman Collection of Agrarian Art*. St. Joseph, Missouri: Albrecht-Kemper Museum of Art, 2021.

Nelson, Mary. "Honoring the People of the Past [Charles Fritz]." *Art of the West* 29:3 (March/April 2016).

O'Brien, Geoffrey. "Hitler in Antarctica." *New York Review of Books* LXVIII:1 (January 14, 20121).

Overman, Wendell P. "Hillside Combine: A Farm Machinery Case History." *Harvester World* 45:11-12 (November-December 1954), 12-16.

Rackham, Oliver. *The History of the Countryside*. London: Phoenix Press, 1986.

Riha, Thomas, ed. *Readings in Russian Civilization*, Volume 2: Imperial Russia, 1700-1917. Chicago: University of Chicago Press, 1964.

Roach, Nancy. *Reflections of Faith: Tri-Cities Prep and the St. Thomas Aquinas Chapel*. Richland, Washington: Etcetera Press, 2019.

Service Book and Hymnal of the Lutheran Church in America. Minneapolis: Augsburg Publishing House, 1958.

Scheuerman, Richard D., and Alexander C. McGregor. *Harvest Heritage: Agricultural Origins and Heirloom Crops of the Pacific Northwest*. Pullman: Washington State University Press, 2013.

Scheuerman, Richard D. *Hallowed Harvests: Agrarian Depiction from the Bible, Literature, and Art to Early Modern Times*. Pasco, Washington: Triticum Press, 2024.

Scheuerman, Richard D., and Clifford E. Trafzer. *Hardship to Homeland: Pacific Northwest Volga Germans*. Pullman: Washington State University Press, 2018.

Scheuerman, Richard D. *Illuminated Earth-Sky Connections: The Photography of John Clement*. Kennewick, Washington: John Clement Gallery, 2012.

Scheuerman, Richard D. "'This Wonderful Place': Zane Grey's Northwest and *The Desert of Wheat*." *The Franklin Flyer* 50:4. Pasco, Washington: Franklin County Historical Society (October 2018).

Schleuning, Neala J. Y. "Meridel Le Sueur: Toward a New Regionalism." *Books at Iowa* 33 (November 1980).

Schillinger, William F., and Robert I. Papendick. *Then and Now: 125 Years of Dryland Wheat Farming in the Inland Pacific Northwest*. Pullman: Washington State University Extension Service, 2008.

Schmidt, Peter J. *Back to Nature: The Arcadian Myth in Urban America*. New York: Oxford University Press, 1969.

Schrager, Nick. "Review: *Days of Heaven*." *Slant Magazine*. October 22, 2007.

Scott, Heidi. "Ag as Art [Kim Matthews Wheaton]." *Wheat Life* 61:1 (January 2018).

Signorelli, Anthony. "Creating a Place." *Small Farmer's Journal* 17:4 (Fall 1993).

Smith, Henry Nash. *Virgin Land: The American West as Symbol and Myth* (Cambridge: Harvard University Press, 1950.

Solbrack, Dennis. "History of the Palouse, Hillside Farming, and Hillside Combines." Unpublished manuscript, 2013.

Swanson, Vern, and others. *Springville Museum of Art: History and Collections*. Springville, Utah: Cedar Fort, 2013.

Tallman, Susan. "Who Decides What's Beautiful?" *The New York Review* LXVII:14 (September 24, 2020).

Topping, Gary. "Zane Grey: A Literary Reassessment." *Western American Literature* 13:1 (Spring 1978), 51-64.

Updike, Robin. "An Artist from the Palouse in Tacoma." *The Seattle Times*, June 12, 1995.

USDA Study Team on Organic Farming. *Report and Recommendations on Organic Farming*. Washington, D.C.: United States Department of Agriculture, 1980.

Usher, M. D. *How To Be a Farmer: An Ancient Guide to Life on the Land, A Work of Many Hands*. Princeton: Princeton University Press, 2021.

Vergauwen, David, and Ive De Smet. "Genomes on Canvas: Artist's Perspective on Evolution in Plant-Based Foods." *Trends in Plant Science* 25:8 (August 2020).

Weaver, Matthew. "Agriculture Draws Out the Best." *Capital Press*, April 10, 2010.

Wilson, Jake. "Trash and Treasure: The Gleaners and I." *Senses of Cinema* 23 (December 2002).

Wheaton, Kim Matthews. *Landscapes of Eastern Washington*. Moses Lake, Washington, 2012.

Wood, Barbara. *E. F. Schumacher: His Life and Thought*. New York: Harper & Row, 1984.

Woods, Michael. *Cultivating Soil and Soul: Twentieth-Century Catholic Agrarians Embrace the Liturgical Movement*. Collegeville, Minnesota: Liturgical Press, 2009.

Zabinski, Catherine. *Amber Waves: The Extraordinary Biography of Wheat, from Wild Grass to World Megacrop*. University of Chicago Press, 2020.

Oral Histories

Arreguín, Alfredo. Seattle, Washington. June 24, 2009.
Bafus, Larry. Pasco, Washington. May 2, 2019.
Bailie, Brad. Connell, Washington. September 12, 2014.
Bamburg, Larry. Bozeman, Montana, June 5, 2024.
Beaverson, Tony. Salsbury, Missouri. June 15, 2016.
Blomfield, Anna. Silver City, New Mexico. March 30, 2022.
Blumenschein, Conrad. St. John, Washington. May 5, 1980.
Broeckel, Vicki. LaCrosse, Washington. October 18, 2023.
Colley, Ron. Richland, Washington. January 15, 2022.
DeRuff, Elizabeth. Larkspur, California. July 28, 2016.
Dorsey, Jack. Camano Island, Washington. June 2, 2016.
Druffel, Greg. Colton, Washington. March 12, 2022.
Eastwood, Harland. Moses Lake, Washington. January 7, 2013.
Edstrom, Arvin. Fernwood, Idaho. September 3, 2016.
Fritz, Charles. Billing, Montana. October 19, 2020.
Hager, Roger. Oakesdale, Washington. October 30, 2023.
Henderson, Steve. Dayton, Washington. September 8, 2020.
Holbert, Bruce. Blanchard, Idaho. October 29, 2021.
Hooks, Kathleen. Pasco, Washington. July 30, 2018.
Helga Jansons. Pasco, Washington. July 31, 2022.
Jackson, Chad. Aberdeen, Idaho. June 30, 2023.
Jacobs, Lisa. Pasco, Washington. March 11, 2023.
Kroll, Jim. Colfax, Washington. February 17, 2022.
Kroll, Mike. Colfax, Washington. February 17, 2022.
LePage, Bruce. Pasco, Washington. November 4, 2021.
Lewis, Paul and Bonnie. Bainbridge Island, Washington. October 8, 2021.
Lowry, Mike. Renton, Washington. May 4, 2007.
Luft, Catherine. San Francisco, California. June 16, 1977.
Lust, Mike. Winona, Washington. August 6, 2019.
Mansfield, Judy. Springville, Utah. June 29, 2023.
McGregor, Alexander. Pullman, Washington. February 17, 2022.
McGuire, Lee. Cashup, Washington. February 9, 2022.
Moore, Jenny. Bozeman, Montana. June 5, 2024.
Morasch, Adam and Katherine. Endicott, Washington. December 28, 1973.
Morasch, Larry. Lacey, Washington. May 20, 2019.
Morasch, Mary K. Endicott, Washington. April 23, 1971.
Moseman, Mark and Carol. David City, Nebraska. July 22, 2022.
Nash, Sue. Richland, Washington. July 2, 2019.
Nelson, Katherine. Coeur d'Alene, Idaho. July 12, 2018.
O'Callaghan, David and Michelle Bertrand. Benton City, Washington. January 27, 2024.
Ochs, John and Li. Colfax, Washington. November 15, 2018.
Papendick, Robert. Pullman, Washington. May 18, 2023.
Patawa, Allen. Pullman, Washington. May 31, 2020.

Payne, James. Walla Walla, Washington. May 16, 2023.

Peschel, Rene. Seattle, Washington. October 8, 2012.

Pfaffenroth, Alexander. Endicott, Washington. August 14, 2003.

Pratt, Alan. Newport, Oregon. June 17, 2019.

Rankin, John. Ritzville, Washington. April 29, 2017.

Riebold, Stan. Endicott, Washington. September 22, 2022.

Rosenoff, Arlon. Coeur d'Alene, Idaho. December 22, 2019.

Ruark, David. Pomeroy, Washington. September 15, 2023.

Schauble, Al. St. John, Washington. March 20, 2017.

Scheuerman, Don. Endicott, Washington. July 1, 2024.

Sewell, Andy. Spokane, Washington. June 4, 2022.

Shaw, Maja. Richland, Washington. June 20, 2021.

Solbrack, Dennis. Colfax, Washington. May 1, 2020; February 17, 2022.

Strader, Verne and Barbara. Endicott, Washington. March 8, 2023.

Wells, David and Georgia. Endicott, Washington. July 7, 2019.

Weitz, Anna B. Endicott, Washington. March 10, 1972.

White, Pat and Sid. Olympia, Washington. March 6, 1991.

Woolston, Bill. Moscow, Idaho. June 14, 2024.

Worth, Gene. Pasco, Washington. June 12,